OFFSHORE WIND ENERGY

Many countries have plans to expand wind energy to meet CO_2 emissions targets. Lack of available land area and the need for good and stable wind conditions have stimulated the development of offshore wind generation technology, which allows for the development of larger turbines. The offshore environment, however, involves new challenges related to the design, installation, operation and maintenance of the turbines.

Based on graduate-level courses taught by the author, this book focuses on the opportunities and challenges related to offshore wind turbines. It introduces the offshore environment, including wind and wave dynamics, before discussing the aerodynamics of wind turbines, hydrodynamic loading, marine operations and wind farm layout. Featuring examples that demonstrate practical application of the topics covered and exercises to consolidate student understanding, this is an indispensable reference text for advanced students and researchers of environmental science and engineering and for industry professionals working in the wind energy sector.

FINN GUNNAR NIELSEN is Professor at the Geophysical Institute at the University of Bergen and was until 2023 Director of the Bergen Offshore Wind Centre, responsible for coordinating the offshore-wind-related research at the University of Bergen. He has worked in industrial research for more than 40 years. From 2002 he headed the research project developing the world's first full-scale floating wind turbine, the Hywind concept.

OFFSHORE WIND ENERGY

Environmental Conditions and Dynamics of Fixed and Floating Turbines

FINN GUNNAR NIELSEN
University of Bergen

Shaftesbury Road, Cambridge CB2 8EA, United Kingdom

One Liberty Plaza, 20th Floor, New York, NY 10006, USA

477 Williamstown Road, Port Melbourne, VIC 3207, Australia

314–321, 3rd Floor, Plot 3, Splendor Forum, Jasola District Centre, New Delhi – 110025, India

103 Penang Road, #05–06/07, Visioncrest Commercial, Singapore 238467

Cambridge University Press is part of Cambridge University Press & Assessment, a department of the University of Cambridge.

We share the University's mission to contribute to society through the pursuit of education, learning and research at the highest international levels of excellence.

www.cambridge.org
Information on this title: www.cambridge.org/9781009341431

DOI: 10.1017/9781009341455

© Finn Gunnar Nielsen 2024

This publication is in copyright. Subject to statutory exception and to the provisions of relevant collective licensing agreements, no reproduction of any part may take place without the written permission of Cambridge University Press & Assessment.

First published 2024

A catalogue record for this publication is available from the British Library.

Library of Congress Cataloging-in-Publication Data
Names: Nielsen, Finn Gunnar, 1951– author.
Title: Offshore wind energy : environmental conditions and dynamics of fixed and floating turbines / Finn Gunnar Nielsen.
Description: Cambridge ; New York, NY : Cambridge University Press, [2023] | Includes bibliographical references and index.
Identifiers: LCCN 2023028780 (print) | LCCN 2023028781 (ebook) | ISBN 9781009341431 (hardback) | ISBN 9781009341455 (ebook)
Subjects: LCSH: Wind turbines. | Offshore wind power plants.
Classification: LCC TJ828 .N54 2023 (print) | LCC TJ828 (ebook) | DDC 621.31/2136–dc23/eng/20230927
LC record available at https://lccn.loc.gov/2023028780
LC ebook record available at https://lccn.loc.gov/2023028781

ISBN 978-1-009-34143-1 Hardback

Additional resources for this publication at www.cambridge.org/owe.

Cambridge University Press & Assessment has no responsibility for the persistence or accuracy of URLs for external or third-party internet websites referred to in this publication and does not guarantee that any content on such websites is, or will remain, accurate or appropriate.

Contents

Additional resources for this title can be found at www.cambridge.org/owe.

Preface

The inspiration to write this book came from the recent focus on offshore wind energy as part of the solution to convert the global energy supply from fossil fuel to renewables. Several important actors, e.g., the International Energy Agency, have pointed toward the potential large contribution from offshore wind energy in the future energy supply. However, moving wind turbines into the ocean and scaling up the turbine sizes adds new challenges compared to traditional wind turbines on land. To realize the offshore wind energy potential, increased competence in a wide range of professional disciplines is thus required.

In my personal professional carrier, I have dealt with topics such as ship hydrodynamics, propeller design, model and full-scale testing, wave loads, structural loads and marine operations, as well as analyzing and testing fixed and floating wind turbines. I have realized that the design and operation of offshore wind turbines relies upon an understanding of a wide range of scientific and engineering disciplines. Even if no individual involved in planning, design and operation can master all the required disciplines in detail, some basic insight is needed to secure precise and efficient communication between specialists. My hope is that this book can contribute to this. The book covers a wide range of topics but does not cover every topic in great detail. For that purpose, references are given to specialized literature.

Most of the material included in this book is based upon material used in lecturing graduate-level courses in marine operations at the Norwegian University of Science and Technology (NTNU) (Nielsen, 2007) and courses in offshore wind energy at the University of Bergen (UiB). The material also builds upon methods developed and published during my carrier in industry. Having the privilege to lead the research project developing the world's first multimegawatt floating wind turbine, Hywind Demo (installed offshore of Norway in 2009), gave me a unique opportunity to experience the importance of cross-disciplinary insight. Also, having the privilege to head the Bergen Offshore Wind Centre (BOW) at UiB from its inauguration in 2018 has given me the opportunity to further explore and learn the importance of cross-disciplinary work. At BOW, e.g., noise from wind turbines, the erosion of turbine blades, impacts on the marine environment, societal perspectives and legal issues have been covered. These are all important issues for the successful development of offshore wind energy, but are not covered in this book.

My hope is that this book can be used in graduate as well as continued education within offshore wind energy, giving a broad overview of key topics related to the environmental conditions and dynamic response of offshore wind turbines. For deeper insight into, e.g., meteorology, aerodynamics and marine hydrodynamics, specialized courses and literature should be consulted.

This book could not have been written without the encouragement and inspiration of several persons, among them: Christina Aabo of Aabo Energy and a member of the Scientific Advisory Board (SAC) of BOW; Jan-Fredrik Stadaas of Equinor and a member of SAC; Adjunct Associate Professor Marte Godvik, UiB and Equinor; Professor Henrik Bredmose, DTU and member of SAC; and Dr. Kristin Guldbrandsen Frøysa, Energy Director at UiB. Many individuals have also contributed with comments and advice on specific chapters, among them several present and former colleagues: Adjunct Professor Bjørn Skaare, University of Stavanger and Equinor; Dr. Herbjørn Haslum, Equinor; Dr. Rolf Børresen; Adjunct Professor Birgitte Rugaard Furevik, UiB and Norwegian Meteorological Institute; Associate Professor Mostafa Bakhoday Paskyabi, UiB; Professor Cristian Guillermo Gebhardt, UiB; Professor Joachim Reuder, UiB; Associate Professor Etienne Cheynet, UiB; Dr. Astrid Nybø, Odfjell Offshore Wind; and Dr. Ida Marie Solbrekke, NORCE. Thanks to all for their contributions! Thanks also to Anna Therese Klingstedt for creating the illustrations of floating wind turbines.

1
Introduction

1.1 Motivation

Wind energy is an increasingly important source of renewable energy. Presently, most wind turbines are installed on land. However, lack of available land area as well as the need for good and stable wind conditions have stimulated the development of offshore wind turbines. Moving offshore has also allowed for a considerable increase in the size of wind turbines; presently, 10–15 MW turbines are the state of the art, while even larger turbines are under development. Moving offshore involves new challenges, both with respect to the design for the offshore environment as well as for the installation, operation and maintenance of the turbines.

The first offshore wind farm, Vindeby, Denmark, was installed in 1991 in a water depth of approximately 4 m. The 11 turbines had a rated power of 0.45 MW each. The technology used for the early offshore wind turbines was the same as for the land-based turbines, only that the tower was "elongated" to provide support for the structure through the water column. The early development of offshore wind turbines resembles that of the offshore oil and gas industry in the 1940s to 1960s. The technology used offshore was very similar to the technology used on land. To compensate for the "new" wet environment, the equipment was placed upon a simple platform, or substructure. In the beginning, the installations were made in water depth of less than 10 m. As the industry moved, step by step, into deeper waters and the size of the wind turbines increased, the same basic design was applied and the technology was extrapolated. However, a significant shift in technology has occurred through the introduction of floating platforms, which have required a rethinking of the complete design.

As the size of individual wind turbines as well as wind farms increases, and as wind farms are located further from shore, in deeper waters and in harsher environments, new technologies are needed. Modification of land-based technology is not sufficient. To design and operate offshore wind turbines, a real multidisciplinary approach is needed. The disciplines comprise meteorology, aerodynamics, hydrodynamics, structural dynamics, material technology, cybernetics, power electronics and several others related to environmental impact, logistics and economics.

1.2 Content of the Book

The focus of the present book is on some key issues that make offshore wind turbines different from land-based wind turbines.

Understanding the offshore environment – the wind and wave conditions – is key to understanding the dynamics of offshore wind turbines. Therefore, a description of the marine atmospheric boundary layer and classical wave theory is given in Chapter 2. Some basic wave statistics are also included in this chapter.

For the sake of completeness, the aerodynamic principles behind a wind turbine are dealt with in Chapter 3. Classical approaches such as the Betz limit and the beam element momentum (BEM) method are described. Further, the dynamic response of the wind turbine, which becomes more important as the size of the turbine increases, is discussed. In addition, vortex methods are described in some detail, which may represent an intermediate and efficient step between the classical BEM and more complete RANS[1] or LES[2] methods for analyzing the aerodynamics of wind turbines.

In Chapter 4, various offshore support structures are discussed. There has been continuous development of new support structures in the last decades. Still, the monopile is the most frequently used substructure.[3] Various jacket structures and tripods are also used. In more recent years, the development of floating support structures has gained speed. The world's first multimegawatt floating offshore wind turbine, Hywind Demo, was installed off the west coast of Norway, at Karmøy, in 2009. A 2.3 MW turbine was used on this test unit. In 2017, the first floating offshore wind farm was installed off the east coast of Scotland. The wind farm consists of five turbines, each of 6 MW. The support structure is of the Hywind brand.

To provide a background for the dynamic analysis of wind turbines, Chapter 5 is included to summarize some simple classical dynamics of linear systems.

The basic principles for computing wave loads on fixed support structures are covered in Chapter 6. The focus is on slender vertical structures, but the principles behind the theory of computing wave loads on large-volume structures of general shape are also included. Further, some issues related to wave forces in steep waves

[1] Reynolds-averaged Navier–Stokes equations. [2] Large eddy simulation.

[3] See Chapter 4 for a further description of the various concepts as well as definitions of wind turbine components such as the nacelle, tower, support structure, substructure and foundation.

are included. These issues are important to estimate extreme loads as well as fatigue effects due to waves.

Wave loads and wave-induced dynamics of floating support structures are covered in Chapter 7. Here, some classical approaches based upon slender body assumptions and linear theory are outlined in detail. Such methods are computational-efficient and may thus be useful in, for example, optimization processes. To compute the rigid body dynamics of a floating body, the mass, damping and inertia matrices must be established. Each of these matrices are discussed and the restoring effects of mooring lines are examined in some detail. Finally, the need for a motion control system for floating wind turbines is discussed.

Marine operations are an important part of installing as well as operating a wind farm. In Chapter 8, the issues of weather windows and duration statistics are discussed and the dynamics of lifting operations from a floating crane vessel are considered in some detail. The probability of impact during a load transfer operation is discussed, as well as a simple approach for estimating the impact loads during load transfer by a crane. Issues related to wind farms are discussed in some detail in Chapter 9, in particular, simple wake models and the summation of wakes.

This book does not handle in detail the classical aerodynamic analysis of wind turbines. This may be found in textbooks such as those by Hansen (2015) or Manwell et al. (2009). Issues related to choices of material and corrosion in the offshore environment are not covered in this book. The same goes for the generation and transformation of electrical power, as well as electrical cable issues.

1.3 The Design Process

A wind turbine must be designed to fulfil several criteria related to, for example, operational conditions and extreme conditions. The various design criteria will determine which analysis is needed in the design process. In standards such as IEC 61400–3 (2009) and DNV (2021a), the criteria are formulated as limit states. A limit state is "a condition beyond which a structure or structural component will no longer satisfy the design requirements." According to the standard issued by DNV (2021a), four different "limit states" have to be considered in the design of offshore wind turbine structures.

The *serviceability limit state* (SLS) corresponds to the tolerance criteria applicable to normal use, i.e., the SLS design check should ensure that the structure works under normal operating conditions.

The *fatigue limit state* (FLS) corresponds to failure due to the effect of cyclic loading, i.e., the FLS design check should ensure that during the lifetime of the structure, the cumulative effect of cyclic loading should not damage the structure.

The *ultimate limit state* (ULS) corresponds to the maximum load-carrying resistance, i.e., the ULS design check should ensure that the structure is able to withstand the extreme loads typically experienced without damage.

The *accidental limit state* (ALS) corresponds to (1) maximum load-carrying capacity for (rare) accidental loads or (2) post-accidental integrity for a damaged structure, i.e., the ALS design check is to secure that the structure, if exposed to loads beyond the ULS level, does not collapse, and that it may survive even in damaged condition.

DNV (2021a) lists the following examples of limit states within each category.

Serviceability limit states (SLS)

Deflections that may alter the effect of the acting forces.
Excessive vibrations producing discomfort or affecting nonstructural components.
Excessive vibrations affecting turbine operation and energy production.
Deformations or motions that exceed the limitation of equipment.
Durability.
Differential settlements of foundations soils causing intolerable tilt of the wind turbine.
Temperature-induced deformations.

Fatigue limit states (FLS)

Cumulative damage due to repeated loads.

Ultimate limit states (ULS)

Loss of structural resistance (excessive yielding and buckling).
Failure of components due to brittle fracture.
Loss of static equilibrium of the structure, or of a part of the structure, considered as a rigid body, e.g., overturning or capsizing.
Failure of critical components of the structure caused by exceeding the ultimate resistance (which in some cases is reduced due to repetitive loading) or the ultimate deformation of components.
Transformation of the structure into a mechanism (collapse or excessive deformation).

Accidental limit states (ALS)

Structural damage caused by accidental loads (ALS type 1).
Ultimate resistance of damaged structures (ALS type 2).
Loss of structural integrity after local damage (ALS type 2).

Prior to the design process for the support structure,[4] the site-specific environmental condition as well as the design of the rotor-nacelle assembly (RNA) must be

[4] The names of the various components of an offshore wind turbine are shown in Section 1.4 and Chapter 4.

known. Based upon these, the design basis for the support structure is formulated. Several operational cases (design situations) and environmental conditions (load cases) are analyzed. These include:

- normal design situations and appropriate normal or extreme external conditions
- fault design situations and appropriate external conditions
- transportation, installation and maintenance design situations and appropriate external conditions

For each case, the load effects are analyzed and checked as to whether they satisfy the appropriate limit state. If not, the design must be modified.

To check if the structure fulfils the limit states, one has to check if the "capacity" (e.g., material strength) of the structure is sufficient to withstand the loads acting upon it. This is normally done by use of the "partial safety factor method." Formally, this is expressed by requiring that the "resistance" of the structure, R_d, shall exceed the design load effect, S_d:

$$S_d \leq R_d \qquad [1.1]$$

The design load effect, S_d, is obtained from the "characteristic load" effect, S_k, by multiplying by a load factor, γ_f, $S_d = \gamma_f S_k$. For example, the characteristic load effect for a ULS is the load combination that has an annual probability of exceedance of 0.02 or less. That is, the load has a return period of at least 50 years. In an FLS, the characteristic load effect history is defined as the expected load effect time history during the lifetime of the structure.

Similarly as for the load effect, the design resistance is obtained by dividing the characteristic resistance by a material factor, γ_m, $R_d = R_k/\gamma_m$.

The load factor, γ_f, is introduced to account for uncertainties in the methods for estimating the loads as well as to account for statistical uncertainties in the loads, for example, the statistics of waves. Similarly, the material factor, γ_m, shall account for uncertainties in the capacity of the material, such as yield strength, thickness, fatigue capacity and so on. As an example, DNV (2021b) states that γ_f should be set to 1.35 for environmental loads in a ULS check, while the material factor γ_m should be set to 1.10 for the strength of steel tubular members (DNV, 2021a). The load and material factor will vary depending on load case considered and failure mode.

For FLS checks, several issues must be accounted for. The fatigue capacity of a steel material depends upon the steel quality, plate thickness, corrosion protection and weld geometry and quality, as well as accessibility for inspection of critical details. With all these factors included and using a characteristic load time history, the cumulative fatigue damage during the lifetime of the structure should be less than one half to one third of the damage causing failure.

1.4 The Layout of Wind Turbines

The function of a wind turbine is to convert the kinetic energy in the wind to rotational energy. Rotational energy was in the past used for grinding ("windmills") or pumping water. Today, rotational energy is converted to electric energy by an electrical generator. Various design principles of wind turbines exist and some features of the main ones are discussed briefly in the following sections.

1.4.1 Horizontal-Axis Wind Turbines

The horizontal-axis wind turbine (HAWT) is the most frequently used wind turbine. In Figure 1.1 (left), an illustration of the layout of a HAWT is given. Both on land and offshore, a HAWT is mounted upon a slender tower. In most cases the rotor consists of three blades mounted on an almost horizontal axis. The axis may have some tilt to avoid collision between the blades and the tower during rotation. The rotor is usually mounted on the upwind side of the tower. Each blade has the profile of an aerofoil. How this works is described in Chapter 3.

Wind blowing into the rotor makes the rotor rotate. Considering the aerodynamics of large rotors, the optimum rotational speed is too low for a standard electrical generator to be efficient. Therefore, a gear is mounted between the rotor shaft (the

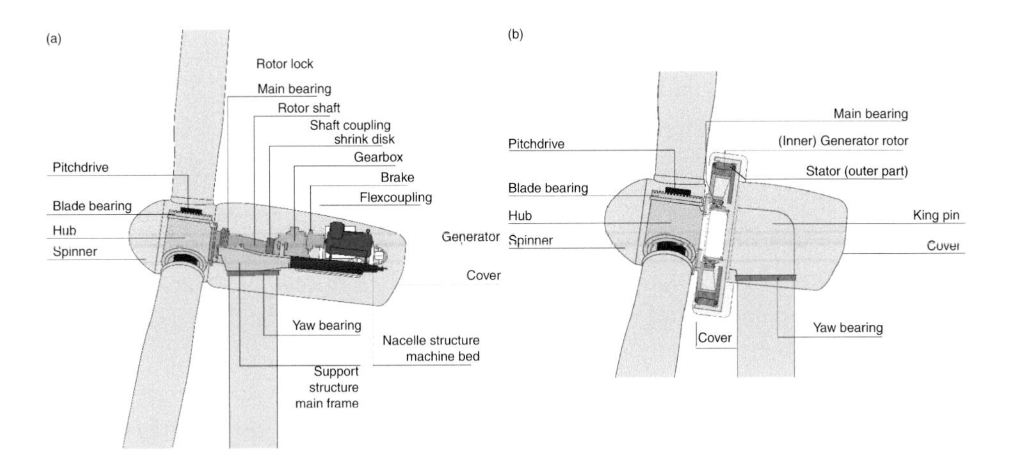

Figure 1.1 Illustration of the main components in a geared wind turbine (left) and a direct-drive turbine (right). Reproduced with permission of Fraunhofer IWES. Source: Wenske (2022).

low-speed shaft) and the generator shaft (the high-speed shaft), with a gear ratio that may be in the order of 100. A four-pole generator should run at 1500 revolutions per min to generate 50 Hz AC power. Thus, if the rotational speed of the turbine is 15 revolutions per min, a gear ratio of 100 is needed. Several multimegawatt turbines are now equipped with so-called *direct-drive* generators, illustrated in Figure 1.1 (right). A direct drive implies that the gear is removed and a generator with a large diameter and many poles along the circumference is used. The generator used is frequently a permanent magnet generator. The many poles secure a sufficiently efficient generation of electrical power even if the rotational speed is low. The frequency of the AC power generated does not normally fit the grid frequency. The AC current is therefore converted to DC current and back to AC with the correct frequency and phase.

The rotor blades are mounted to the rotor hub. In the hub, a mechanism for twisting or pitching the rotor blades is located. The pitching of blades is used to control the power and reduce the loads on the rotor at high wind speeds. The shafts, gear, generator, control system and power converter systems are assembled in a housing denoted the nacelle. The rotor and nacelle assembly, frequently denoted the RNA, is located on top of the tower. A weathervane is mounted on the top of the nacelle, by which the direction of the wind is measured. The RNA may rotate on top of the tower so that the rotor heads into the wind; this is accomplished by the yaw mechanism. The operation of a modern wind turbine relies upon several systems that control rotational speed, blade pitch and yaw angle.

As discussed in Chapter 3, the three-bladed upwind turbine illustrated in Figure 1.1 has many advantages. However, two-bladed designs exist as well. A two-bladed rotor is lighter and has a higher optimum rotational speed than a three-bladed turbine. Therefore, the required gear ratio is lower. A three-bladed rotor has a beneficial dynamic property as the mass moment of inertia is independent of which axis in the rotor plane is considered. This contrasts with a two-bladed rotor, which has very large difference in the mass moment of inertia about the axis coinciding with the blade axis and the axis normal to the blade axis. This difference in inertia may trigger unfavorable dynamic loads.

Downwind rotors also exist. By locating the rotor on the downwind side of the tower, the collision issue between blades and tower is solved. However, every time a blade passes the leeward side of the tower, it moves into a wind shadow, causing severe dynamic loads. Downwind rotors may, under certain conditions, be self-correcting with respect to wind direction. In that case, the yaw mechanism required for the upwind turbine may be avoided.

1.4.2 Horizontal-Axis Multirotors

The size of horizontal-axis wind turbines has steadily increased. In the 1980s, wind turbines with a rated power of 50 KW and a rotor diameter of 15 m were common. Currently, turbines with a rated power of 15 MW and more and a rotor diameter in the order of 250 m are the state of the art. The power that can be extracted by a wind turbine is proportional to the area swept by the rotor, i.e., proportional to the diameter squared. However, the bending moment in the blade root will increase by approximately the cube of the rotor diameter. The same goes with the mass. It may thus be assumed that increased mass, large deformations and large and heavy units (blades, generator) to be handled during installation and maintenance will at some point limit the size of turbines. So far, this has not happened, but alternative designs have been proposed to address the upscaling challenge. Wind turbine designs with two or more rotors mounted on the same tower have been built (van der Laan et al., 2019). Also, offshore, floating wind turbine concepts with serval rotors mounted upon the same hull have been proposed (Jamieson, 2017). These designs span from using two rotors to large arrays using in the order of 100 rotors. The arguments for having many smaller rotors rather than one large rotor are related to costs of energy and ease of handling. It may also be argued that the complete production and assembly line is simpler and faster.

1.4.3 Vertical-Axis Turbines

As the name states, the blades of a vertical-axis wind turbine (VAWT) rotate around a vertical axis, as shown in Figure 1.2. In addition to the two principles shown in Figure 1.2, designs with helical-shaped blades exist. The cross-section of each blade is shaped like an aerofoil, thus creating a lift force on the blade section, driving the rotation. Drag-based systems also exist. The differences between lift-based (aerofoil) systems and drag-based systems are discussed in Chapter 3. An important difference is that drag-based systems are far less efficient than lift-based systems.

The main advantages of a VAWT are that most of the technical equipment, including the generator, may be placed at ground level, and that the turbine works independent of the wind direction. Generally, a VAWT has lower efficiency than a HAWT. The reason for this is mainly related to the angle of attack for the blades. This angle varies during the rotation, thus, the angle of attack most of the time is not optimum. Further, when the blade is moving at the downwind side of the rotor, it is moving in the wake of the upwind blade, causing reduced efficiency and larger dynamic loads due to the increased turbulence. However, for smaller turbines – e.g., “rooftop” versions – a VAWT may be an attractive solution.

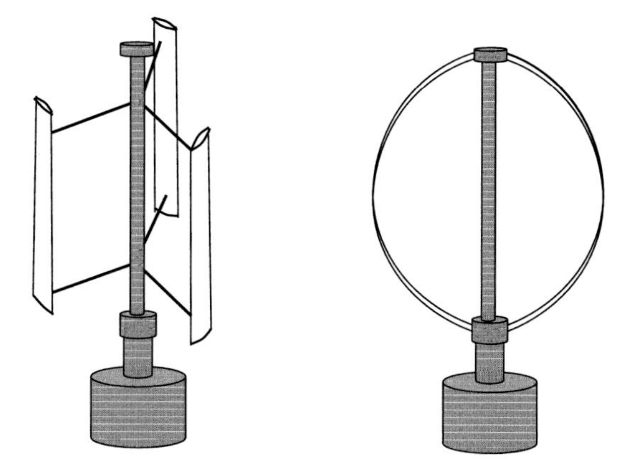

Figure 1.2 Vertical-axis wind turbines. "H-shaped" type (left) and Darrieus ("eggbeater") type (right).

1.4.4 High-Altitude Wind Power Devices

High-altitude wind power devices (HAWP) use wind forces to stay high above ground level and at the same time extract energy. The systems are inspired by kites. A kite, or a wing, is connected via a line to a drum on the ground. As the kite is flying, it is forced into a special motion pattern, for example, a figure-eight pattern, as illustrated in Figure 1.3, or a circular pattern. By allowing the drum to rotate, the line is little by little released from the drum and the combination of the tension in the line and rotational velocity of the drum creates mechanical power, driving an electrical generator. When the full length of the line is out of the drum, a "home-flying" mode is activated; the kite descends and the line is wound up on the drum under low tension. Power is thereby produced as the kite flies up, while a small amount of power is used when it flies back "home."

As an alternative to extracting power when the kite flies up and tensions the line, other concepts equip the kite, or rigid wing, with small wind turbines that extract wind power while the system is flying at a steady pattern at high altitude. The Makani system[5] is an example. The power is transferred via the line down to ground level. Such a system can in principle produce power continuously.

The main idea behind HAWP devices is to utilize the strong and steady wind of high altitudes. These devices operate several hundred meters above ground level. However, at the present stage of development the systems are not commercial and do not deliver power at the megawatt scale.

[5] The Moonshot Factory. n.d. "Makani: Harnessing Wind Energy with Kites to Create Renewable Electricity." www.x.company/projects/makani/ (accessed November 3, 2021).

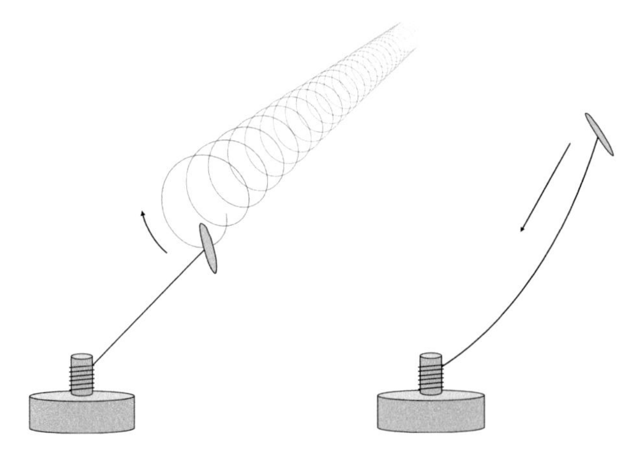

Figure 1.3 Simplistic view of a kite connected to a drum, with an indication of the flying pattern when power is produced (left) and when flying back (right).

Exercises Chapter 1

1. Describe the main differences between HAWT and VAWT. Why can the rotor blade of a VAWT not work at an optimum angle of attack all the time?
2. Consider a H-shaped VAWT, as illustrated in Figure 1.2. Assume the angular velocity of the blades is given by ω and the incident wind speed is U. Derive an expression for the angle of attack as a function of the angular position of the rotor blade. (See Chapter 3 for definition of the angle of attack.) Ignore induced velocities.
3. Consider a Cartesian coordinate system with one axis oriented along the axis of rotation of the rotor and the two other axes in the rotor plane. Show that the mass moment of inertia of a three-bladed wind turbine is independent of which axis in the rotor plane is considered.

2
The Offshore Environment

In this chapter the properties of wind and ocean waves will be described, focusing on issues important to the extraction of wind energy and the computation of loads from wind and waves.

The main difference between land-based and offshore wind turbines is exposure to the offshore environment. In addition to wind, an offshore wind turbine is exposed to waves and currents.

The main characteristics of wind are similar over land and over the ocean. However, there are some important differences. These are, among others, related to the sea surface. In contrast to wind over land, waves on the sea surface represent a boundary for the wind, with surface roughness depending upon wave height. Further, the temperature difference between sea and air follows a different pattern over time than the temperature difference between land and air. The consequence is that the wind field offshore differs from that on land. Thus, experience gained from land-based wind turbines has some limitations when it comes to offshore turbines.

Ocean waves represent a new environmental parameter to consider when moving wind turbines offshore. A proper description of ocean waves is thus needed to establish proper estimates of fatigue and extreme loads on structures. For floating wind turbines, wave-induced motion is also an important design consideration. How much detail is needed for the description of wave kinematics depends upon the geometry of the support structure, the load cases considered and, e.g., the relative importance of wind versus wave loads. For some applications, a linearization of the kinematics and loads is sufficient, while in other cases a careful description of the kinematics of the extreme waves should be considered.

Ocean currents in most cases do not themselves represent critical loads. However, currents modify waves and may thus modify wave-induced loads. For bottom-fixed support structures, a speed-up of currents close to the seabed may cause the transport of sediments away from the immediate vicinity of the wind turbine foundation in a phenomenon called scouring. Scouring may deteriorate the fixture of the wind turbine to the sea floor.

In the following, some basic principles for describing the wind field and ocean waves are presented. The description of the theoretical background and the level of detail are limited. To get a deeper insight into the material presented, please see the references given.

2.1 Wind

2.1.1 Introduction

For offshore wind turbines, the wind represents the energy resource. However, the wind also causes loads on the turbine structure. These loads must be accounted for in the design. In applications related to offshore wind farms, the wind field must be understood and described across a wide range of length and timescales. In the design of rotor blades, the turbulent structures of the wind field in the range of meters and seconds are important in the assessment of dynamic loads, while to understand the flow inside and in the vicinity of a wind farm length scales in the range of hundreds of meters to several kilometers and time scales in the range of a few minutes to hours must be considered. Prediction of the energy production hours ahead requires an understanding of the wind field at meso-scale range (tens of kilometers in length and hours in time).

In the design of offshore structures, simplistic descriptions of the wind field have been used in most cases. This is because the focus has been on the estimation of extreme loads with probability of occurrence in the range 10^{-2}–10^{-4} per year. For offshore wind turbines the needs are different. Still, extreme as well as operational loads are important factors in design. However, as the largest rotors exceed 200 m in diameter and are very slender and flexible structures, a detailed understanding of the structure of the wind field is very important, from both a power extraction and a structural design point of view.

A thorough description of the turbulent wind field over the ocean will not be given here. For a detailed description of the meteorology of the atmospheric boundary layer, reference is made to special textbooks, e.g., Lee (2018) and Stull (1988). The following description will restrict the discussion to some of the main parameters used to describe the wind field suitable for the design of wind turbines. It will also address some of the challenges encountered, such as with increased rotor sizes. Relevant time and length scales are illustrated in Figure 2.1.

2.1.2 The Marine Atmospheric Boundary Layer

The atmospheric boundary layer[1] is the region between ground or sea surface and the "free" atmosphere. In the free atmosphere the effect of the surface friction on

[1] The general discussion of the atmospheric boundary layer is valid for boundary layers both over sea and land. When the marine atmospheric boundary layer is considered, the abbreviation MABL is used.

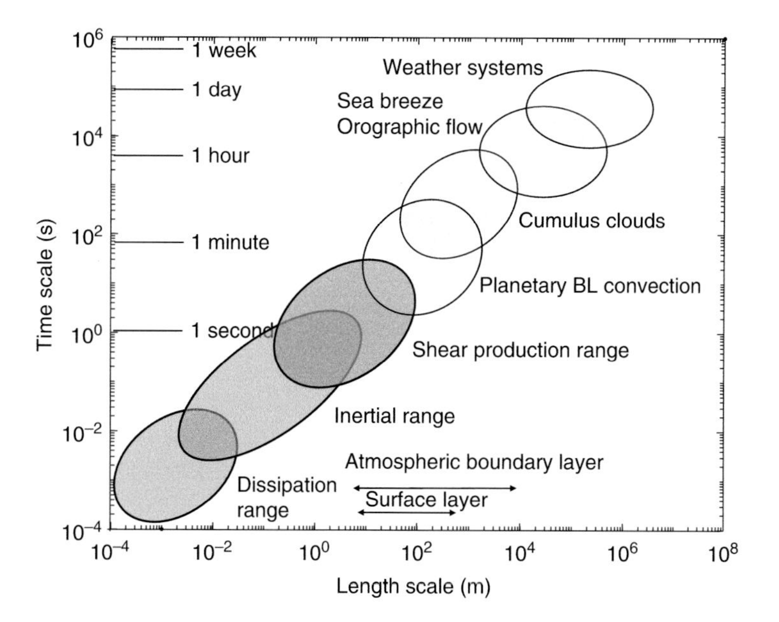

Figure 2.1 Illustration of the various time and length scales involved in the atmospheric flow. The three ranges considered for wind turbines are marked in bold. Based upon an original figure in Busch et al. (1978). Reproduced with permission of Springer eBook.

the wind field can be ignored. Between the free atmosphere and the boundary layer there frequently exists a capping layer, which limits the exchange of air between the boundary layer and the free atmosphere. The elevation of this capping or inversion layer is highly dependent upon the weather conditions. The various layers of the atmospheric boundary layer are illustrated in Figure 2.2. A typical order of magnitude of the elevation of the capping layer during neutral and unstable atmospheric conditions, also denoted as convective conditions, may be around 1 km. However, this may vary a lot and during stable atmospheric conditions the elevation of the capping layer can be as low as a few hundred meters and even below 100 m. Atmospheric stability is discussed in more detail in Section 2.1.2.2. In the upper part of the boundary layer, also denoted as the Ekman layer, the direction of the mean wind is strongly influenced by the combined effect of the Coriolis forces and the vertical gradient of the mean wind speed, causing a continuous shift of mean wind direction down through the boundary layer. The lower part of the boundary layer is called the surface layer. The thickness of the surface layer is in the order of 10% of the boundary layer, and traditionally it has been assumed that wind turbines operate in this layer. However, large offshore wind turbines may operate above this height. In the surface layer the turbulent mixing due to vertical velocity shear is

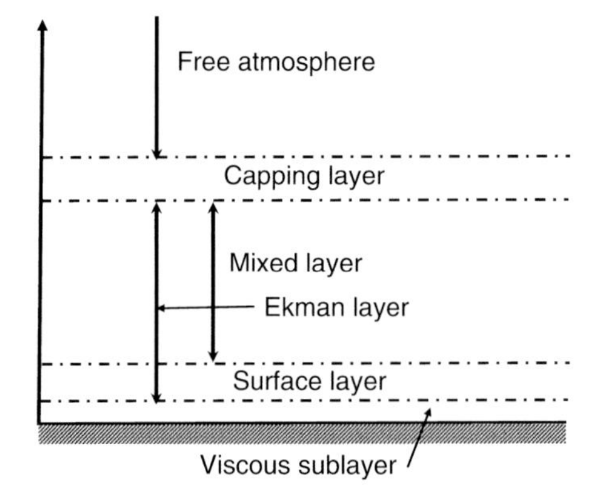

Figure 2.2 Various parts of the atmospheric boundary layer. Not to scale. The mixed layer is also denoted as the convective layer.

significant, but also temperature (buoyancy) effects may be important for the turbulence. In the immediate vicinity of the ground or sea, there is a viscous sublayer. We will focus here on the surface layer. The considerations are somewhat simplified as, e.g., the Coriolis forces are neglected, as is frequently done in wind energy applications.

2.1.2.1 Mean Velocity Profile

To describe the flow field in the atmospheric boundary layer, the Navier–Stokes equations are invoked. In Appendix A, a brief summary of the two-dimensional boundary layer equations for an incompressible flow is given. Here, some classical relations between the mean velocity profile, the fluctuating velocity components and the shear are given. These relations are utilized in describing some key characteristics for the MABL.

A Cartesian coordinate system is used, with the x-axis horizontal and positive in the mean wind direction and the z-axis vertical, positive upwards and zero at the sea level. The velocity in x-direction can then be written as a mean value plus the turbulent fluctuation, $u = \overline{u} + u'$. The mean transverse and vertical velocities are assumed to be zero, while the fluctuating components are denoted as v' and w' respectively. The variation in mean wind direction with height due to the Coriolis effect is thus disregarded. In solving the Navier–Stokes equations for the mean flow, a closure problem exists. Terms denoted Reynolds stresses of the form $\rho_a\overline{u'w'}$, $\rho_a\overline{u'v'}$ and $\rho_a\overline{v'w'}$ appear in the equations, where ρ_a is the density of air. The Reynolds stresses are related to characteristics of the mean flow. Based upon empirical evidence, the Reynolds stresses are assumed proportional to the vertical velocity gradient in the boundary layer. This is according to the Prandtl mixing length theory; see, for example, Curle and Davies (1968). In a two-dimensional flow, the Reynolds stress is written as:

$$\rho_a \overline{u'w'} = -K_m \rho_a \frac{\partial \overline{u}}{\partial z}. \quad [2.1]$$

K_m is denoted as the eddy diffusivity. In the absence of heat fluxes, the boundary layer is denoted as neutral. Under such conditions and close to a smooth surface but above the viscous sublayer (see Appendix A), K_m may be parameterized as:

$$K_m = k_a z u_*. \quad [2.2]$$

k_a is the von Kármán constant, based upon experimental data found to be approximately 0.40. u_* denotes the friction velocity and is given from the relation $u_* = \sqrt{\tau/\rho_a}$, where τ is the shear stress at the surface. In the surface layer it is assumed that the turbulent momentum fluxes are constant in the vertical direction. Thus, by combining Equations 2.1 and 2.2, and integrating from the bottom of the surface layer, z_0 to z, the velocity profile is obtained as:

$$\overline{u}(z) = \frac{u_*}{k_a} \ln\left(\frac{z}{z_0}\right). \quad [2.3]$$

z_0 is frequently denoted as the surface roughness length scale. Charnock (1955) found a relation between z_0 and u_* by measuring the wind field over a water reservoir. The vertical profile of the lowest 8 m over the water surface fitted the logarithmic profile in Equation 2.3 well. The mean velocity profile in the surface layer thus gives $\overline{u}(z_0) = 0$, even if the mean velocity at top of the friction layer has some small but finite value. Charnock (1955) found the empiric relation between the roughness length scale and the friction velocity as:

$$z_0 = \alpha_0 \frac{u_*^2}{g}. \quad [2.4]$$

This is called the Charnock relation. Over open sea, with fully developed waves, the Charnock constant α_0 has, according to DNV (2021c), values in the range of 0.011–0.014, and near coastal sites it may be 0.018 or higher.

For offshore conditions, z_0 and u_* depend upon the wind speed, upstream distance from land and wave conditions. DNV (2021c) proposes z_0 in the range 0.0001 to 0.01 for open sea conditions. The lowest value corresponds to calm water and the highest to rough wave conditions. Onshore, higher values for z_0 apply.

The friction velocity may be related to the mean velocity at a certain vertical elevation and a surface friction coefficient. Frequently, the 10 min mean wind velocity at a height z_r over the ground or sea surface, denoted $U_{10}(z_r)$, is used as a reference. By combining the above relations and defining a surface friction coefficient $\kappa = \tau/\rho_a U_{10}^2(z_r)$, the following expression for the friction velocity and friction coefficient is obtained:

$$u_* = \sqrt{\kappa} U_{10}(z_r)$$
$$\kappa = \frac{k_a^2}{\left(\ln\left(\frac{z_r}{z_0}\right)\right)^2}. \qquad [2.5]$$

Frequently, a reference height $z_r = 10m$ is used. With the above range of z_0, κ- values in the range 0.0012 and 0.0034 are obtained.

The above formulations rely upon the assumption of a near neutral boundary layer and constant turbulent momentum fluxes in the vertical direction. Above the surface layer and in stable or unstable boundary layers the assumptions are not valid. Below the surface layer, in the viscous sublayer with a vertical extent in the order of mm, the flow is mainly laminar and is, as the name indicates, dominated by viscous effects. Viscous effects do not play an important role above this layer. Some more details related to the two-dimensional boundary layer equations are outlined in Appendix A.

In engineering applications, [2.3] is frequently replaced by a simpler power function, written as:

$$U_{10}(z) = U_{10}(z_r)\left(\frac{z}{z_r}\right)^{\alpha}. \qquad [2.6]$$

Here, $U_{10}(z)$ is the 10 min mean wind velocity at vertical level z. By requiring the mean wind speed at the vertical level z as obtained by [2.3] and [2.6] to be equal, a relation between z_0 and α is obtained:

$$\alpha(z) = \frac{\ln\left(\frac{\ln(z/z_0)}{\ln(z_r/z_0)}\right)}{\ln(z/z_r)}. \qquad [2.7]$$

The result is plotted in Figure 2.3 (left) for three different values of the roughness length, z_0. It is observed that the proper α value depends upon at which height the two expressions for the mean velocity are required to fit. In Figure 2.3 (right), the relation between z_0 and α for three different fitting heights is plotted. A reference height of $z_r = 10\ m$ is applied in the graphs.

As indicated above, the mean wind velocity depends upon the time of averaging. When considering extreme wind velocities, the extreme mean velocity with a

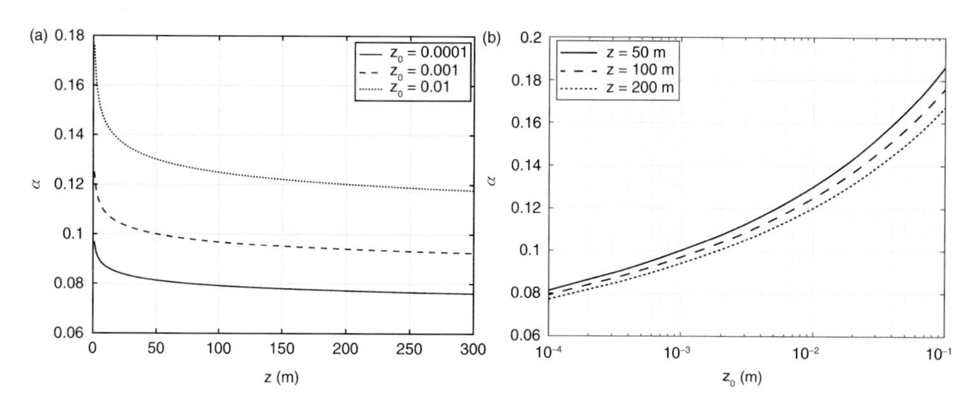

Figure 2.3 Relation between roughness length, z_0 and the exponent α in the power law formulation of the mean wind profile. Left: height dependence of α for three different z_0. The logarithmic and exponential profile gives the same mean wind speed at height z. Right: α versus z_0 for three different heights. Mean wind speed fits at vertical level z. Reference height $z_r = 10\ m$ is used.

certain return period, for example 10 or 50 years (corresponding to a yearly probability of occurrence of 0.1 and 0.02, respectively), increases if the time of averaging is reduced, e.g., from 1 h to 10 min. The extreme wind velocity also increases with increased turbulence intensity.

The above wind profiles should be used with care in heights above approximately 100 m as the assumption of constant turbulent fluxes above this level may be doubtful. Few wind measurements of the mean wind profile above this level exist. An example on data beyond this level is *rawinsonde* data, as discussed by Furevik and Haakenstad (2012).[2] Another option to measure the wind speed at high elevations is by using Lidars; see Section 2.1.5.3.

2.1.2.2 Stability

The mean velocity profiles discussed in the previous section are obtained assuming constant turbulent fluxes in vertical direction and neutral stability of the atmosphere. During unstable and stable atmospheric conditions, the profiles must be modified.

The atmospheric stability is related to the vertical temperature profile in the air. In simple terms, the stability condition may be understood by considering a volume of air (an air parcel) at a certain vertical level. In the initial position of the parcel, the temperature inside the parcel is equal to the temperature of the surroundings. Moving this parcel upwards causes the volume to expand due to the reduced pressure. The expansion implies work is done. Assuming the expansion is adiabatic, i.e., there is no

[2] A rawinsonde is a balloon equipped with instruments that is released from the ground or a ship. It measures parameters such as pressure, temperature, relative humidity and position (by GPS) on its way up to altitudes of greater than 30 km. Sometimes a rawinsonde is called a radiosonde; however, a radiosonde normally does not measure position (see the National Weather Service website, www.weather.gov/; accessed December 2022).

heat exchange with the exterior, the temperature in the parcel is reduced. If this new temperature is equal to the surrounding temperature, the density of the air in the parcel is equal to the density of the surrounding air and there is an equilibrium, or neutral stability. If, on the other hand, the temperature inside the expanded parcel is lower than the external temperature, the density of the air in the parcel is higher than the density of the surrounding air. The parcel will thus tend to sink back down to its original position, i.e., the conditions are stable. However, if the temperature of the air parcel in the new vertical position is higher than the surrounding temperature, the air parcel will tend to move further upward, i.e., the conditions are unstable.

Assuming dry air, the adiabatic expansion or compression of the air parcel will behave according to the ideal gas law:

$$\begin{aligned} pV &= nRT \\ pV^{\gamma} &= \text{Constant.} \end{aligned} \quad [2.8]$$

Here, n is the number of moles in the volume, R is the universal gas constant, V is the volume considered, p is the absolute pressure, T is the absolute temperature and $\gamma = c_p/c_v$ is the gas constant for air. c_p is the specific heat under constant pressure, while c_v is the specific heat under constant volume. For dry air, $\gamma = 1.40$. Combining the expressions in [2.8], the temperature T_1 at pressure p_1 is obtained as:

$$T_1 = T_0 \left(\frac{p_0}{p_1} \right)^{\frac{1-\gamma}{\gamma}}. \quad [2.9]$$

T_0 and p_0 define an initial state of the gas.

Vertical Temperature Variation

Assume the temperature at ground level is $20^{o}C$ or $293\ K$. *Normal air pressure* $p_0 = 1013\ hPa$. *With an air density of* $1.225\ kg/m^3$, *the pressure difference at 100 m versus at ground level becomes* $\Delta p = -\rho_a g \Delta z = -1.225^{*}9.80665^{*}100 = -1201.3\ Pa$. *The temperature at 100 m elevation, assuming adiabatic expansion, becomes:*

$$T_{(z=100m)} = T_{(z=0m)} \left(\frac{p_0}{p_0 + \Delta p} \right)^{\frac{1-\gamma}{\gamma}} = 292.00\ K.$$

I.e., under the assumption of dry air, the temperature is lowered by 1°C per 100 m of increased elevation. The requirement of "dry air" is strict; the relation holds also if no phase change takes place and there is no heat transfer by radiation. Note that the change in density with pressure has been ignored as we are considering small pressure differences.

If the air contains a lot of water vapour, condensation of water may take place as the air is cooled. The condensation releases heat, resulting in a lower temperature decay. The effect is temperature- and pressure-dependent, but in most cases the lapse rate is in the range of 0.5–1.0°C per 100 m.

In studying the atmospheric boundary layer, it is convenient to use the potential temperature, θ, rather than the sensible (measured) temperature, T. To obtain the potential temperature, assume that the air at any $z-$ level is moved adiabatically downwards to the ground level. The temperature the air then will obtain is the potential temperature. Assuming dry air, the potential temperature becomes:

$$\theta = T_z \left(\frac{p_z}{p_0} \right)^{\frac{1-\gamma}{\gamma}}. \quad [2.10]$$

Thus, in a neutral stratified atmosphere, the potential temperature is constant with height.

Over land, the ground is heated during the day and cools at night. Over sea, the diurnal variation vanishes, as the incoming radiative energy is efficiently distributed over a large volume of water. Therefore, over the sea, the synoptic weather situation, e.g., cold air advection with northerly winds (in the northern hemisphere) over relatively warm water, or warm air advection with southerly winds over relatively cooler water, and seasonal effects control the stability of the marine atmospheric boundary layer. The first example, cold air over warm water, happens most frequently during autumn and winter and causes an unstable or convective situation, while the case of warm air over cold water occurs more frequently during spring and summer, causing a stable situation. In Figure 2.4, a simplistic illustration of the vertical profile of the potential temperature in the case of stable, neutral and unstable (convective) conditions is given. Figure 2.6 illustrates the vertical variation of the mean wind speed for the unstable, neutral and stable conditions. Here the simple exponential velocity profile, [2.6] is assumed.

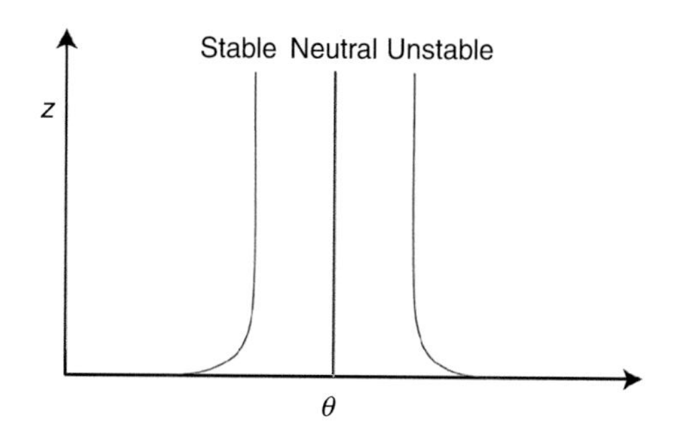

Figure 2.4 Simplistic illustration of the vertical distribution of the potential temperature in the surface layer. Stable, neutral and unstable conditions.

During unstable, or convective, conditions, the air close to the sea surface will be warmed by the sea and tend to move upwards, causing vertical mixing due to the buoyancy effect. The lower part of the boundary layer, in the region with a negative gradient of the potential temperature, the combined effect of turbulent shear stresses and buoyancy effects gives good vertical mixing. Above the surface layer, in the mixed layer, the potential temperature is almost constant. In the mixed layer, the vertical gradient of the mean velocity is low. The mixed layer may extend to about 1 km above sea level. Above the mixed layer, normally an inversion or capping layer exists, where the potential temperature increases and thus partly blocks the mixing of air into the free atmosphere. During unstable conditions, the surface layer may become fairly thick, in the order of 100–200 m. An illustration of the potential temperature through the various layers is given in Figure 2.5 (left).

During stable conditions the potential temperature increases from the sea level upwards. As the velocity gradient is large close to the sea surface, the vertical mixing is efficient in this region. Further up, however, the mixing is suppressed by the positive gradient of the potential temperature and the mixing diminishes. A vertical distribution of potential temperature through the various layers is illustrated in Figure 2.5 (right). During stable conditions, the surface layer is thinner than during unstable conditions, perhaps even less than 50 m. Above the surface layer, a residual layer is present where the potential temperature has a slightly positive gradient and the turbulent mixing is low.

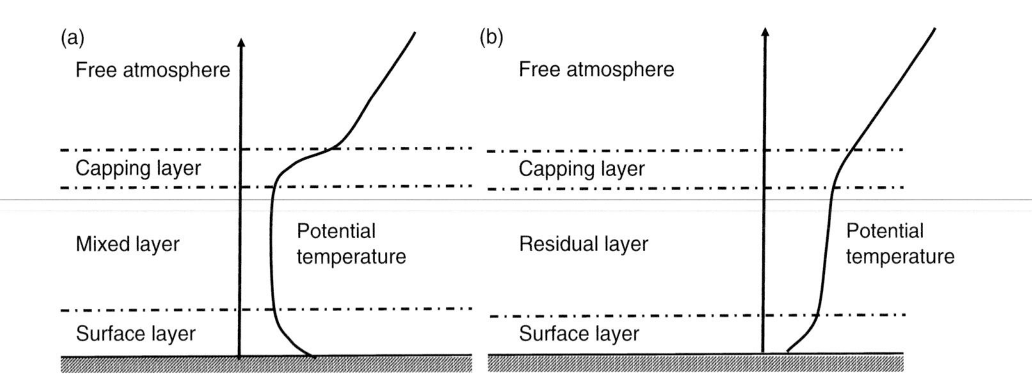

Figure 2.5 Illustration of the potential temperature variation through the various layers of the atmospheric boundary layer during unstable (convective; left) and stable (right) conditions. Based on Lee (2018). Reproduced with permission of Springer eBook.

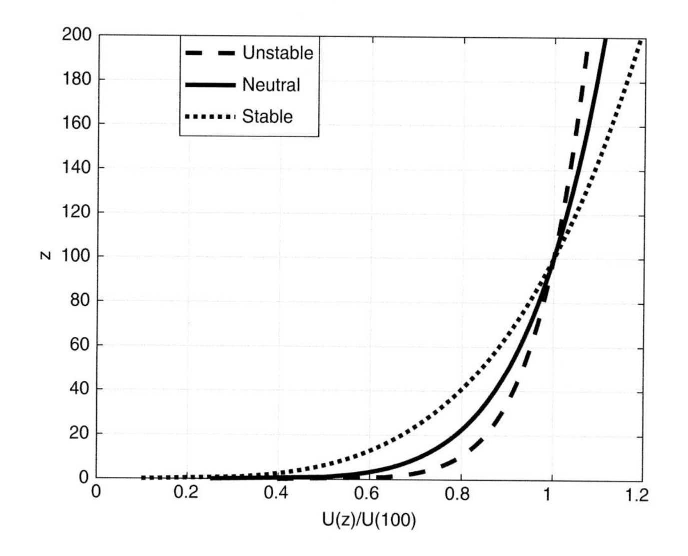

Figure 2.6 Illustration of mean wind profiles in the surface layer at different atmospheric stability conditions, assuming same mean velocity at height 100 m.

Coriolis forces act on the wind field. As the mean wind speed in general increases with height, Coriolis forces causes the wind direction to change with height. In the northern hemisphere, the wind direction at ground level is directed to the left relative to the wind direction above the boundary layer, the geostrophic wind.[3]

There are various ways to characterize the atmospheric stability conditions. A simple, qualitative assessment of the static stability can be performed by considering the gradient of the potential temperature.[4] $\partial\overline{\theta}/\partial z < 0$ corresponds to unstable conditions, $\partial\overline{\theta}/\partial z = 0$ corresponds to neutral conditions, while $\partial\overline{\theta}/\partial z > 0$ corresponds to stable conditions.

A physical, consistent way to quantify the degree of stability of the boundary layer is to compare the contribution by velocity shear and buoyancy effects to the production of turbulent kinetic energy. The buoyancy effect may, depending upon the sign of the surface heat flux, both increase and decrease the kinetic turbulent energy. The flux Richardson number – see, e.g., Lee (2018) – expresses the ratio between the buoyancy and velocity shear effects in the production of turbulent kinetic energy. The flux Richardson number is written as:

[3] Above the boundary layer, the wind direction is determined from a balance between the pressure gradient force and the Coriolis force. The Coriolis force deflects the wind direction away from the direction of the pressure gradient (toward the right in the northern hemisphere) until the wind direction is parallel to the isobars. This is denoted as geostrophic wind.

[4] In the calculation of potential temperature, dry air is assumed. Frequently, the virtual potential temperature is used. In the virtual potential temperature, the humidity of the air is accounted for.

$$R = \frac{\text{buoyancy production}}{\text{shear production}} = \frac{\frac{g}{\theta}\overline{w'\theta'}}{\overline{u'w'}\frac{\partial \overline{u}}{\partial z}}. \qquad [2.11]$$

Shear production is the shear stress times the gradient of the mean velocity. θ' denotes the turbulent potential temperature fluctuations, similar to the velocity fluctuations. The relation is derived by using the two-dimensional Navier–Stokes equation. In deriving the relation for the buoyancy production, the vertical pressure gradient set to $-\rho g$ and the relation between the density and the temperature is according to the ideal gas law. The minus sign in the shear production term (see Appendix A) has been omitted. $R = 0$ corresponds to no buoyancy effects and thus a neutral condition. $R < 0$ corresponds to positive buoyancy effects and thus an unstable condition. At $R = -1$ the turbulence production from buoyancy effects and the velocity shear effect are equal. $R > 0$ indicates a stable condition. Theoretically, $R = 1$ represents an upper limit at which the destruction of turbulence caused by the buoyancy balances the production due to the vertical shear.

The flux Richardson number is height-dependent. This is not explicitly shown in [2.11]. However, as discussed above, in the surface layer, where constant vertical flux of turbulence may be assumed, the velocity gradient may be expressed by [2.3]. Further, using that $-\overline{u'w'} = u_*^2$ in the surface layer, the flux Richardson number may be written as:

$$R = \frac{\frac{g}{\theta}\overline{w'\theta'}}{-\frac{u_*^3}{k_a z}} = \frac{z}{-\frac{u_*^3}{k_a \frac{g}{\theta}\overline{w'\theta'}}} = \frac{z}{L}. \qquad [2.12]$$

Here, L denotes the Obukhov length (also denoted the Monin–Obukhov length). As the Obukhov length is expressed by the surface friction and the vertical turbulent heat flux, it is independent of height (in the region of validity of the assumptions applied). It is therefore frequently used to classify the stability of the boundary layer. Van Wijk et al. (1990) propose the following ranges for characterization of stability.

Very stable:	$0\ m < L < 200\ m$
Stable:	$200\ m < L < 1000\ m$
Near neutral:	$\lvert L \rvert > 1000\ m$
Unstable:	$-1000\ m < L < -200\ m$
Very unstable:	$-200\ m < L < 0\ m$

A discussion of measured offshore wind speeds, turbulence and stability is found in Nybø et al. (2019) and Nybø et al. (2020). In Figure 2.7, the various regions of stability are shown as a function of the inverse of the Obukhov length.

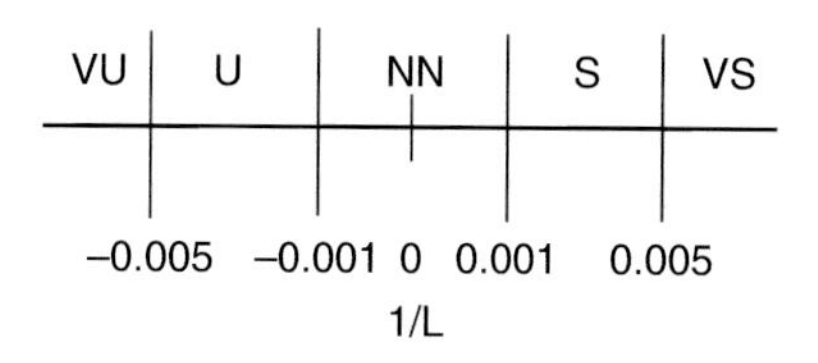

Figure 2.7 Regions of stability as a function of the inverse of the Obukhov length, $1/L$. VU: very unstable; U: unstable; NN: near neutral; S: stable; VS: very stable.

As the quantities used in the flux Richardson number are not readily available, an alternative is to use the gradient Richardson number. Here, the ratio between the gradient of the potential temperature and the gradient of the square of the mean wind speed is considered. The flux Richardson and gradient Richardson numbers are related via the eddy diffusivities for eddies and heat; for details, see Lee (2018).

2.1.2.3 Shear Exponent and Stability

Over land, the shift between stable and unstable conditions in the atmospheric boundary layer frequently follows a diurnal period. As mentioned in Section 2.1.2.2, the heat exchange between sea and atmosphere differs from that between land and atmosphere. This affects the stability condition of the atmospheric boundary layer and thus the vertical shear profile of the mean wind speed. Figure 2.8 illustrates how the shear exponent α in [2.6] may vary over the year at an offshore location in the North Sea (Myren, 2021). The results are obtained by considering 11 years of hindcast data in the NORA3 dataset (Haakenstad et al., 2021). In Figure 2.8 it is also observed that the variation shear exponent closely follows the temperature difference between 100 m elevation and sea level, $\Delta T = T_{100} - T_0$. With this difference less than -1°C, stable atmospheric conditions may be assumed. In the case shown in Figure 2.8, this happens most frequently during autumn and winter. The study by Myren (2021) showed that the variations over the year were less pronounced as the location moved farther north and farther offshore. This may be explained by lower variation between sea and air temperature in these regions.

It should be noted that the shear exponents in Figure 2.8 in general are lower than recommended in many standards. DNV (2021c) recommends a general α- value over open sea with waves offshore equal to 0.12. This value does not account for atmospheric stability. However, DNV (2021c) gives advice on how the shear profile may be adjusted considering the Obukhov length. The IEC standard 64100-3 (2009) recommends the "normal wind profile" over sea $\alpha = 0.14$, using the reference height equal to hub height. For extreme wind speeds averaged over 3 s and a return period of 50 years, IEC 64100-3 (2009) recommends $\alpha = 0.11$ and use of a gust factor of 1.1. No correction for stability conditions is included in this case.

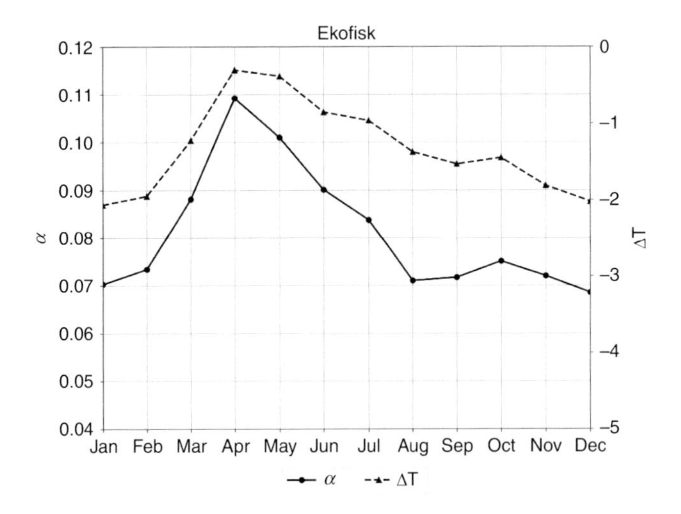

Figure 2.8 Hourly mean shear exponent, α, at the Ekofisk area (left axis). α obtained from NORA3 hindcast data (Haakenstad et al., 2021). The dashed line (right axis) shows the mean temperature difference between 100 m above sea level and sea surface. Time period 2004–2015. Courtesy of Myren (2021).

Furevik and Haakenstad (2012) studied a large number of offshore wind conditions where both measured and hindcast wind profiles were available. They defined the stable, near neutral and unstable conditions from the temperature difference ΔT_{150} between the temperature 150 m above sea level and the sea temperature. The unstable, near neutral and stable conditions were defined by $\Delta T_{150} < -1$, $-1 \leq \Delta T_{150} \leq 0$, $\Delta T_{150} > 0$. ΔT_{150} in °C. The corresponding average α-values were found as 0.04, 0.05 and 0.09. These α-values are, as the hindcast data above, lower than recommended by the standards. In all standard wind energy applications an increasing wind speed with height has been assumed. Furevik and Haakenstad (2012) observed frequently (in more than 1500 of 8700 cases) a decreasing wind speed with height. In general, these cases had unstable atmospheric conditions. The observations were made in the North Atlantic on the weather ship *Polarfront*.[5]

2.1.2.4 Turbulence

As discussed above, the wind speed at a specific point may be characterized by the mean value plus a stochastic variation, $u = \overline{u} + u'$. The turbulence level is characterized by the standard deviation of the wind speed, $\sigma_u = \sigma_{u'}$. Also, the transverse and the vertical components of the velocity fluctuations are of importance. However, traditionally, and partly due to the measurements available, the wind

[5] *Polarfront* was a Norwegian weather ship located in the North Atlantic at 62° north, 2° east until the end of 2009. See "Polarfront" in *Store norske leksikon*, https://snl.no/Polarfront (accessed November 2021).

energy community has considered fluctuations in the mean wind direction only. The turbulence intensity is thus normally written as:

$$TI = \frac{\sigma_u}{\bar{u}}. \quad [2.13]$$

The numerical value of σ_u is sensitive to the length of the record considered. Traditionally, records of 10 min duration have been used in the wind industry. However, using longer records, e.g., 1 h, increases the computed value of σ_u as the wind spectra contain significant energy at low frequencies. Also, real wind data are never really stationary. Nybø et al. (2019) discuss ways to handle real wind data for use in the analysis of wind turbine dynamics.

Considering a short period of time, e.g., from 10 min to 1 h, the time history of the wind speed may be considered a stationary process. Several formulations of wind spectra are proposed. The most common spectra used for offshore conditions are the Kaimal, von Kármán, Davenport and Harris spectra; see, e.g., DNV (2021c). For example, the Kaimal spectrum may, according to IEC 61400 (2005), be written as:

$$S_k(f) = \sigma_k^2 \frac{A \frac{L_k}{U_{ref}}}{\left(1 + B \frac{fL_k}{U_{ref}}\right)^{5/3}}. \quad [2.14]$$

f is the frequency and L_k is a length scale parameter. U_{ref} is the mean wind velocity at a reference height, z_r, e.g., the hub height. The index k refers to the velocity components: $k = 1$ mean wind direction, $k = 2$ lateral direction and $k = 3$ vertical direction. IEC 61400 (2005) recommends the values given in Table 2.1 for the parameters, depending upon direction.

The length scale parameter is given as:

$$\Lambda_1 = \begin{cases} 0.7\, z_r & \text{for } z_r < 60\, m \\ 42\, m & \text{for } z_r \geq 60\, m \end{cases}. \quad [2.15]$$

A = 4 and B = 6 are recommended values. Note that in all wind spectral formulations, the high-frequency tail of the spectrum should be proportional to $f^{-5/3}$.

Table 2.1 *Parameters to be used in the Kaimal spectrum according to IEC 61400 (2005)*

Direction, k	**1**	**2**	**3**
Standard deviation, σ_k	σ_1	$\sigma_2 \geq 0.7\sigma_1$	$\sigma_3 \geq 0.5\sigma_1$
Integral length scale, L_k	$L_1 = 8.1\Lambda_1$	$L_2 = 2.7\Lambda_1$	$L_3 = 0.66\Lambda_1$

The wind spectrum may be divided into three ranges: a low-frequency range where large-scale turbulence is generated, denoted as the production range; a medium-frequency range where there is a balance between energy gained from the larger eddies and the energy lost to even smaller eddies, denoted as the inertial subrange; and a high-frequency range where the eddies are so small that the energy is lost in viscous dissipation, denoted as the Kolmogorov dissipation range. The order of magnitude length and time scales for each of these ranges are indicated in Figure 2.1. Experience shows that spectra such as, e.g., the Kaimal formulation represent the measured spectra well in the inertial subrange, at least in near neutral and unstable conditions. The length and time scales in the dissipation range are so small that they do not have any practical impact on wind turbines. However, the energy content in the lower range of the inertial subrange and in the production range may not be well represented by the "standard" spectra. This is a frequency range important to floating structures due to the low natural frequencies for such structures.

Figure 2.9 gives examples of the Kaimal spectrum. For frequencies above about 0.1 Hz, the slope of the spectrum is close to $f^{-5/3}$. Observe that changing the turbulence intensity while keeping the mean velocity causes a vertical shift in the spectral curve.

The formulation in [2.14] represents the spectrum in the main wind direction and uses the standard deviation in that direction, σ_u. For the lateral and vertical velocity components, the design standards propose the use of the same spectral formulation, but with modified standard deviations (see Table 2.1).

The turbulence intensity, as defined by [2.13], is in general assumed to decrease with increasing mean wind velocity. Nybø et al. (2019) and Nybø et al. (2020) investigated

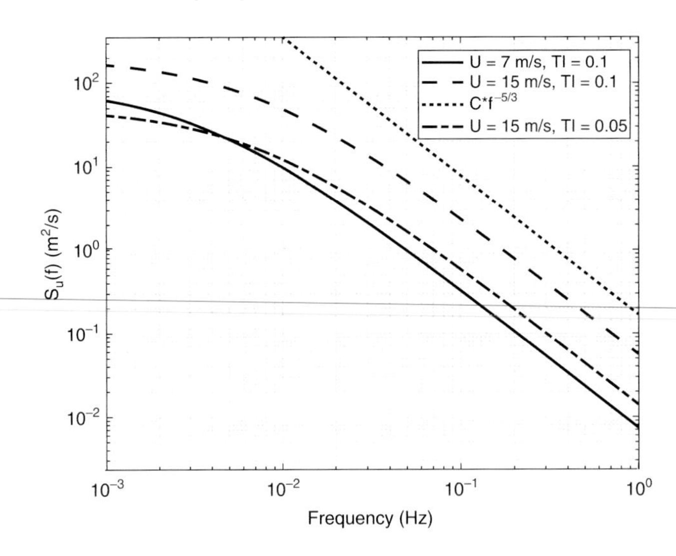

Figure 2.9 The Kaimal spectrum for the velocity in the mean wind direction according to IEC 61400 (2005). Reference height $z_r = 100\,m$. Double logarithmic scale.

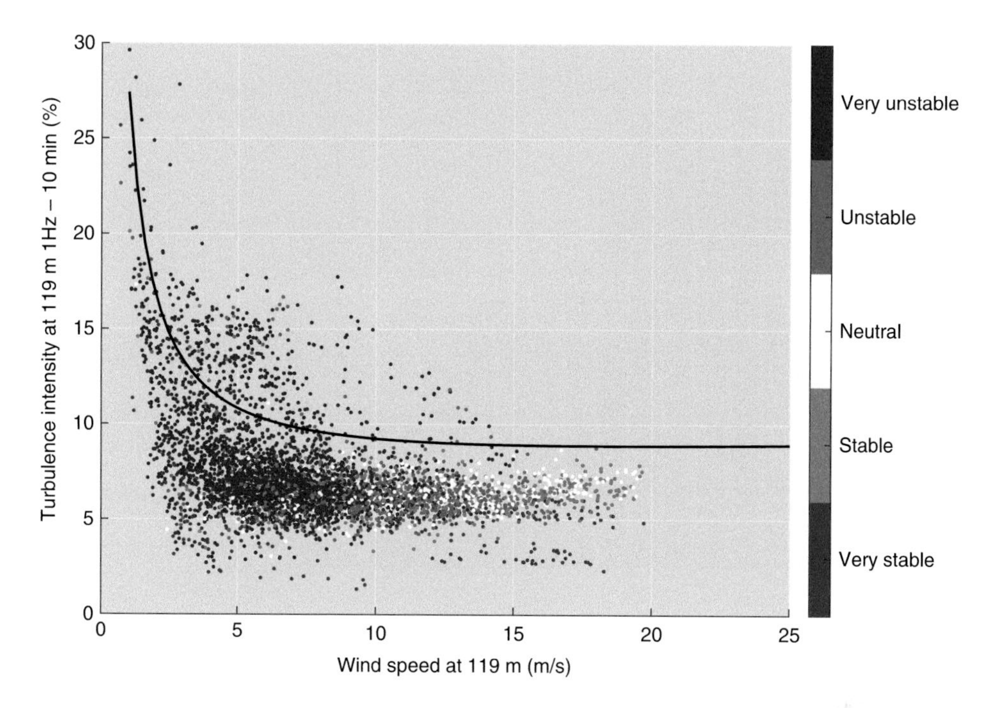

Figure 2.10 The turbulence intensity (TI) as a function of mean wind speed, from Nybø et al. (2020). The solid black line represents the 90th percentile as given in the IEC (2009) standard for offshore conditions with reference TI equal to 0.12. Reproduced from Nybø et al. (2020) under Creative Common Attribution License No. 5460641106526.

about one year of wind data from the offshore meteorological measurement platform FINO1 (www.fino1.de/en/; accessed July 2023). Careful quality control of the data was performed prior to the analysis. The quality control implies removing erroneous data due to spikes, missing samples, mast shadow effects etc., as well as implementing requirements related to the stationarity of the records. In Figure 2.10, turbulence intensities for the FINO1 data at 119 m above sea level are plotted for a large number of 60-min records. The TI values are averaged over six consecutive 10-min records sampled at 10 Hz. As shown in the figure, the turbulence intensity is sensitive to the stability of the atmospheric boundary layer. Nybø et al. (2020) used the classification of stability based upon the Obukhov length as given in Section 2.1.2.2. Figure 2.10 also includes turbulence intensities as recommended in the IEC standard 61400-3 (2009) for offshore wind turbines. These values are supposed to represent the 90th percentile of measured turbulence intensities. The 90th-percentile curves as given in the IEC standard decay monotonically with increasing wind speed. In Figure 2.11 the distribution of stability conditions in the data of Nybø et al. (2020) are plotted as function of mean wind speed. At very low and very high wind speeds, the few occurrences introduce large uncertainties in the distribution.

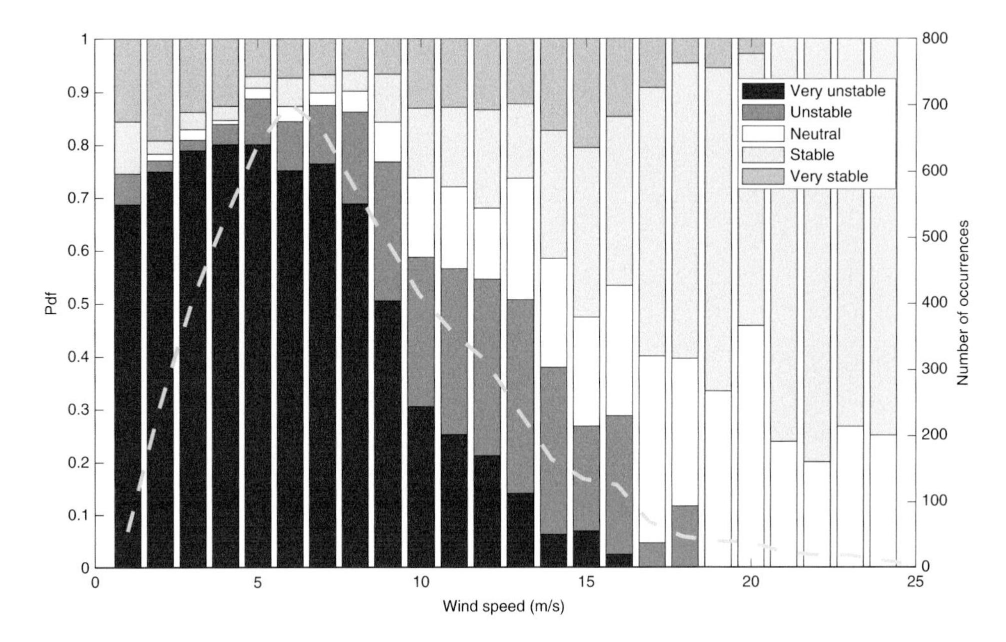

Figure 2.11 Atmospheric stability as a function of mean wind speed at 80 m above sea level. FINO1 data as processed by Nybø et al. (2020). The number of occurrences within each wind speed interval is given by the solid black line (right axis). The stability limits given in Section 2.1.2.2 are used. Copied from Nybø et al. (2020) under Creative Common Attribution 3.0 License No. 5460641106526.

In contrast to onshore conditions, the surface roughness offshore increases with wind speed. The IEC standard assumes the surface roughness can be solved iteratively from the expression $z_0 = A_c/g[\kappa\overline{U}_{hub}/\ln(z_{hub}/z_0)]^2$. Here, A_C is the Charnock parameter and κ is von Kármán's constant. The standard deviation of the wind in the mean wind direction is written as $\sigma_1 = \overline{U}_{hub}/\ln(z_{hub}/z_0) + 1.84 I_{15}$. Here, I_{15} is the turbulence intensity at hub height at 15 m/s wind speed. In Figure 2.10, the IEC turbulence intensity curve is included, using $A_C = 0.011$, $\kappa = 0.4$ and $I_{15} = 0.12$, corresponding to Class C wind turbines. The surface roughness length varies in this case from 4.8E-07 at 1 m/s wind speed to 7.9E-04 at 25 m/s wind speed. The measurements reveal a large scatter in turbulence intensity for the various records analyzed. The turbulence intensity during unstable atmospheric conditions is in general higher than during neutral and stable conditions.

Classical theory describes how the wind shear over the ocean surface creates surface waves, starting from short, capillary waves. As the duration of the wind and or the fetch length increase, the waves become longer, ending up as long-periodic swells with periods beyond 15 s. The understanding of how waves cause variation of the wind speed over an ocean surface is less developed. The waves interact with the wind field both by representing a wavy boundary condition and by a time-varying drag force. The drag force between the air and water is related to the relative velocity between the two media. As the water particles moves back and forth, a time-varying drag force will result. Kalvig (2014) studied the effect of the moving boundary and found that long swells could cause variation in the wind speed at vertical levels corresponding to the rotor of an offshore wind turbine. However, more research is needed to fully understand the impact of the waves on the wind field.

2.1.2.5 Coherence

The above discussion on the turbulence spectra is confined to the velocity variation in time observed in one point. To compute the wind loads on a wind turbine, the velocity variations in space are also of importance. To describe the spatial variations, the coherence is used. The coherence of a wind field tells how the variation in wind speed at one point correlates to the wind speed at another point. Consider the rotor plane of a wind turbine. If the rotor diameter is small, it may be assumed that the wind speed is fully correlated over the rotor plane. This holds at least for the frequency ranges important for estimating the power production and dynamic loads. As seen from the power spectra in Figure 2.9, high frequencies correspond to low energy content.

As the diameter of the rotor is increased to beyond 200 m, the issue of correlation of the turbulence over the rotor plane beomes increasingly significant. The variation in the $u-$ velocity over the rotor plane is obviously important. Due to the shear in the mean velocity profile, it is to be expected that the correlation in the u-velocity for points separated in the horizontal direction differs from the correlation between points separated in the vertical direction. Variations in the v and w turbulent velocity components over the rotor plane have had less attention but do also influence the rotor blade loads. In the following, the discussion is limited to the variation in the u-velocity component.

Consider two points, 1 and 2, separated by a distance r_{12}. The auto-spectra for the wind velocity at points 1 and 2 may be denoted $S_{11}(f)$ and $S_{22}(f)$. Similarly, the cross-spectrum for the velocities at points 1 and 2 may be denoted $S_{12}(f)$. The coherence may now be written as:

$$\gamma(f, r_{12}) = \frac{S_{12}(f)}{\sqrt{S_{11}(f)S_{22}(f)}}. \qquad [2.16]$$

The auto-spectra are real and positive, and in most cases almost equal in the two points considered. However, the cross-spectrum is in the general case a complex quantity. The coherence is therefore also complex, i.e., it contains information about the phases or the time shift between the two time series considered. The real and imaginary parts of $\gamma(f,r)$ are denoted the co-coherence and quad-coherence respectively. In most wind energy applications the absolute value of [2.16] is used as the coherence, i.e., the coherence is written as follows (Burton et al., 2011):

$$coh(f,r) = |\gamma(f,r_{12})| = \frac{|S_{12}(f)|}{\sqrt{S_{11}(f)S_{22}(f)}}. \qquad [2.17]$$

In most wind standards an exponential coherence function is proposed, e.g., IEC (2005) proposes a coherence on the form:

$$coh(f,r) = \exp\left[-12\left(\left(\frac{fr}{U}\right)^2 + \left(0.12\frac{r}{L_k}\right)^2\right)^{0.5}\right]. \qquad [2.18]$$

Here, U is the mean wind velocity, to be taken at hub height. $L_k = 8.1\Lambda_1$ is the coherence scale parameter; see [2.15]. [2.18] is supposed to be valid for the $x-$ component of the velocity and describes the coherence between two points located in the same $x-y$ plane (e.g., the rotor plane) at a distance r. From [2.18] it is observed that the coherence according to this formulation is always real and positive. It is also observed that the coherence tends to $\exp(-1.44r/L_k)$ as the frequency tends to zero. For other coherence models, e.g., the Davenport model (Davenport, 1962), the coherence converges toward unity for zero frequency.

The above formulation of the coherence is to be combined with a model of the point spectrum of the wind, e.g, the Kaimal turbulence spectrum [2.14]. Using the "Sandia" method (Veers, 1988), a complete wind field satisfying the point spectrum as well as the coherence function may be generated. Further details are given below. Frequently, only the variation of the $u-$ component of the wind is considered in these models.

2.1.2.6 Mann's Turbulence Model

Based upon a linearized version of the Navier–Stokes equations, Mann (1994) developed a tensor-based model for the spatial structure of the turbulence in the surface layer of the atmosphere. This is the original version of Mann's turbulence model. The effect of the earth's rotation, i.e., Coriolis forces, is ignored. To come up with a model for all three turbulent components of the wind field, several important simplifying assumptions are made. Key assumptions are that the air is assumed incompressible; further, a neutral stability of the atmosphere and a uniform shear of the mean velocity are assumed, i.e.:

$$\bar{u}(z) = z\frac{d\bar{u}}{dz}, \qquad [2.19]$$

with $d\bar{u}/dz$ constant. The model uses von Kármán's turbulence spectrum with isotropic turbulence as a starting point. The effect of shear is included by the so-called "rapid distortion theory" (RDT). The RDT describes how the sheared wind field distorts eddies. Eddies with rotation are either stretched or compressed along the axis of rotation depending upon the direction of rotation. Further, large eddies are assumed to live for longer than small eddies. Small eddies are assumed to be isotropic, simplifying the spectral tensor considerably. Large eddies are assumed to be anisotropic, causing the three components of the velocity fluctuations to differ and fulfil the relation:

$$\sigma_u > \sigma_v > \sigma_w. \tag{2.20}$$

Further, the ensemble average of the product of the horizontal and vertical turbulent velocities, $\langle u'w' \rangle$, becomes negative, as expected from the 2D boundary layer equations; see Appendix A.

If the fluctuation of the wind speed is measured at a fixed point in space, a time-history of the wind speed is obtained and, by spectral analysis, a frequency spectrum is obtained, as shown in Figure 2.9. In the spectral formulation by Mann (1994), a wave number spectrum is used rather than a frequency spectrum. The instantaneous wind speed variations along the x-axis at a certain point in time are considered. Computing the spectrum of these wind speeds, $u(x)$, $v(x)$, $w(x)$, a spectrum based upon wave number is obtained. Invoking the Taylor's frozen turbulence hypothesis implies that the turbulent structures are assumed to be moving downstream with the average wind speed, i.e.:

$$\mathbf{u}(x + \bar{u}\Delta t, y, z, t + \Delta t) = \mathbf{u}(x, y, z, t). \tag{2.21}$$

Bold types denote a vector. Denoting the spectrum based upon wave number $F(k)$, the relation between the frequency and wave number representation is given by $S(f)df = F(k_1)dk_1$ with $k_1 = 2\pi f/\bar{u}$. Here, k_1 is the wave number along the $x-$ axis. The wave numbers along the y and z axes are independent of the mean velocity.

The model uses the covariance tensor as a basis, denoting the covariance tensor as:

$$R_{ij}(\mathbf{r}) = \langle u_i(\mathbf{x}) u_j(\mathbf{x} + \mathbf{r}) \rangle, \tag{2.22}$$

where $\mathbf{x} = (x, y, z)$ is the position vector, $\mathbf{r}$ is a distance vector and u_i are the turbulent velocity components, and $\langle \rangle$ denotes ensemble average. In a homogeneous flow field, the covariance tensor depends upon the absolute value of the distance only. Similarly, as the ordinary frequency spectrum is obtained from the

Fourier transform of the auto-correlation function, a spectral tensor is obtained from the Fourier transform of the covariance tensor, i.e.:

$$\Phi_{ij}(\mathbf{k}) = \frac{1}{(2\pi)^3} \iiint R_{ij}(\mathbf{r}) e^{(-i\mathbf{k}\cdot\mathbf{r})} dr_1 dr_2 dr_3. \qquad [2.23]$$

The integrals are taken from $-\infty$ to $+\infty$. $\mathbf{k} = (k_1, k_2, k_3)$ denotes the wave number vector. The spectral tensor is transformed to an orthogonal process, and the velocity field is then obtained from the Fourier transform of this process. Mann (1994) shows how the spectral tensor and $F_i(k_i)$ can be obtained assuming an isotropic turbulence, using von Kármán's energy spectrum. Then, introducing a vertical shear, the spectral tensor is modified, and an anisotropic flow is obtained. The degree of anisotropy is determined by a parameter controlling the length of life of the eddies.

Three key parameters are used in Mann's (1994) formulation of the wind field: a length parameter L to describe the characteristic size of eddies; a parameter Γ to characterize the length of life of eddies; and a viscous dissipation rate for the turbulent kinetic energy.

Starting out with von Kármán's turbulence spectrum, the point spectrum for the $u-$ velocity is obtained as:

$$F_u(k_1) = \frac{9}{55} \alpha \varepsilon^{\frac{2}{3}} \frac{1}{\left(L^{-2} + k_1^2\right)^{5/6}}. \qquad [2.24]$$

Here, $\alpha\varepsilon^{2/3}$ is the parameter characterizing the viscous dissipation rate for the turbulent kinetic energy. k_1 is the wave number in $x-$ direction. ε is the specific turbulent dissipation as given from the kinematic viscosity and the kinetic turbulent energy $\varepsilon = \nu\overline{\left(du_i/dx_j\right)}$ (see Appendix A). α is an empirical constant equal to approximately 1.7.

From [2.24] it is observed that the characteristic length, L, may be obtained from $F_u(0)$. However, Mann (1994) comments that the value of the spectrum at low wave numbers (low frequencies) is strongly influenced by non-stationarity and large-scale meteorological phenomena. He therefore suggests using the maximum of $k_1 F(k_1)$ as reference. Considering the $u-$ component of the velocity, it is found that:

$$L \simeq \frac{1.225}{k_{1\max}}, \qquad [2.25]$$

where $k_{1\max}$ corresponds to the maximum value of $k_1 F(k_1)$.

The linearization of the Navier–Stokes equations causes unrealistic behavior of the flow and the eddies to be distorted beyond what is physical realistic. Therefore, the parameter Γ, limiting the length of life of the eddies, is introduced. Eddies are supposed to break up as time goes on and small eddies break up faster than large eddies. Mann (1994) applies an assumption that the length of life of the eddies may be written in the form:

$$\tau(k) \propto k^{-\frac{2}{3}}\Lambda \propto \begin{cases} k^{-2/3} & \text{for} \quad k \to \infty \\ k^{-1} & \text{for} \quad k \to 0 \end{cases}. \qquad [2.26]$$

Λ is related to the hypergeometric function;[6] for details, see Mann (1994). The limit of τ as $k \to \infty$ is assumed valid in the inertial subrange. In the inertial subrange, [2.26] may be rewritten as a nondimensional lifetime as:

$$\beta(k) \equiv \frac{d\bar{u}}{dz}\tau(k) = (kL)^{-2/3}\Gamma. \qquad [2.27]$$

Γ is to be determined empirically. For $\Gamma = 0$, an isotropic flow is obtained, i.e., $\sigma_u = \sigma_v = \sigma_w$. As Γ increases, the differences between the three standard deviations increase. In Figure 2.12, the relation between Γ and the variances in the three

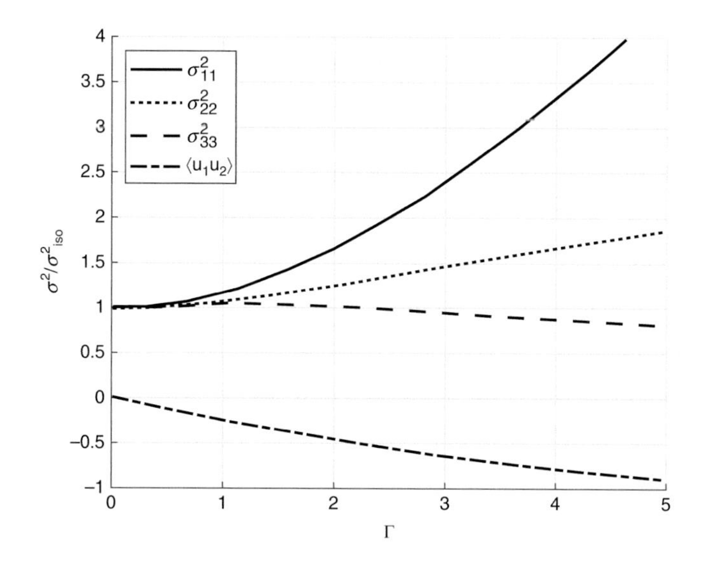

Figure 2.12 Relation between the parameter Γ controlling the length of life of eddies and the variances of the turbulence in the three directions as well as the covariance between horizontal and vertical velocity component. Reproduced from Mann (1994) with permission from the *Journal of Fluid Mechanics*, License No. 5431500056035.

[6] Hypergeometric functions represent solutions of a special group of second-order differential equations expressed as a series expansion; see, e.g., Wolfram MathWorld, https://mathworld.wolfram.com/HypergeometricFunction .html (accessed July 2023).

directions of the flow are shown. Further modifications to the model have been made by introducing a blocking effect, accounting for zero vertical velocity component at ground level.

In Mann (1998) and (1994), the above model was tested and the parameters estimated based upon full-scale measurements and for various model wind spectra. Cheynet, Jakobsen and Obhrai (2017) found, based upon offshore data at the FINO1 platform for wind velocities at 80 m above sea level and a wind speed range of 14–28 m/s, average values $\Gamma = 3.7$, $\alpha\varepsilon^{2/3} = 0.04\ m^{4/3}\ \mathrm{s}^{-2}$ and $L = 70\ m$. The estimated values for Γ and L are approximately constant over the range of velocities considered, while $\alpha\varepsilon^{2/3}$ shows a marked increasing trend, from about $0.02\ m^{4/3}\ \mathrm{s}^{-2}$ at the lowest velocity to $0.07\ m^{4/3}\ \mathrm{s}^{-2}$ at the highest velocity.

To account for non-neutral atmospheric stability, Chougule et al. (2018) introduce a modification of the original Mann model. A linear vertical variation of the potential temperature is introduced. One or alternatively two extra parameters are needed for this model.

When applying the Mann model for wind field simulations, the tensor description of the wind field in the wave number domain is transformed to an instantaneous wind field in the (x, y, z) domain for the three turbulent components (u, v, w). This is done by first transforming the spectral tensor, $\Phi_{ij}(\mathbf{k})$, into a set of orthogonal processes (functions). By Fourier transform and summation of these orthogonal functions, the wind field is obtained. Consistent with the frozen turbulence hypothesis, the wind turbine may be moved through this field to obtain a time-dependent inflow.

2.1.3 Numerical Generation of Wind Fields

Several options exist to generate wind fields that can be applied for analyzing wind turbines. First, an undisturbed wind field approaching a wind farm must be considered. Next, to analyze the wind field inside a wind farm, the effect of wakes has to be accounted for. Here, the generation for the free wind field is discussed briefly. The effect of wind turbine wakes is discussed in Chapter 9.

In Mann (1998), an efficient algorithm is given for computing the wind field according to the principles outlined above. One or more components of the turbulence may be generated assuming neutral atmospheric stability and constant vertical gradient of the mean wind speed. The result is a "wind field box" of specified size in (x, y, z). Realistic spectral shape and variances for the three components of the wind speed as well as the covariance of the horizontal and vertical velocity components are obtained.

Veers (1984) and (1988) describes the "Sandia method" for generating a wind field. In this method the three turbulent components are assumed to be uncorrelated and can thus be generated independently. The time histories of the wind speed at several points in a plane perpendicular to the mean wind direction are considered. The plane may coincide with the rotor plane. The time histories of the wind speed in each point are supposed to be Gaussian and the point spectrum of the wind in each point (point number n of total N) is given by a frequency spectrum, $S_n(f)$. In most cases the spectrum is assumed equal for all the points considered. The subscript n is thus omitted in the following. The spectrum is represented by a summation of M frequency components, each with amplitude $\sqrt{2S(f_m)\Delta f}$. Here, Δf is the frequency increment and a single-sided spectrum is assumed. The wind speeds at two points are not independent. The degree of correlation depends upon the distance between the points. Consider the rotor plane and denote two points in the plane, j and k, with the coordinates (x, y_j, z_j) and (x, y_k, z_k). The cross-spectrum for the wind speed between the two points can be written as:

$$S_{jk}(f_m) = \gamma(f_m, \Delta r_{jk})\sqrt{S_{jj}(f_m)S_{kk}(f_m)}. \quad [2.28]$$

$S_{jj}(f_m)$ is the auto-spectrum (point spectrum) in point j. As mentioned above, $S_{jj}(f_m) = S_{kk}(f_m)$ is normally assumed. Δr_{jk} is the distance between the two points. Note that the coherence depends upon the distance between the points and does not distinguish between vertical and horizontal separation. Taylor's frozen turbulence hypothesis is applied, i.e., the turbulent eddies, the velocity variations, are moved downstream with the mean wind velocity. Veers (1984) assumed that the coherence function is real and positive. This assumption was justified by the fact that within the typical rotor sizes considered, the phase shift in the velocity components over the rotor, even in vertical direction, is small. The assumption simplifies the computations significantly. The assumption of a real and positive coherence function contrasts with the results that are obtained by Mann's formulation, which may result in negative cross-spectra and an imaginary part of the cross-spectra different from zero.

The time history of the wind speed is to be generated in N points in the rotor plane. Each time history is assumed to be a filtered white noise process, obtained by summation of the M frequency components. Further, the correlation between the time series shall satisfy [2.28]. For each frequency, the spectral matrix $\mathbf{S}(f_m)$ contains the spectral value for all combinations of j and k altogether N^2 values. As the distance between two points is equal in both directions, $\mathbf{S}(f_m)$ is symmetric. In general, $\mathbf{S}(f_m)$ may be rewritten as the product between a complex transformation matrix and the transposed of its complex conjugate, i.e.:

$$\mathbf{S}(f_m) = \mathbf{H}(f_m)\mathbf{H}^{*T}(f_m). \quad [2.29]$$

As $\mathbf{S}(f_m)$ is assumed real, $\mathbf{H}(f_m)$ is real and equal to its complex conjugate. It is now assumed that $\mathbf{H}(f_m)$ is a lower triangular matrix. The non-zero values may then be determined by a set of recursive equations (see Veers, 1984), i.e.:

$$\begin{aligned} H_{11} &= S^{1/2} \\ H_{21} &= S_{21}/H_{11} \\ H_{22} &= \left(S - H_{21}^2\right)^{1/2} \\ H_{31} &= S_{31}/H_{11} \\ &\vdots \\ H_{jk} &= \left(S_{jk} - \sum_{l=1}^{k-1} H_{jl}H_{kl}\right)/H_{kk} \\ H_{kk} &= \left(S - \sum_{l=1}^{k-1} H_{kl}^2\right)^{1/2} \end{aligned} \quad [2.30]$$

This matrix is unique for each frequency, f_m. Veers (1988) states that "the elements of $\mathbf{H}$ may be thought of as weighting factors for the linear combinations of N independent, unit-magnitude, white noise inputs that will yield N correlated outputs with the correct spectral matrix. Each row of $\mathbf{H}$ gives the contribution of all the inputs to the output at point k."

The harmonic components along the diagonal of $\mathbf{H}$ at each frequency f_m are assumed to be uncorrelated. This is obtained by assuming each component has an independent, random phase θ_{jm} between 0 and 2π. A unit amplitude diagonal matrix $\mathbf{X}$ is thus introduced with the elements:

$$X_{jj}(f_m) = e^{i\theta_{jm}}. \quad [2.31]$$

The harmonic component of the velocity with frequency f_m in point n may thus be written as a linear sum of the contribution from the point itself and its neighbors, accounting for the lower diagonal structure of $\mathbf{H}$:

$$V_n(f_m) = \sum_{k=1}^{n} H_{nk}(f_m)X_{kk}(f_m) = \sum_{k=1}^{n} H_{nk}(f_m)e^{i\theta_{km}}. \quad [2.32]$$

Having established all the frequency components for each point, the time-history is obtained by a Fourier transform of V_n. For details and an example of usage, see Veers (1988).

In Nybø et al. (2021), the coherence of the u-component of the velocity as computed by various methods is compared. The results are illustrated in

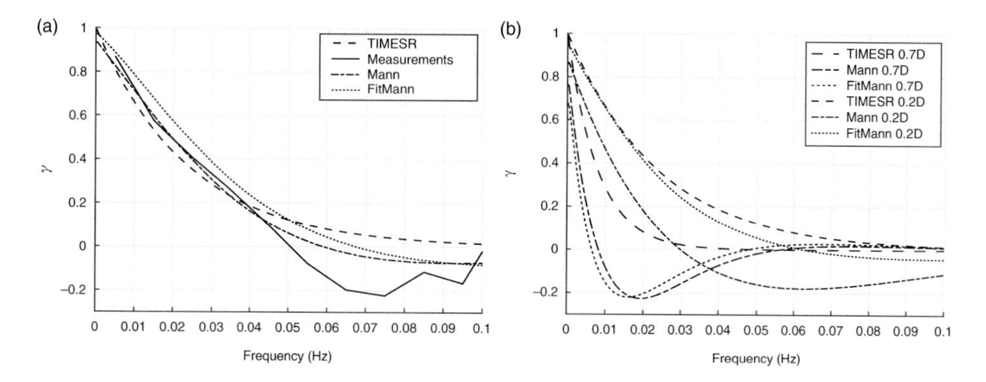

Figure 2.13 Computed coherence based upon offshore measurements and various wind field models. Left: vertical co-coherence of the *u*-component computed for 40 m vertical separation. Right: horizontal co-coherence of the *u*-component computed for 36 m (0.2D) and 125 m (0.7D) horizontal separation. Copied from Nybø et al. (2021) under Creative Commons Attribution 3.0 license.

Figure 2.13. The wind fields are computed by Mann's method, as described, using standard parameters as recommended from the design standards and fitting the turbulence level to measurements. Further, a version of Mann's model with fitting of the spectral shape is employed, known as "FitMann" (Cheynet, 2019). "TIMESR" uses the Sandia method, as described, with input from wind measured at three different vertical levels. The Davenport exponential coherence function is used with parameters fitted to the measurements. The specific condition considered in Figure 2.13 has close to neutral atmospheric stability. The vertical coherence is also computed directly from the measurements. The measurements are from an offshore meteorological mast, FINO1 (Nybø et al., 2019; Nybø et al., 2020).

From Figure 2.13 it is observed that the measured vertical co-coherence is negative in some ranges of frequencies. The observed negative co-coherence is partly captured by the two Mann's models. This is not the case for the Davenport coherence model, in which the co-coherence is forced to be positive. For the horizontal coherence, no measurements are available. However, Mann's model still predicts negative co-coherence in some ranges of frequencies. Measurements also show quad-coherence different from zero, in particular in the vertical direction. A vertical quad-coherence different from zero is also obtained by Mann's model. However, present practice in wind turbine design frequently ignores the quad-coherence.

In addition to the methods discussed above, numerical methods based upon solving the Navier–Stokes equations are used. Still, most of the methods available are too computationally demanding to be used directly in the design process.

However, for calibrating more simplistic methods and in studies of special flow phenomena, these methods are very useful.

The simplest version is the so-called Reynolds-averaged Navier–Stokes (RANS) solvers. As the name indicates, these methods average out all turbulent variations and give a picture of the time-averaged flow only. The effect of turbulence is parameterized to ensure proper shear and dissipation of energy in the flow. Such methods are of limited value in considering wind turbines.

The most relevant Navier–Stokes solver methods are the large-eddy simulation (LES) methods. In this approach the turbulent structures with sizes above a certain limit are resolved in the numerical model, while the smaller-scale turbulence is parameterized to ensure proper turbulent dissipation of energy. Two frequently used implementations are the Parallelized Large-Eddy Simulation Model (PALM; Maronga et al., 2015) and the Simulator fOr Wind Farm Application (SOWFA; see, e.g., Churchfield et al., 2012). These computational tools typically start from a meso-scale wind field that is refined by using a finer-grid scale in the wind farm area, and this locally refined flow model is coupled with a wind turbine model to compute power production, loads and wakes behind the turbines.

The starting point for the models is the three-dimensional momentum equation driven by a horizontal pressure gradient and accounting for Coriolis forces due to the rotation of the earth. The air may be considered incompressible, but with the buoyancy effect included by assuming that the vertical variation of the density of the air follows the variation in potential temperature, the so-called Boussinesq approximation. The density of the air at some vertical level, z, may thus be written as:

$$\rho(z) = \rho(z_r)\left[1 - \frac{\overline{\theta}(z) - \overline{\theta}(z_r)}{\overline{\theta}(z_r)}\right]. \qquad [2.33]$$

Here, z_r is some vertical reference level.

The turbulent fluctuations of the velocities are split into resolved variations and sub-grid variations. The closure of the sub-grid problem is not trivial and several closure methods exist (Maronga et al., 2015). Also, coupling the coarse meso-scale[7] model to a finer micro-scale model represents a challenge. This may be performed by a one-way or a two-way coupling.

[7] The meso-scale has a range from a few kilometers to several hundred kilometers. The micro-scale ranges from a few hundred meters to a few kilometers.

Nested Computational Domains in Numerical Simulation of Wind Fields

To obtain a stationary flow field, a "precursor" simulation may be performed. The simulation is run with a given horizontal pressure gradient, a surface roughness and, for example, a constant heat transfer from the ground. At the upper boundary, a "free slip" condition is normally used. To simulate the capping inversion layer at the top of the atmospheric boundary layer, a fixed positive gradient of the potential temperature may be applied in the upper part of the computational domain. In the precursor run, the flow will gradually change from the initial flow (e.g., a homogeneous flow) to a flow with proper shear, turbulence and temperature profile. The result from this precursor run may then be used as an input to a more refined model with wind turbines. An example of a precursor run is shown in Figure 2.14. Here, a flow field with an extent of 5.12 km x 5.12 km x 1.28 km is used. Thus, about 34 million grid cells are used in the domain. To obtain the wanted turbulence intensity at nacelle level, in this case approximately 100 m above sea level, a surface roughness length of 0.0001 m is used. Neutral atmospheric stability is obtained by using zero heat flux from the ground. To include turbines in the flow, a refined grid is needed in the vicinity of the turbines and a stepwise reduction in the size of the grid size is used. In Figure 2.15, an example of a such stepwise reduction of the grid size is shown. The placement of the turbines is also illustrated. The technique of including more refined grids inside a coarser grid is called nesting. In the present example, the inner domain has a grid size of 2.5 m. By using nesting as illustrated in Figure 2.15, the total number of grid cells will increase to approximately 100 milion.

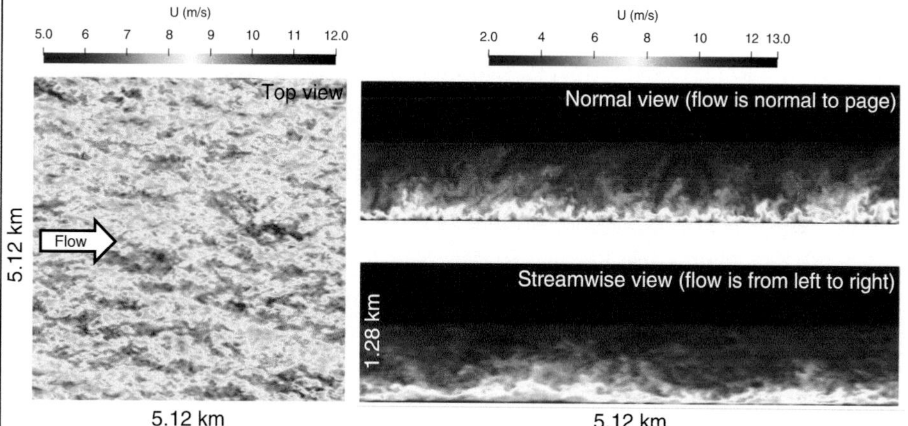

Figure 2.14 Example of result from a precursor run of an offshore wind field. Left: instantaneous flow velocities 140 m above sea. Right: instantaneous flow velocities in stream-wise and normal directions relative to the flow. The boundary layer is neutral and stable capped. Grid resolution 10 m. Courtesy Matt Churchfield, National Renewable Laboratory (NREL).

(cont.)

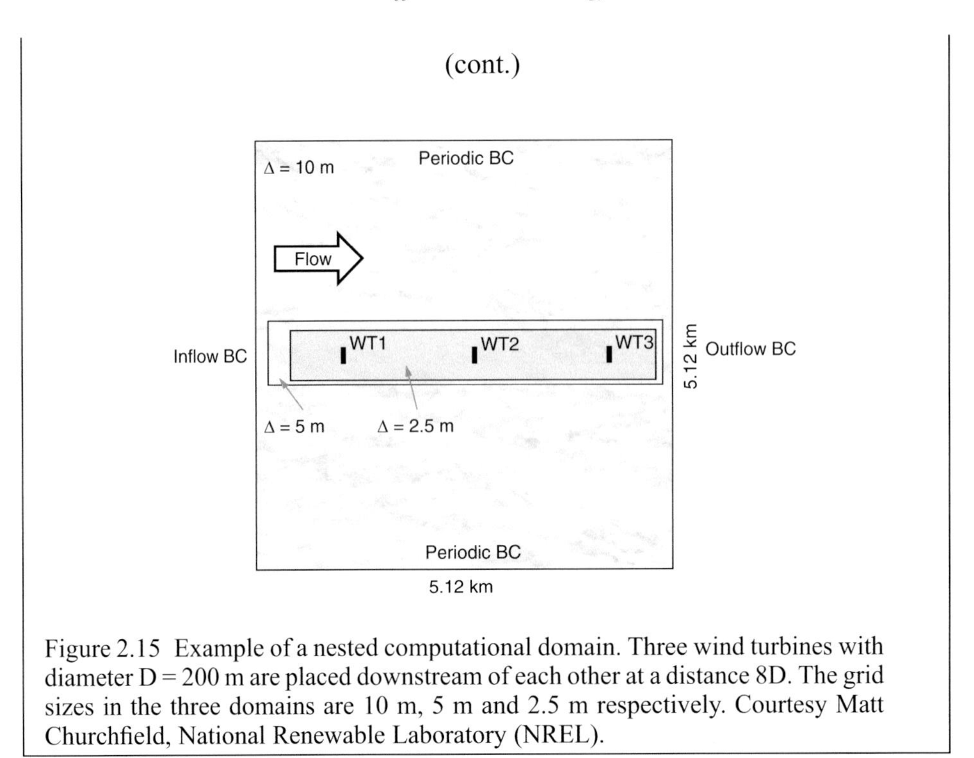

Figure 2.15 Example of a nested computational domain. Three wind turbines with diameter D = 200 m are placed downstream of each other at a distance 8D. The grid sizes in the three domains are 10 m, 5 m and 2.5 m respectively. Courtesy Matt Churchfield, National Renewable Laboratory (NREL).

2.1.4 Long-Term Wind Statistics

In the design of wind turbines and to estimate the expected energy production, a long-term statistic of the mean wind speed is required. For several locations around the world, historical time series for the key meteorological data are available for several years. The data are given on a horizontal grid and at several vertical levels. Examples of such datasets are NORA10 and its updated version NORA10EI (Haakenstad et al., 2020) and NORA3 (Haakenstad et al., 2021). Both NORA10 and NORA3 cover significant parts of the North Sea, Norwegian Sea and Barents Sea. NORA10 has a temporal resolution of 3 h and a horizontal grid of approximately 10 km x 10 km. NORA3 has a finer temporal and spatial resolution of 1 h and 3 km x 3 km. The datasets cover about 50 and 30 years respectively. The models are based upon downscaling from larger-scale meteorological models. Such datasets are called hindcast data as they are based upon numerical simulation of historical weather with observations assimilated into the simulations. In addition to wind data, pressure, wave and temperature information is available.

Figure 2.16 shows an example of the long-term distribution of the 10 min mean wind velocity 100 m over the sea level at a specific location in the North Sea (the proposed wind field area "Utsira North"). The data were obtained over 15 years using the NORA3 database. The three-parameter Weibull probability distribution is normally used to

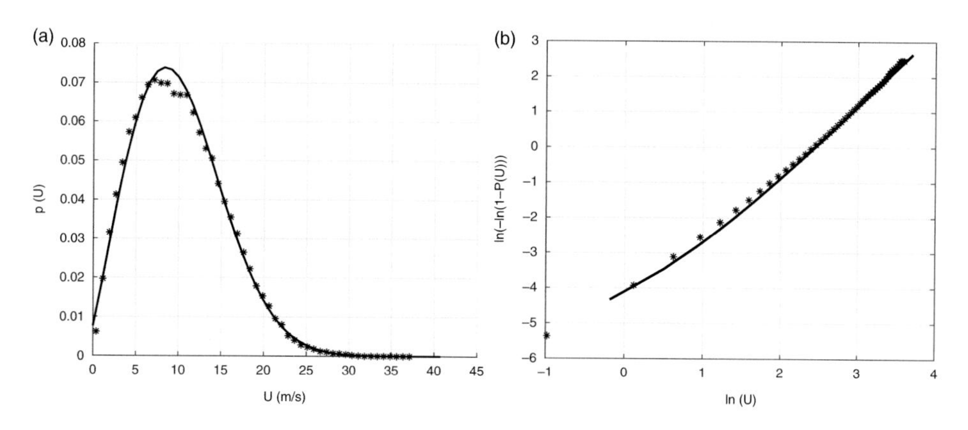

Figure 2.16 Example of distribution of hourly 10 min average wind speeds 100 m above sea level in the North Sea. Fifteen years of data are used. Left: probability distribution. Stars: binned data, 50 bins; solid line: fitted three-parameter Weibull distribution. Right: cumulative distribution plotted on Weibull scale. Data from the NORA3 database (Haakenstad et al., 2021).

represent the long-term distribution of the mean wind velocity. The three-parameter Weibull probability density and cumulative distributions for a variable x are written as:

$$\begin{aligned} p(x) &= \frac{\beta}{\alpha}\left(\frac{x-\gamma}{\alpha}\right)^{\beta-1} e^{-\left(\frac{x-\gamma}{\alpha}\right)^{\beta}} \\ P(x) &= 1 - e^{-\left(\frac{x-\gamma}{\alpha}\right)^{\beta}} \end{aligned} . \qquad [2.34]$$

Here, $\alpha\,(>0)$ is the scale parameter, used to normalize the variable; $\gamma\,(\leq x)$ is the location parameter, used to define a lower threshold for the variable; and $\beta\,(>0)$ is the shape parameter, defining the shape of the distribution. For the data shown in Figure 2.16, the parameters are estimated as $\alpha = 12.31$, $\gamma = -0.848$ and $\beta = 2.18$. The distribution may be used to estimate extreme values with certain return periods; for details, see Section 2.3.2. By combining the long-term distribution of the mean velocity with the power production characteristics of the wind turbine, the statistics of the power production at a certain location are obtained. The power characteristic is discussed in Section 3.8. Similarly to the long-term yearly distribution of the wind speed, seasonal distributions are frequently used.

The instantaneous wind velocity differs over the ocean area. Figure 2.17 illustrates the 10 min mean wind speed during the extreme weather event "Dagmar" in 2011. Large spatial variations in wind speed are observed. Such spatial variations are also present during more normal weather situations. Knowledge about the spatial correlation of the mean wind velocities may be utilized to mitigate the intermittent nature of wind power; see, e.g., Solbrekke et al. (2020).

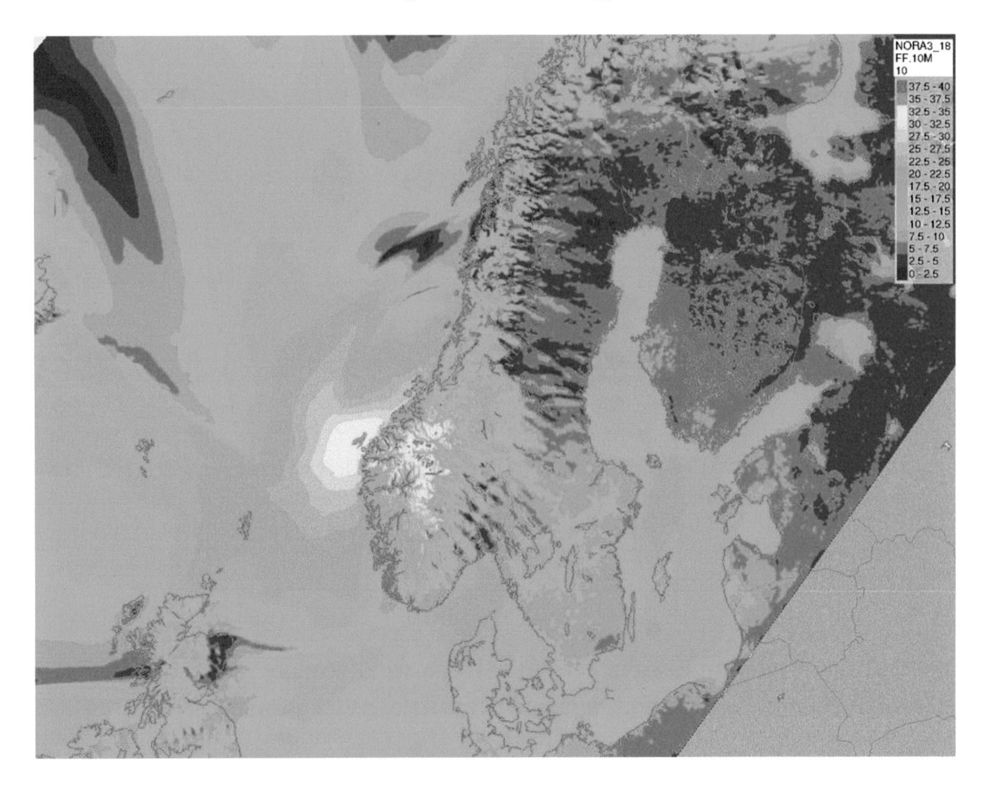

Figure 2.17 Wind speed during the extreme weather event "Dagmar," December 25, 2011. 10 min mean wind speed in m/s at 21:00 UTC. Courtesy the Norwegian Meteorological Institute.

2.1.5 Wind Measurements

Several techniques exist for measuring wind speed. They differ in the ability to measure wind direction, resolve high-frequency wind components and measure the different directional components of the turbulent wind. Some techniques also average the wind speed over a volume. In the following sections, some of the most common techniques are discussed in brief.

2.1.5.1 Cup Anemometers

Cup anemometers (Figure 2.18) are widely used for wind measurements. The main principle relies upon the fact that the drag force of a conical cup is larger for flow into the cup than for flow from the back of the cup. In the "Risø" version of the cup anemometer, three cups are mounted on a vertical axis. The rotational resistance in the bearings is very low, causing the rotation to start at very low wind speeds. The rotational speed of the cup anemometer is close to proportional to the wind speed. The relation between wind speed and rotational speed is discussed in some detail in

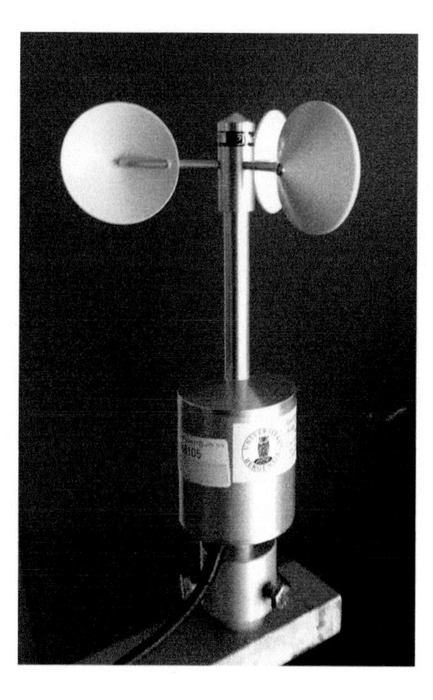

Figure 2.18 Cup anemometer with three cups. Photo by Stephan Kral, University of Bergen.

Section 3.5, where a cup anemometer is considered in a power production context. Cup anemometers are considered accurate but have some disadvantages, e.g., they measure the total horizontal wind speed only and give no information about wind direction. To obtain the wind direction, a separate wind vane must be used. As the rotor system has a certain inertia, high-frequency wind speed variations will not be measured. In turbulent wind, the average wind speed tends to be overestimated. This is because the drag characteristic of the cups will cause the rotor to rapidly speed up during a wind speed increase, but the rotational speed will reduce more slowly during wind speed decrease. The cup anemometer must be mounted on a mast. Care must thus be taken to avoid shadow or speed-up effects due to the mast and mounting system. Wind directions affected by the mast and mounting system should be removed from the subsequent data analysis.

2.1.5.2 Sonic Anemometers

Sonic anemometers use ultrasonic senders and receivers. Consider the two combined sender and receiver sensors illustrated in Figure 2.20. The time a sound signal requires to move from one sensor head to the other is given by:

$$T_{AB} = \frac{L}{c + U_{AB}}. \qquad T_{BA} = \frac{L}{c - U_{AB}} \qquad [2.35]$$

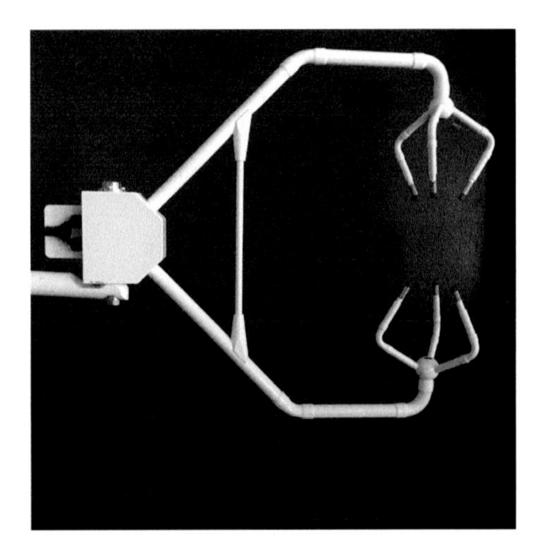

Figure 2.19 Example of a three-component sonic anemometer. Photo by Stephan Kral, University of Bergen.

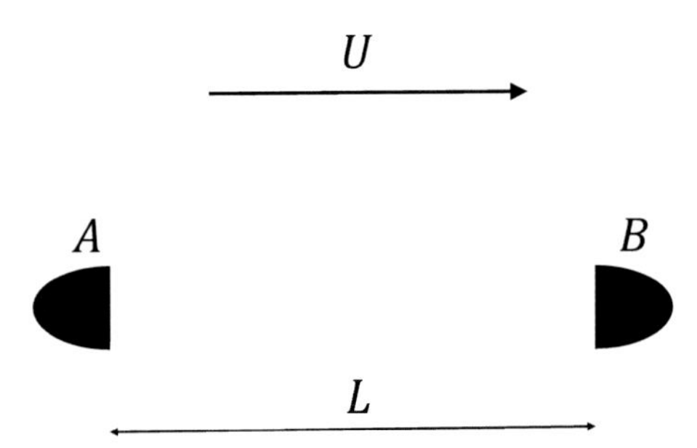

Figure 2.20 Principle of a sonic anemometer. Sensor heads A and B are located a distance L apart.

Here, U_{AB} is the component of the wind velocity along a straight line between the sensors A and B. The distance between the sensors is L. The speed of sound is c. Rearranging [2.35], the following relations are obtained:

$$\begin{aligned} U_{AB} &= \frac{L}{2}\left(\frac{1}{T_{AB}} - \frac{1}{T_{BA}}\right) \\ c &= \frac{L}{2}\left(\frac{1}{T_{AB}} + \frac{1}{T_{BA}}\right) \end{aligned} . \qquad [2.36]$$

It is observed that the expression for the wind speed is not dependent upon the speed of sound and that the speed of sound is obtained as well. By locating sensors in three orthogonal directions, all three components of the wind vector can be measured. Figure 2.19 shows an example of a three component sonic anemometer. Compared to the cup anemometer, the sonic anemometer has the advantage of a very high frequency resolution and the possibility of measuring all components of the wind. As for the cup anemometer, one must be aware of possible shadow effects of the mast and mounting arrangement. Experience shows that the sonic anemometer may give erroneous results during rainy weather; see, e.g., Nybø, et al. (2019).

2.1.5.3 Lidar

The LIDAR measurement technique is a remote sensing technique. The name LIDAR originates from an abbreviation of "light-based radar" or "light detection and ranging." A simplistic view of the basic principle is illustrated in Figure 2.21. A narrow beam of laser (monochrome) light is sent out from the LIDAR unit. The light is scattered by aerosols in the air and some of the light is back-scattered into the LIDAR's receiver system, which utilizes the Doppler effect. By detecting the change of frequency in the back-scattered light, the speed of the aerosol particles, i.e., the wind speed in the direction of the laser beam, U_S, is obtained. As the light beam has a finite angle and the duration of the analyzed scattered signal has a finite duration, the velocity obtained is a weighted average over a volume, as illustrated in Figure 2.21. Typical averaging lengths in the direction of the beam are in the order of tens of meters. Assuming no mean vertical component of the wind speed, the

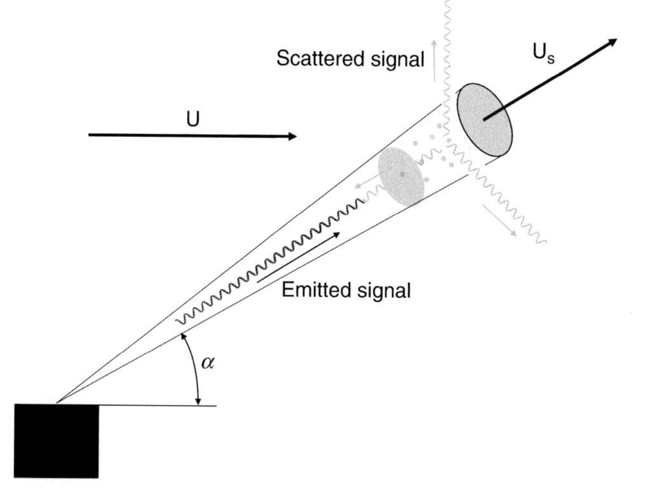

Figure 2.21 Basic principle of LIDAR. Some of the emitted light is back-scattered from the aerosol particles in the air between the two planes. From the frequency shift in the back-scattered signal, the component of the wind velocity in the direction of the beam, U_S, may be determined.

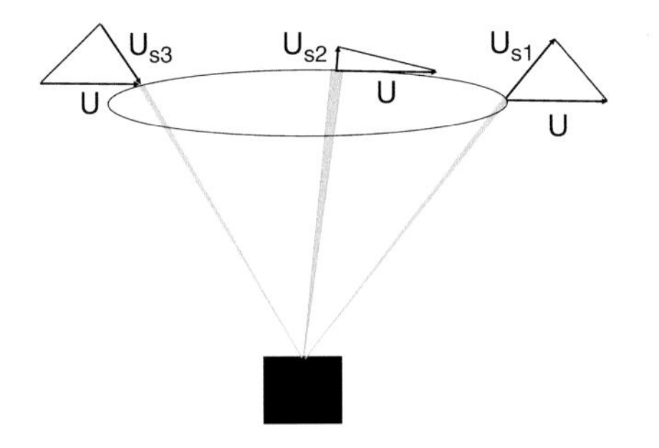

Figure 2.22 Example of circular sweeping pattern of the laser beam to obtain both components of the horizontal mean wind speed. U is the total horizontal wind speed; U_{si} is the measured component of the wind speed in azimuthal angle β_i.

horizontal component of the mean wind speed in the vertical plane described by the laser beam is obtained as $U_\beta = U_s/\cos\alpha$. U_β is the horizontal wind speed component in direction β and α is the angle between the laser beam and the horizontal plane.

To obtain the total horizontal wind speed, at least two orthogonal measurements must be performed. In practice, to obtain the total horizontal velocity, a circular sweeping pattern of the laser beam can be used, as illustrated in Figure 2.22. The laser beam measures the velocities U_{si} at a number of azimuthal angles β_i. The relation between the mean horizontal wind velocity, $\overline{U}$, and the measured speed in the direction of the laser beam is thus:

$$U_{si} = \overline{U}\cos(\beta_i - \beta_0)\cos\alpha. \qquad [2.37]$$

Here, β_0 is the mean direction of the wind. By measuring for several values of β_i, the mean wind speed and the mean direction are obtained. A key assumption used to obtain the mean wind speed by this approach is that the wind field is homogeneous over the sweeping area.

Other remote measurement techniques utilize sound (SODAR) and microwaves (RADAR) to obtain the wind velocity. The sound signals are scattered by temperature differences in the air. Assuming that such temperature structures are advected with the mean wind speed, the wind speed in the direction of the sound wave is obtained by considering the frequency shift between emitted and reflected sound (for details, see, e.g., Lang and McKeogh, 2011). Microwave radars may be used in combination with SODAR. The radar signals may interfere with the sound waves through the so-called Bragg effect. This makes it possible to determine temperatures in addition to velocities.

2.2 Ocean Waves

2.2.1 Introduction

This section will give the classical linear description of gravity waves in the ocean. This is the most common wave description used in the computation of wave loads on marine structures. To compute the loads on marine structures, we must know about the surface elevation of the waves as well as the velocity and acceleration fields in the water below the free surface. An important feature of the linear description is the principle of superposition. This makes it possible to describe a complex sea state by a summation of harmonic components. Some issues related to the nonlinearity of waves will also be addressed. Fixed offshore wind structures are normally located at water depths where the sea bottom "is felt" by the waves. Such finite water depth conditions are important for the wave kinematics.

More details on the modeling of ocean waves may be found in World Meteorological Organization (2018), Mork (2010), Faltinsen (1990) and DNV (2021c).

2.2.2 Assumptions

Ocean gravity waves may be modeled under the assumptions that water is incompressible and that capillarity effects and viscous effects may be ignored. In the following, constant density is also assumed. This implies that we are not considering internal waves. Such waves may occur in cases with water of low salinity on top of more saline water. However, internal waves are normally not considered in the design of offshore wind turbines.

To derive the governing equations for linear-gravity waves, it is assumed that the velocity field can be described by a velocity potential ϕ and that the fluid velocities are obtained from the gradients of the velocity potential:

$$\mathbf{u} = \nabla\phi. \qquad [2.38]$$

Here, $\mathbf{u} = (u, v, w)^T$ is the velocity vector given by the components of the velocity in the x, y and z directions. The $(x, y, 0)$ plane is at the mean free surface. z is vertical, zero at the mean free surface and positive upwards; see Figure 2.23. Under the condition that the fluid is incompressible and irrotational, the potential will satisfy Laplace equation throughout the fluid:

$$\nabla^2\phi = \frac{\partial^2\phi}{\partial x^2} + \frac{\partial^2\phi}{\partial y^2} + \frac{\partial^2\phi}{\partial z^2} = 0. \qquad [2.39]$$

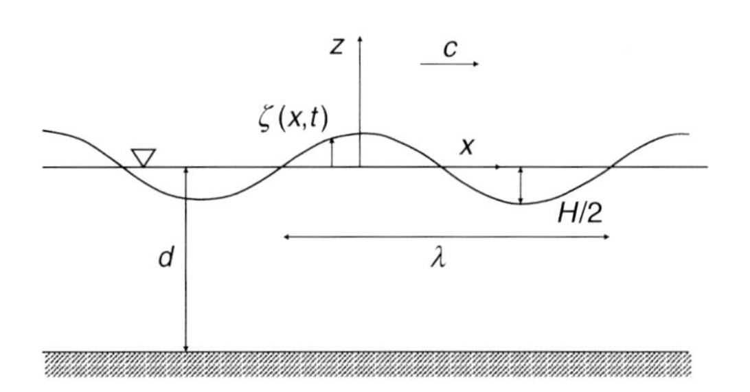

Figure 2.23 Notations used in describing the linear ocean gravity waves. The waves propagate in positive x–direction. Surface elevation $\zeta(x,t)$ is shown at $t = 0$.

2.2.3 Solution

To solve the above Laplace equation, proper boundary conditions must be imposed. To simplify the problem, it is assumed that the waves propagate in positive x-direction, so the velocities in y-direction are zero, $\frac{\partial \phi}{\partial y} = 0$. The following boundary conditions are then to be imposed at the sea bottom and at the free surface.

Under the condition that the bottom is horizontal, zero vertical velocity must be required at the bottom, i.e.:

$$\frac{\partial \phi}{\partial z} = 0 \quad \text{at} \quad z = -d. \qquad [2.40]$$

At the free surface, two conditions are to be imposed: one kinematic and one dynamic. The kinematic condition implies that particles, once located at the free surface, will remain there, i.e., the particles on the free surface must follow the motion of the free surface. This may be expressed as:

$$\frac{\partial \zeta}{\partial t} = \frac{\partial \phi}{\partial z} - \frac{\partial \phi}{\partial x}\frac{\partial \zeta}{\partial x} \quad \text{at z} = \zeta. \qquad [2.41]$$

Here, ζ is the free surface elevation. The first term is the vertical velocity of the free surface at a specific x-position. This vertical velocity must equal the two terms on the right-hand side, the first one being the vertical velocity of the water and the second being the horizontal velocity of the water multiplied by the slope of the surface. This condition is to be satisfied at the instantaneous free surface level.

The dynamic free surface condition is the requirement of constant pressure on the free surface, equal to the atmospheric pressure. Invoking Bernoulli's equation, we may write:

$$\rho\frac{\partial\phi}{\partial t}+\frac{\rho}{2}\left[\left(\frac{\partial\phi}{\partial x}\right)^2+\left(\frac{\partial\phi}{\partial z}\right)^2\right]+\rho g\zeta=C \text{ at } z=\zeta. \qquad [2.42]$$

The first term is the pressure related to the acceleration of the fluid, the second term is due to the velocities and the third term is the hydrostatic contribution.

The kinematic and dynamic boundary conditions on the above forms imply a nonlinear solution of the problem. Exact analytical solutions of this problem do not exist. However, several approximate solutions are derived, in particular the class of Stokes wave solutions that are found by a series expansion of the problem. Classic solutions are Stokes second-order and Stokes fifth-order waves; see, e.g., Sarpkaya and Isaacson (1981). These classical solutions assume an infinite train of equal waves. Each wave is also symmetric in the sense that the front and back slope of the wave crest are equal. This is not in accordance with the geometry of real steep waves.

To obtain a linear solution of the above boundary value problem, the following approximations are introduced. It is assumed that the slope of the waves is small and that the boundary conditions may be satisfied at $z=0$ rather than at $z=\zeta$.

The small slope approximation implies that the second term on the right-hand side of [2.41] is much smaller than the first term and is thus ignored. Similarly, the velocity-squared terms in [2.42] are ignored. [2.41] and [2.42] are in linearized form thus written as:

$$\begin{aligned}&\frac{\partial\zeta}{\partial t}-\frac{\partial\phi}{\partial z}=0 \text{ at } z=0\\&\frac{\partial\phi}{\partial t}+g\zeta=0 \text{ at } z=0\end{aligned}. \qquad [2.43]$$

The constant in the Bernoulli equation may be set to zero. By taking the time derivative of the dynamic boundary condition and combining it with the kinematic condition, the two equations may be combined into one, which constitutes the linearized free surface boundary condition:

$$\frac{\partial^2\phi}{\partial t^2}+g\frac{\partial\phi}{\partial z}=0 \text{ at } z=0. \qquad [2.44]$$

The linearized free surface elevation is obtained as:

$$\zeta=-\frac{1}{g}\left(\frac{\partial\phi}{\partial t}\right)_{z=0}. \qquad [2.45]$$

It is further assumed that the solution shall represent a harmonic progressive wave. The Laplace equation [2.39] is combined with the linear boundary conditions in [2.40] and [2.45] and the solution may be written as:

$$\phi = i\frac{gH}{2\omega}\frac{\cosh[k(z+d)]}{\cosh(kd)}e^{i(\omega t-kx)}. \qquad [2.46]$$

Here, $k = 2\pi/\lambda$ is the wave number and λ is the wavelength. $\omega = 2\pi/T$ is the wave angular frequency and T is the wave period. d is the water depth. H is the double amplitude for the wave, equal to twice the single amplitude, ζ_A, under the assumption of linear waves. For nonlinear waves, however, the height of the wave crest and the depth of the wave trough differs, and it is convenient to use the distance from trough to crest, the wave height, as a measure. It is implicit that it is the real part of the quantities which has physical meaning. The free surface elevation is obtained as:

$$\zeta = \mathrm{Re}\left\{\frac{H}{2}e^{i(\omega t-kx)}\right\} = \zeta_A\cos(\omega t - kx). \qquad [2.47]$$

As the water depth tends to infinity, the potential may be written as:

$$\phi = i\frac{g\zeta_A}{\omega}e^{kz}e^{i(\omega t-kx)} \quad \text{as} \quad d\rightarrow\infty. \qquad [2.48]$$

A frequently used assumption is that the deep-water formulation may be used if the water depth is greater than half the wavelength. Another result is the so-called dispersion relation, the relation between the wave number and the wave (angular) frequency:

$$\omega = \sqrt{kg\tanh(kd)}. \qquad [2.49]$$

In the limits $d\rightarrow 0$ and $d\rightarrow\infty$, the dispersion relation becomes:

$$\begin{aligned} &\omega\rightarrow k\sqrt{gd} \ \text{as} \ \mathrm{d}\rightarrow 0 \\ &\omega\rightarrow\sqrt{kg} \quad \text{as} \ \mathrm{d}\rightarrow\infty \end{aligned}. \qquad [2.50]$$

The speed of the wave, the wave celerity or the phase speed is given as $c = \lambda/T = \omega/k$. In the general case, the wave celerity is given as:

$$c = \frac{\omega}{k} = \sqrt{\frac{g}{k}\tanh(kd)}. \qquad [2.51]$$

In the shallow and deep-water cases, the following limiting values are obtained (for more details, see Section 2.2.5):

$$c \to \sqrt{gd} \quad \text{as } d \to 0$$
$$c \to \frac{g}{\omega} \quad \text{as } d \to \infty \qquad [2.52]$$

We observe that the speed of the waves becomes independent of wave period as the water depth tends to zero. This is the case as long waves approaches a shallow beach. In deep waters, waves with long periods move faster than those with short periods. If we observe the wave field in the open ocean, we see that wave crests exist for a short period of time and then seems to disappear. This is because the crests we observe is composed of many wave components with different wave periods. As the long waves move faster than the short waves, the positive summation of components that formed the crest at a certain instant in time no longer exists after a short while. It should also be noted that the energy flux in the wave is slower than the wave celerity. The energy moves with a speed called the group velocity. In deep water, the group velocity is half the celerity.

The above equations summarize the key features of linear-gravity waves. From these equations, the velocities, accelerations etc. can be derived. Table 2.2 presents some key relations. Derivation of higher-order solutions, e.g., Stokes waves of orders two and five, may be found in Sarpkaya and Isaacson (1981) and Fenton (1985).

Table 2.2 gives an expression for the average energy density in the wave. This density is per unit free surface area. It can be shown that in deep water the average energy density has equally large contributions from the free surface elevation, the hydrostatic part and the kinematic part, i.e., the velocity squared term integrated from the bottom to the free surface. Note, however, that in wave energy applications, it is not the energy density per unit surface area that matters, but the energy flux in the direction of wave propagation. The energy flux per unit width of the wave crest is equal to the energy density multiplied by the group velocity of the waves.

The kinematics according to linear wave theory is valid for $z \leq 0$ only. To estimate the kinematics above the mean water level, various approximations are used in practical applications. These are discussed in some detail in Section 2.2.7.

2.2.4 *Waves in Shallow Water*

In deep water, the wave steepness is the only parameter important to the validity of the linear wave theory discussed here. In shallower water, the ratio of the water depth to the wavelength also becomes an important parameter. The following three parameters may be used to characterize waves in limited water depth (see DNV, 2014a).

Table 2.2 *Linear gravity on finite and infinite water depth.* $\theta = (\omega t - kx\cos\chi - ky\sin\chi)$, *where* χ *is direction of wave propagation relative to the positive x-axis;* $\zeta_A = H/2$ *is the wave amplitude*

Quantity	Finite water depth	Deep water
Potential	$\phi = i\frac{g}{\omega}\zeta_A\frac{\cosh[k(z+d)]}{\cosh(kd)}e^{i\theta}$	$\phi = i\frac{g}{\omega}\zeta_A e^{kz+i\theta}$
Surface elevation	$\zeta = \zeta_A e^{i\theta}$	$\zeta = \zeta_A e^{i\theta}$
Wave period $T = 2\pi/\omega$	$T = \frac{2\pi}{\sqrt{kg\tanh(kd)}}$	$T = \frac{2\pi}{\sqrt{kg}}$
Wavelength $\lambda = 2\pi/k$	$\lambda = cT = \frac{2\pi}{\omega}\sqrt{\frac{g}{k}\tanh(gd)}$	$\lambda = 2\pi\frac{g}{\omega^2}$
Phase velocity $c = \omega/k$	$c = \sqrt{\frac{g}{k}\tanh(kd)}$	$c = g/\omega$
Dynamic pressure	$p = -\rho\frac{\partial\phi}{\partial t} = \rho g\zeta_A\frac{\cosh[k(z+d)]}{\cosh(kd)}e^{i\theta}$	$p = \rho g\zeta_A e^{kz+i\theta}$
Particle velocity in x-direction	$u_x = \frac{\partial\phi}{\partial x} = \frac{kg\cos\beta}{\omega}\zeta_A\frac{\cosh[k(z+d)]}{\cosh(kd)}e^{i\theta}$	$u_x = \omega\cos\beta\zeta_A e^{kz+i\theta}$
Particle velocity in y-direction	$u_y = \frac{\partial\phi}{\partial y} = \frac{kg\sin\beta}{\omega}\zeta_A\frac{\cosh[k(z+d)]}{\cosh(kd)}e^{i\theta}$	$u_y = \omega\sin\beta\zeta_A e^{kz+i\theta}$
Particle velocity in z-direction	$u_z = \frac{\partial\phi}{\partial z} = i\frac{kg}{\omega}\zeta_A\frac{\sinh[k(z+d)]}{\cosh(kd)}e^{i\theta}$	$u_z = i\omega\zeta_A e^{kz+i\theta}$
Particle acceleration in x-direction	$a_x = \frac{\partial\phi}{\partial x\partial t} = ikg\cos\beta\zeta_A\frac{\cosh[k(z+d)]}{\cosh(kd)}e^{i\theta}$	$a_x = i\omega^2\cos\beta\zeta_A e^{kz+i\theta}$
Particle acceleration in y-direction	$a_y = \frac{\partial\phi}{\partial y\partial t} = ikg\sin\beta\zeta_A\frac{\cosh[k(z+d)]}{\cosh(kd)}e^{i\theta}$	$a_y = i\omega^2\sin\beta\zeta_A e^{kz+i\theta}$
Particle acceleration in z-direction	$a_z = \frac{\partial\phi}{\partial z\partial t} = -kg\zeta_A\frac{\sinh[k(z+d)]}{\cosh(kd)}e^{i\theta}$	$a_z = -\omega^2\zeta_A e^{kz+i\theta}$
Group velocity	$c_g = \frac{1}{2}c\left(1 + \frac{2kd}{\sinh(2kd)}\right)$	$c_g = c/2$
Average energy density	$E = \frac{1}{2}\rho g\zeta_A^2$	$E = \frac{1}{2}\rho g\zeta_A^2$
Average energy flux	$P = Ec_g$	$P = Ec_g = E\frac{c}{2}$

$$\begin{aligned}\text{Wave steepness parameter}: \quad & S = \frac{H}{\lambda_0} = 2\pi\frac{H}{gT^2}\\ \text{Shallow water parameter}: \quad & \mu = \frac{d}{\lambda_0} = 2\pi\frac{d}{gT^2}\\ \text{Ursell}^8\text{ parameter}: \quad & U_r = \frac{H}{k_0^2 d^3} = \frac{1}{4\pi^2}\frac{S}{\mu^3}\end{aligned} \qquad [2.53]$$

Here, λ_0 and k_0 are the linear deep-water wavelength and wave number corresponding to the wave period T. The Ursell parameter expresses the ratio between the second-order amplitude and the first-order amplitude in a Stokes wave formulation (Sarpkaya and Isaacson, 1981) and is used to classify the range of validity for various wave descriptions. The Stokes second-order contribution to the surface elevation may be written as:

[8] The Ursell number is also written as $U_R = 4\pi^2 U_r$ (DNV, 2021c).

$$\zeta^{(2)} = \frac{\pi H^2}{8\lambda} \frac{\cosh(kd)}{\sinh^3(kd)} [2 + \cosh(2kd)] e^{2i\theta}. \quad [2.54]$$

Here, H is the wave double amplitude, including both the first- and second-order contribution.

For the linear wave theory to be valid, the wave steepness should be small, $S << 1$. Further, the second-order component of the wave amplitude should be much less than the first-order amplitude. This is obtained if the Ursell parameter is much less than 1, $U_r \ll 1$. It is observed that if the shallow-water parameter, μ, is reduced, then the Ursell parameter increases. This demonstrates that the Stokes wave representation of the waves breaks down in very shallow water.

The linear approach, also called Airy waves, has a fairly large range of validity with respect to wave steepness in deep water, while the range of validity is reduced as the water depth is reduced. Deep water may usually be assumed if $\mu > 0.5$. It is common to assume that the limiting steepness, or breaking limit, for regular waves in deep water is S=1/7. This is, however, far beyond the validity of linear theory. In very shallow water a breaking limit of H/d = 0.78 is commonly assumed. For water that is not so shallow, the limit is lower (Grue et al., 2014).

Various recommendations exist with respect to the applicability of linear wave theory and various nonlinear formulations. Such recommendations may be found in various design standards, e.g., DNV (2021c). However, the applicability of the various approximations depends upon the use, e.g., if an accurate estimate of crest height is important, if local structural loads close to the free surface are to be estimated, if the loads are inertia- or drag-dominated etc.

From the above relations, it is observed that when waves move from deep water and into shallower water, the wavelength is shortened and the wave celerity slows down. As waves move from deep water, where they do not necessarily move perpendicular to the isobaths, into shallower water, the direction of the waves will change, tending to move perpendicular to the isobaths (see Figure 2.24). This phenomenon is called refraction. In Figure 2.24, we observe how the waves are directed away from deeper area focuses toward the shallower area. This causes wave energy to concentrate in the shallow area, increasing wave steepness. The refraction effect is strongest for long waves. If the coastline is not a shallow beach, but rather a steep cliff, waves will be reflected from the cliffs, causing a wave train that moves away from land. The combined refraction and reflection effects may thus cause very "confused" sea in the area. Details on wave refraction are found in Sarpkaya and Isaacson (1981).

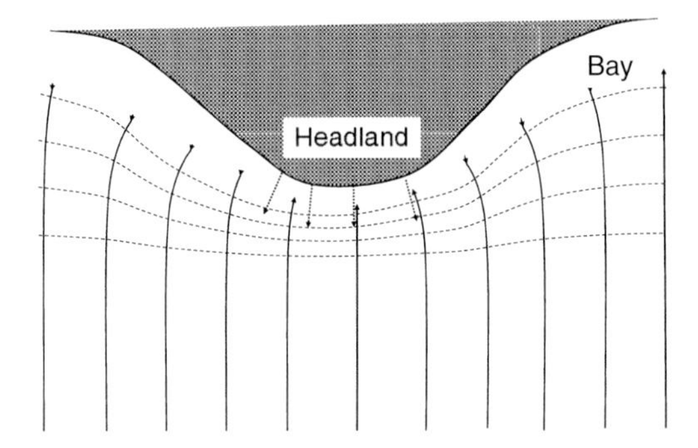

Figure 2.24 Refraction of waves approaching a shore. The dashed lines represent the isobaths (constant depth curves). The solid arrow represents the direction of wave propagation, while the dotted arrow indicates reflection from a steep headland. Based upon World Meteorological Organization (2018).

2.2.5 *Energy in Waves*

As in Table 2.2, the energy density per unit surface area for a linear-gravity wave is given by:

$$E = \frac{1}{2}\rho g \zeta_A^2. \qquad [2.55]$$

This is independent of water depth and wave period. The energy is equally distributed between potential energy, given from the free surface elevation, and kinetic energy, given from the particle velocity throughout the water column.

The energy flux through a surface of unit width and perpendicular to the direction of wave propagation is given as:

$$P = c_g E. \qquad [2.56]$$

c_g is the group velocity of the waves. The group velocity may be illustrated by considering the summation of two regular waves of almost equal wave numbers, k and $k + \delta k$. The surface elevation is then given by:

$$\begin{aligned} \zeta(x,t) &= \zeta_{a1} e^{i\left(\omega(k)t - kx\right)} + \zeta_{a2} e^{i\left(\omega(k+\delta k)t - (k+\delta k)x\right)} \\ &\approx e^{i\left(\omega(k)t - kx\right)} \left[\zeta_{a1} + \zeta_{a2} e^{i\left(\delta k \frac{\partial \omega}{\partial k} t - \delta k\, x\right)} \right] \end{aligned} . \qquad [2.57]$$

Here, it is the real part of the expression that has physical meaning. δk represents the difference in wave number between the two waves. It is assumed that $\delta k \ll k$. $\omega(k+\delta k)$ is the corresponding wave frequency. The last term in [2.57] represents a slowly oscillating term with frequency $\delta k \frac{\partial \omega}{\partial k}$ and wave number δk. It is observed that the speed of this slowly oscillating term, the "group speed," becomes:

$$c_g = \frac{\delta k \frac{\partial \omega}{\partial k}}{\delta k} = \frac{\partial \omega}{\partial k}. \qquad [2.58]$$

Thus, the last term corresponds to a wave which propagates with the speed $\partial \omega / \partial k$. This is the speed of the "group" in the example shown in Figure 2.25. In the general case the group speed is obtained by derivation of the dispersion relation (see [2.49]):

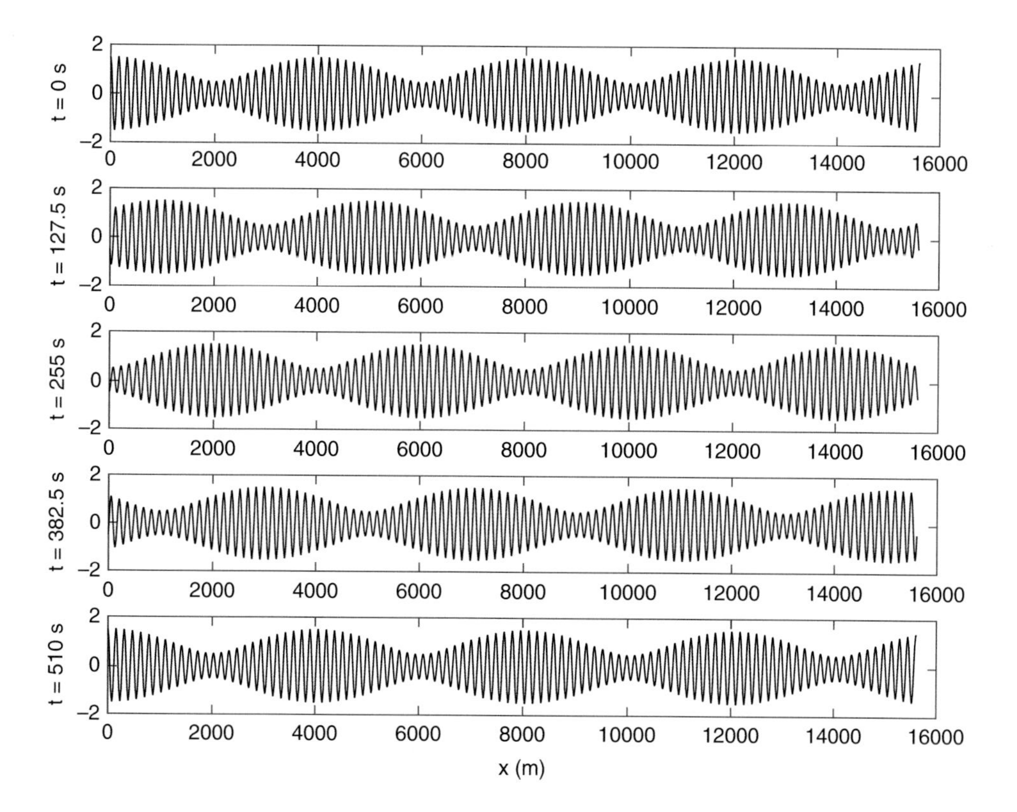

Figure 2.25 Example of summation of two waves of almost equal period (T = 10 s and 10.2 s) in deep waters. The amplitudes are 1 and 0.5 m respectively. The group period becomes $T_g = 1/(1/T_1 - 1/T_2) = 510$ s. The plots are given at five equidistant time instants. The time interval between the plots is $T_g/4$. Each individual wave crest moves with a velocity $c = \omega/g$, about 15.6 m/s, while the peak of the group moves with a speed $c_g = \omega/2g$, about 7.8 m/s.

$$c_g = \frac{\partial}{\partial k}\sqrt{kg\,\tanh(kd)} = \frac{1}{2}\left(\frac{g}{k}\tanh(kd)\right)^{1/2}\left(1 + kd\frac{1-\tanh^2(kd)}{\tanh(kd)}\right). \qquad [2.59]$$

Here, the relations $\frac{d}{dx}\tanh(x) = \frac{1}{\cosh^2 x}$ and $\cosh^2 x = \frac{1}{1-\tanh^2 x}$ have been utilized. The expression for the group speed may be rewritten as:

$$c_g = \frac{1}{2}\left(\frac{g}{k}\tanh(kd)\right)^{1/2}\left(1 + \frac{2kd}{\sinh(2kd)}\right) = \frac{1}{2}c\left(1 + \frac{2kd}{\sinh(2kd)}\right), \qquad [2.60]$$

where c is the phase speed of the wave; see [2.51].

The deep-water and shallow-water approximations for the group speed are obtained as:

$$c_{g(d\to\infty)} = \left.\frac{\partial\omega}{\partial k}\right|_{d\to\infty} = \frac{\partial}{\partial k}\sqrt{kg} = \frac{1}{2}\sqrt{\frac{g}{k}} = \frac{1}{2}\frac{g}{\omega} = \frac{1}{2}c_{(d\to\infty)} \qquad [2.61]$$

$$c_{g(d\to 0)} = \left.\frac{\partial\omega}{\partial k}\right|_{d\to 0} = \frac{\partial}{\partial k}k\sqrt{gd} = \sqrt{gd} = c_{(d\to 0)}. \qquad [2.62]$$

It is observed that in deep water, the group speed is half the phase velocity, while in the shallow water limit, the group and phase speeds are equal. Figure 2.25 illustrates how the sum of two deep-water waves of approximately the same period propagates. It is observed that the peak of the group moves with half the phase speed, as derived in [2.61].

An example of the group speed as a function of water depth is given in Figure 2.26. The general picture is that the group speed is reduced as the water depth is reduced. However, in a range where the water depth is about 0.15–0.2 times the wavelength, the group speed increases.

Consider waves moving from deep water toward shallow water. The energy flux through every cross-section perpendicular to the direction of wave propagation must be constant; i.e., according to [2.56], it is obtained that the wave height has to vary as $H(d)/H_{d\to\infty} = \sqrt{c_{g(d\to\infty)}/c_g(d)}$. The resulting wave height is shown in Figure 2.26, together with the shallow-water approximation.

2.2.6 Superposition of Waves: Wave Spectrum

A real sea state cannot be modeled as a regular wave. Sometimes swells may almost resemble a regular wave with constant period and amplitude. However, under the assumption of linearity, wind-generated gravity waves may be modeled as a summation of regular waves with different wave periods and

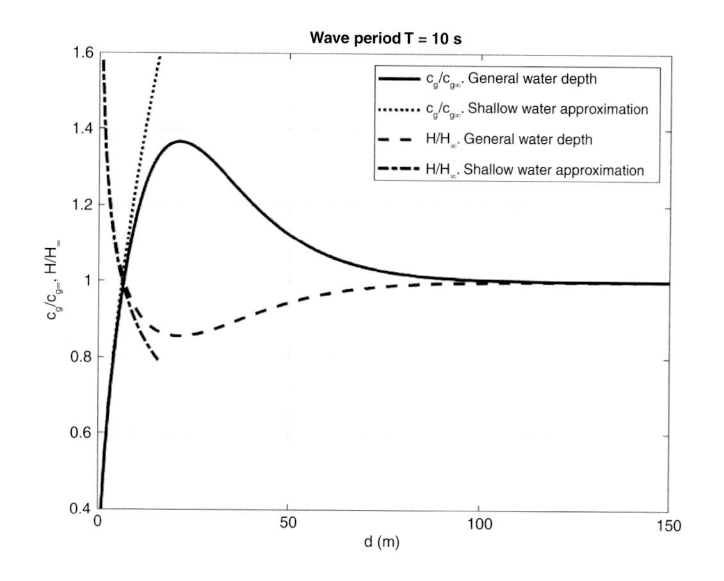

Figure 2.26 Group speed and wave height for a 10 s wave as a function of water depth. The group speed and wave height are normalized with the deep-water values.

amplitudes; i.e., assuming unidirectional waves, the wave elevation may be written as:

$$\zeta(t,x) = \mathrm{Re}\left\{\sum_{j=1}^{N} A_j e^{i\left(\omega_j t - k_j x + \varphi_j\right)}\right\} = \sum_{j=1}^{N} A_j \cos(\omega_j t - k_j x + \varphi_j). \qquad [2.63]$$

Here, A_j is the wave amplitude of wave component j with corresponding wave number k_j and angular frequency ω_j . φ_j is the phase of component j. If $N \to \infty$, the wave field may be described by a continuous wave spectrum, or a wave energy spectrum. In the description of ocean waves, a one-sided spectrum is most commonly used, i.e., the spectrum is given for positive frequencies only. Examples of such spectra are given in Figure 2.27. The wave energy spectrum contains information of the amplitudes of the wave components, or rather the energy, as a function of wave frequency. Information of the phase is not present. If the continuous spectrum is divided into finite frequency intervals, $\Delta\omega = 2\pi\Delta f$, the amplitude of the wave component corresponding to a specific interval is given by:[9]

[9] Note that $S(f) = 2\pi S(\omega)$, so $S(f)\Delta f = S(\omega)\Delta\omega$.

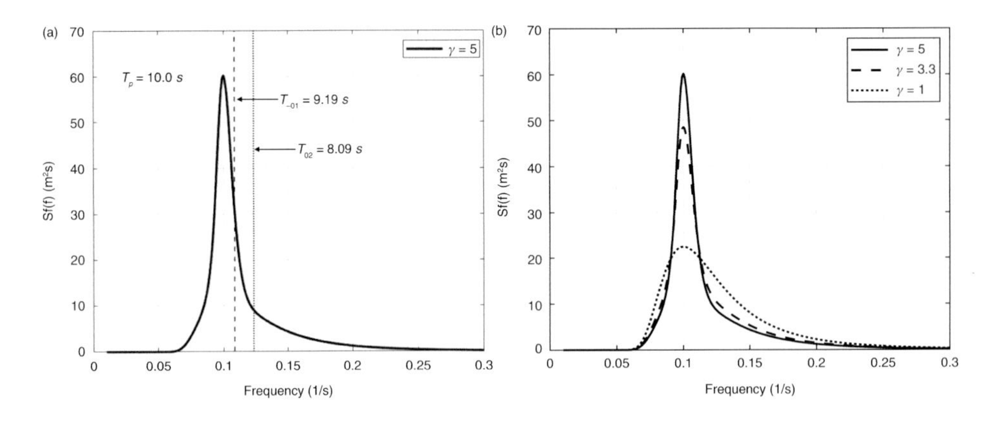

Figure 2.27 The Jonswap wave spectrum. Left: example on a spectrum with $H_s = 5.0\,m$ and $T_p = 10$s. Right: spectra with different γ values. $H_s = 5m$ and $T_p = 10$s.

$$A_j = \sqrt{2S(\omega_j)\Delta\omega_j}. \tag{2.64}$$

This means that the area of the spectrum over this frequency interval is proportional to the energy in a corresponding regular wave component. The wave spectrum may be obtained from measurements, by taking the Fourier transform of the measured wave elevation at a specific point in the sea. For design purposes several standard spectra exist. These are developed to be representative of typical average sea states. Two frequently used spectral representations of ocean waves are the Pierson–Moskowitz (PM) spectrum and the Jonswap spectrum. According to DNV (2021c) the PM spectrum may be written as:

$$S_{PM}(\omega) = \frac{5}{16} H_s \frac{\omega_p^4}{\omega^5} e^{\left[-\frac{5}{4}\left(\frac{\omega}{\omega_p}\right)^{-4}\right]}. \tag{2.65}$$

Here, H_s is the significant wave height. Traditionally this corresponded to the visually observed wave height but now has a more stringent definition related to the energy in the sea state or the area of the wave spectrum (see text following [2.67]). Sometimes, the significant wave height has also been directly related to the time history of the wave record and denoted $H_{1/3}$, which is the average of the 1/3 highest waves (double amplitudes) in the sea state (World Meteorological Organization, 2018). $\omega_p = 2\pi/T_p$ is the peak angular frequency of the spectrum with corresponding period T_p (see Figure 2.27). A wave spectrum is defined for a stationary wave condition, a hypothetical situation where all parameters as well as the shape of the spectrum are constant over time. As sea states are always changing as a response to the weather, a truly stationary sea state cannot exist. However, for design purposes, one normally considers a sea state to be stationary for 3 h. The PM spectrum is considered a reasonable description of a fully developed sea state, i.e., when the wind has been blowing for a long time with constant strength and direction over a large sea area (fetch). A more realistic assumption is that the sea is under development and that the fetch is limited. For such cases the Jonswap spectrum is a reasonable description. The Jonswap spectrum is an γ-adjusted version of the PM spectrum:

$$S_J(\omega) = A_\gamma S_{PM}(\omega)\gamma^{\exp\left[-0.5\left(\frac{\omega-\omega_p}{\sigma\omega_p}\right)^2\right]}. \qquad [2.66]$$

Here, γ is a nondimensional peak shape parameter; σ is a spectral width parameter with $\sigma = \sigma_a$ for $\omega \leq \omega_p$ and $\sigma = \sigma_b$ for $\omega > \omega_p$; $A_\gamma = 1 - 0.287\ln(\gamma)$ is a normalizing factor. Normally $\sigma_a = 0.07$ and $\sigma_b = 0.09$ are used. An average value of γ is 3.3; however, values in the range $\gamma = [1,5]$ may be encountered. $\gamma = 1$ corresponds to the PM spectrum. In Figure 2.27 (right), a Jonswap spectrum is illustrated using different γ-values.

Given a wave spectrum $S(\omega)$, various parameters may be defined. The nth spectral moment is defined by:

$$m_n = \int_0^\infty \omega^n S(\omega)d\omega. \qquad [2.67]$$

The significant wave height is then defined by $H_s = 4\sqrt{m_0}$. m_0 is equal to the square of the standard deviation of the surface elevation. For a "narrow-banded" spectrum the average zero up-crossing period is obtained as $T_{02} = 2\pi\sqrt{m_0/m_2}$. In Figure 2.27 (left), a Jonswap spectrum with $H_s = 5.0\ m$ is illustrated. Here, the peak frequency $f_p = 1/T_p$ as well as the zero up-crossing frequency $f_{02} = 1/T_{02}$ and energy mean frequency $f_{-01} = m_0/(2\pi m_{-1}) = 1/T_{-01}$ are shown. The energy mean period is the period of a regular wave with the same flux of energy as the mean flux of energy in the the wave spectrum, assuming deep water.

The term "narrow-banded" can be understood in the sense that the wave elevation record, and thus the wave spectrum, is narrow-banded if there is only one maximum or minimum value (wave crest or wave trough) between every crossing of the average value. On the contrary, a broad-banded process has several maxima and minima between every crossing to the average value.

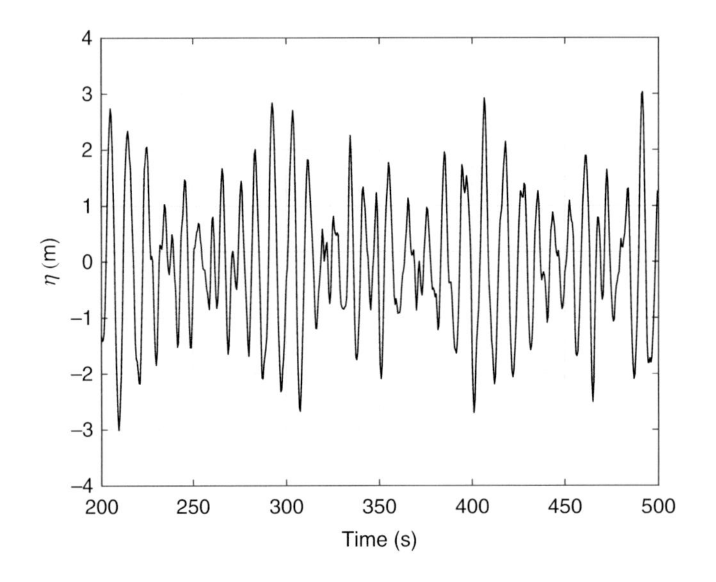

Figure 2.28 Example of a realization of a time history of a Jonswap spectrum with $H_s = 5.0\ m$, $T_p = 10.0\ s$ and $\gamma = 5$. $T_{02} = 8.1\ s$.

This is the case when small, short waves ride on top of long, large waves. Figure 2.28 shows an example of a realization of the wave spectrum in Figure 2.27 (left). The time history is obtained by using a fast Fourier transform of the wave spectrum, using about 1000 frequency components with random phases in the range 0 – 2 Hz. It is observed that the "narrow-band" criterion of only one extremum between each zero-crossing is almost fulfilled. To illustrate a broad-banded process, the rectangular spectrum in Figure 2.29 is used. A realization of a time history based upon this spectrum is shown in Figure 2.30. Here, several extreme values are observed between each zero-crossing. More details on spectral formulation and stochastic processes may be found in, e.g., Naess and Moan (2013).

In the above description of the waves, it is assumed that all wave components are progressing in the same direction. However, within the framework of linearity, a summation of wave components progressing in different directions works as well. We then obtain so-called "short-crested" waves. A common way of writing the directional wave spectrum is by introducing a directional weight function, i.e., the wave spectrum is written as:

$$S(\omega, \chi) = S(\omega) D(\chi). \qquad [2.68]$$

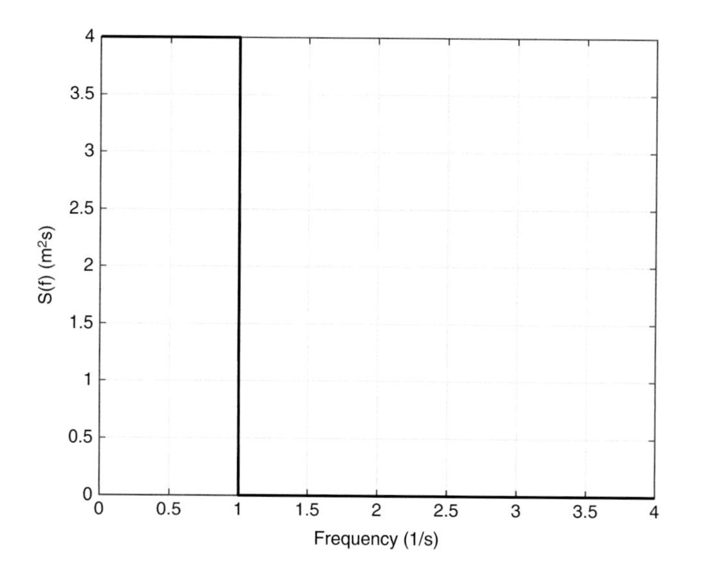

Figure 2.29 Rectangular, broad-banded spectrum with area equal to 4 m^2.

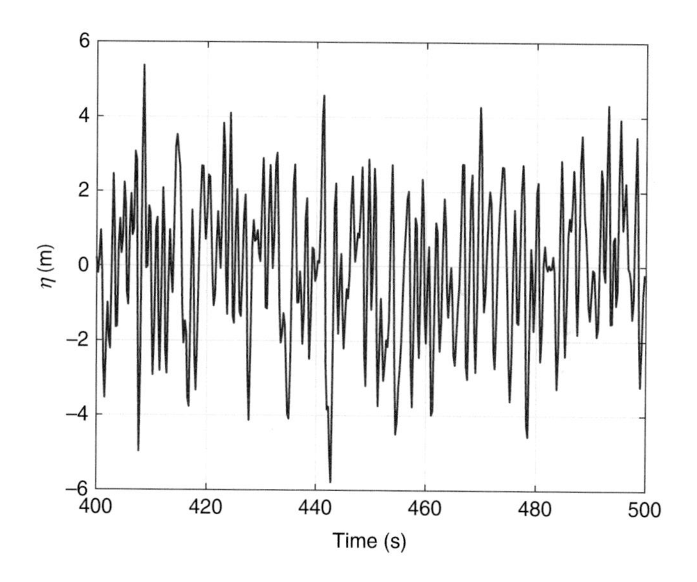

Figure 2.30 Extract of a realization of a time history of the spectrum in Figure 2.29.

Here, $D(\chi)$ is the directional weight factor, which has the property $\int_{\theta} D(\chi)d\chi = 1$. The integral is to be taken over all wave directions. A commonly used directional function (see DNV, 2021c) is:

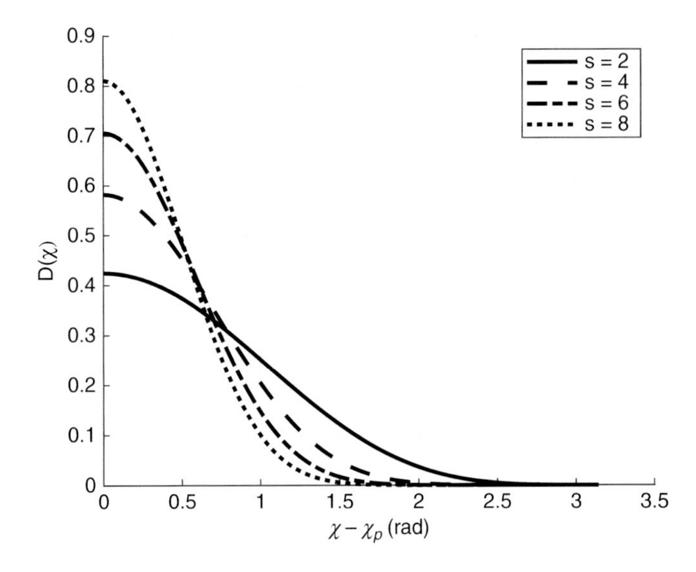

Figure 2.31 The directional spreading function, $D(\chi)$, for different values of s.

$$D(\chi) = \frac{\Gamma(s+1)}{2\sqrt{\pi}\,\Gamma(s+1/2)} \cos^{2s}\left(\frac{1}{2}\left(\chi - \chi_p\right)\right). \qquad [2.69]$$

Here, the first term contains Gamma-functions, which are present to secure that the integral over all directions becomes unity. χ_p is the prevailing wave direction and $|\chi - \chi_p| \leq \pi$. For wind sea, typical values of s are in the range 5–15, while swells may have $s > 15$. Examples of the directional spreading function are shown in Figure 2.31. In the above formulation of the directional spreading, all wave components get the same directional spreading. This is normally not the case in real seas. Thus, a more realistic formulation is to introduce frequency-dependent spreading, i.e., $D = D(\chi, \omega)$. To find a proper function suitable for design is, however, a challenge.

2.2.7 *Wave Kinematics in Irregular Waves*

When estimating wave kinematics, particle velocities and accelerations, in an irregular sea, we may as a first approximation use linear superposition, as for the wave elevation. This works well below the mean free surface level. However, above the mean free surface level we have a challenge as linear theory is not valid here. Several engineering approaches exist for how to estimate the wave kinematics in wave crests. It is of particular importance to have good estimates on the velocities in the water under the highest wave crests. These velocities may in many cases determine the extreme loads on marine structures. Thus, to obtain reliable designs, reliable velocity estimates are of vital importance. The following three methods are frequently used for estimating kinematics in wave crests (NORSOK, 2017).

- Wheeler stretching
- Second-order kinematic models
- CFD techniques

Which method to use depends upon the accuracy required in the estimates as well as the design conditions to be considered. Wheeler stretching is frequently used in engineering applications as the method is simple and it is straightforward to implement. The method is based on the assumption that a wave elevation record is available, e.g., from measurements. The idea is to use the sum of the linear velocities from each spectral component, "re-computed" using a water depth from the actual free surface $d' = d + \zeta$. The computed velocities are assumed valid from the actual free surface level downwards. The principle is illustrated in Figure 2.32. Using direct extrapolation of the linear velocity profile above the mean surface level will usually overestimate the fluid velocities, in particular for the short wave components. In steep waves the Wheeler stretching method may underestimate the fluid velocities.

More consistent methods based upon second-order perturbations have been developed; see, e.g., Stansberg (2011) and Johannessen (2011). Birknes et al. (2013) have compared various approaches. In general, all such perturbation approaches tend to fail for very steep waves. Methods based upon second-order perturbations also have limited applicability in shallow water. Thus, for very steep waves and shallow water, methods using computational fluid dynamics (CFD) techniques or model testing seem to be the most appropriate approaches to obtain proper wave kinematics. However, to study the local kinematics in steep, breaking waves, methods based upon potential theory have also been developed. For

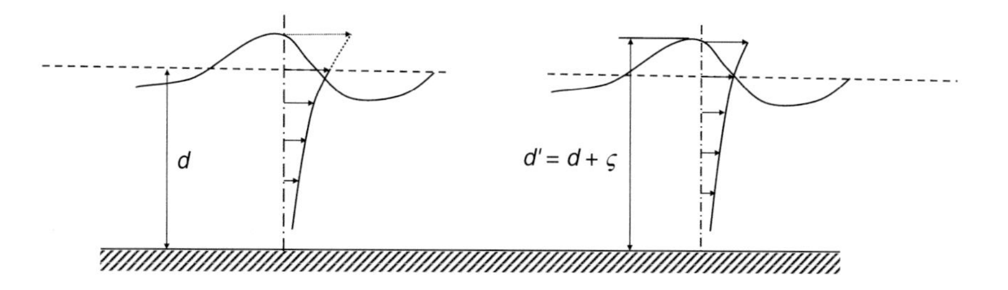

Figure 2.32 Illustration of Wheeler stretching versus pure extrapolation. Left: velocity profile under wave crest, extrapolated linear profile indicated by dotted line. Right: Wheeler stretching.

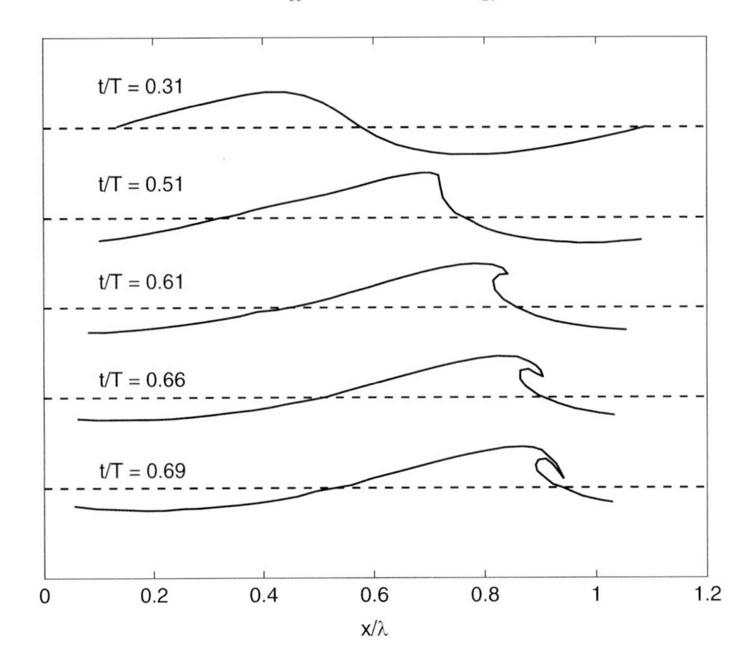

Figure 2.33 Wave profile of a plunging breaker in deep water; T is the wave period. The initial state is a steep sinusoidal wave. As computed by Vinje and Brevig (1980). Reproduced with permission by SINTEF OCEAN.

example, Vinje and Brevik (1980) developed a boundary element method based upon potential theory to compute the development of steep, breaking waves. The method assumes a train of equal waves and breaks down when the wave crest hits the free surface. Results from the method are illustrated in Figure 2.33. The computations indicate maximum fluid velocities in the crest beyond the phase velocity of the wave and accelerations in excess of the acceleration due to gravity.

2.3 Wave Statistics

In wave statistics a distinction is made between short- and long-term statistics. Short-term statistics assumes a stationary sea state, i.e., all statistical parameters are constant in the period considered. The process is also assumed to be ergodic. In ocean wave applications the definition of "short-term" frequently is in the range of 1–3 h. The period chosen is often a compromise between the stationarity of the sea state and the require length to achieve proper response statistics of the structure considered.

Long-term statistics considers the statistics of key parameters as significant wave height and mean zero-crossing period over time intervals of month to many years.

2.3.1 Short-Term Statistics

During a short period of time, the wave condition may be considered stationary. Assuming stationarity, adhering to the linearity assumption and assuming a narrow-banded process, some simple statistics may be derived for the wave heights. We may assume that the surface elevation is represented by a Gaussian process. The statistics of wave heights (trough to crest) may then be modeled by a Rayleigh distribution, see Figure 2.34, i.e.:

$$P_H(h) = 1 - e^{\left[-\left(\frac{h}{\alpha_H H_s}\right)^2\right]}. \quad [2.70]$$

Here, $P_H(h)$ is the cumulative probability of the wave height, being less than h. α_H is a parameter related to the spectral width, for an infinitely narrow-band process $\alpha_H = \frac{1}{2}\sqrt{2}$. Næss (1985) found that for real sea states, α_H could be expressed by $\alpha_H = \frac{1}{2}\sqrt{1-\rho}$. ρ is related to the bandwidth of the wave spectrum. For a Jonswap spectrum with $\gamma = 3.3$ the value of ρ is obtained as −0.73 (DNV, 2021c). Næss (1985) finds that for most sea states with some severity and without significant swell, $-0.75 < \rho < -0.65$; see Næss and Moan (2013).

During a period of time t (e.g., 3 h), the number of waves passing a fixed point is $N = t/T_{02}$. If the random process is repeated many times, and assuming a narrow-banded Rayleigh distribution, the mean values of the maxima in each of the repeated records will tend toward :

$$H_{\max,mean} = \left[\sqrt{\frac{1}{2}\ln N} + \frac{0.2886}{\sqrt{2\ln N}}\right] H_s. \quad [2.71]$$

The most probable highest wave in a realization (the mode) is similarly given by:

$$H_{\max,\text{mod}} = \left[\sqrt{\frac{1}{2}\ln N}\right] H_s. \quad [2.72]$$

Thus, if we have a wave condition with zero up-crossing period $T_{02} = 10\,s$, then 1080 waves will pass during a period of 3 h. The most probably largest wave during a realization will then be $1.87H_s$, while the average maximum wave during several realizations will be $1.95H_s$. Care should be shown in using the Rayleigh distribution in shallow water. Here, the wave height will typically be limited to 0.78 times the water depth and the tail of the distribution will be distorted. Alternative distributions for the wave heights exists for shallow water; see, e.g., DNV (2021c).

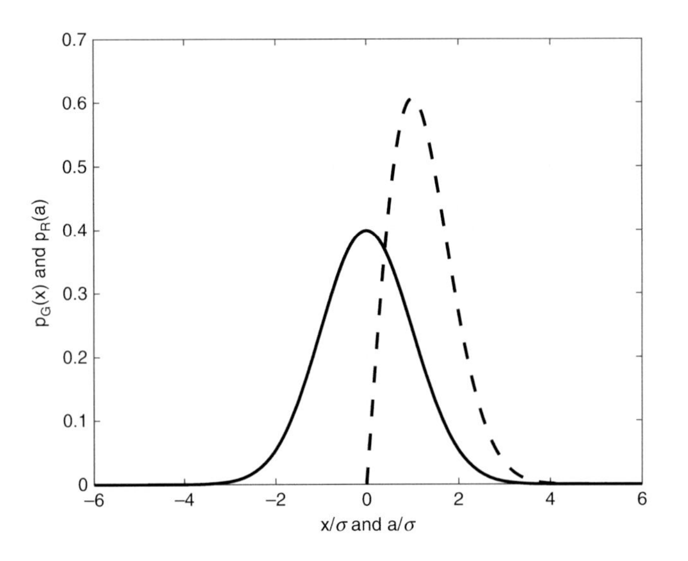

Figure 2.34 Illustration of the probability distribution of a Gaussian free surface elevation x (solid line) and the corresponding Rayleigh distribution of the amplitudes a (dashed line) for a narrow-band process with $\alpha_H = 0.5\sqrt{2}$.

For many applications it is more important to have a good estimate of the extreme crest height rather than the wave height. The Forristall distribution may be used for that purpose (DNV 2021c).

2.3.1.1 Confidence Limits for Short-Term Extreme Values

In analyzing wave time series, very often the extreme value is of interest. It is therefore of interest to estimate the confidence limit of an extreme value obtained from a single realization of a process, either by measurement or simulation. This can be done as follows.

Assume a narrow-band process, $x(t)$, with zero mean, i.e., there is only one positive peak between every zero up-crossing. The peaks are denoted x_a and are assumed independent, i.e., there is no correlation between two neighboring peaks. The cumulative distribution of peaks is denoted $P(\xi) = prob(x_a < \xi)$. As the peaks are assumed independent, the probability of having N peaks less than ξ is given by:

$$prob(x_{a1}, x_{a2}, \ldots, x_{aN} < \xi) = [P(\xi)]^N. \qquad [2.73]$$

The desired confidence limits are the positive and negative extreme values given by:

$$[P(\xi)]^N = \begin{cases} 1-\varepsilon \\ \varepsilon \end{cases}, \quad [2.74]$$

where $\varepsilon \ll 1$ is assumed. Assume further that the amplitudes are Rayleigh-distributed with probability density and cumulative probability given by:

$$\begin{aligned} p(\xi) &= \frac{\xi}{\sigma_x^2} e^{-\frac{\xi^2}{2\sigma^2}} \\ P(\xi) &= 1 - e^{-\frac{\xi^2}{2\sigma^2}} \end{aligned} . \quad [2.75]$$

To establish the probability that the extreme value for a realization of the process stays within the interval given by [2.74], one has to compute $\alpha_1 = P(\xi_1) = \varepsilon^{1/N}$ and $\alpha_2 = P(\xi_2) = (1-\varepsilon)^{1/N}$. Inserting for the Rayleigh distribution, the following is obtained:

$$\begin{aligned} P(\xi) &= 1 - e^{-\frac{\xi^2}{2\sigma^2}} = \alpha \\ \frac{\xi}{\sigma} &= \sqrt{-2\ln(1-\alpha)} \end{aligned} . \quad [2.76]$$

Confidence Limits for Maximum Values

Assume the amplitude distribution obeys the Rayleigh distribution. Set $\varepsilon = 0.05$ *and assume the number of oscillations to be* $N = 200$. *The* $100(1-2\varepsilon) = 90$*-percentile interval is thus given from the probability levels:*

$$P(\xi_1) = \varepsilon^{1/N} = 0.05^{1/200} = 0.98510 = \alpha_1$$

$$P(\xi_2) = (1-\varepsilon)^{1/N} = 0.95^{1/200} = 0.9974 = \alpha_2$$

Inserting into [2.76] the lower and upper limit for the 90th-percentile interval is found as

$$\frac{\xi_1}{\sigma} = 2.90, \quad \frac{\xi_2}{\sigma} = 4.07,$$

while the most probable extreme value in a Rayleigh distribution is given from $\frac{x_{\max,\text{mod}}}{\sigma} = \sqrt{2\ln N} = 3.26$.

Some nonlinear response processes induced by, for example, slow-drift wave loads or low-frequency wind forces may have a statistic of peaks that significantly deviates from the Rayleigh distribution. Sometimes the response statistics are closer to an exponential distribution, i.e.:

$$P(\xi) = 1 - e^{-\frac{\xi}{\sigma}}. \qquad [2.77]$$

Following the procedure above, the confidence limit for the exponential distribution is found from:

$$\frac{\xi}{\sigma} = -\ln(1 - \alpha). \qquad [2.78]$$

In Figure 2.35, the most probable maximum amplitudes and the corresponding 90th-percentile interval for a Rayleigh and an exponential distribution are shown. It is clearly observed that the exponential process has higher expected extreme values as well as a larger confidence interval.

2.3.2 Long-Term Wave Statistics

Assume that during a long period of time, say, one year, the wave conditions may be described by a large number of stationary wave conditions of duration, for example, of 3 h. Each of these stationary conditions may be characterized by key spectral parameters such as significant wave height and spectral peak period. Each of the 3 h periods are assumed independent. In an average year, there will be 2922 such

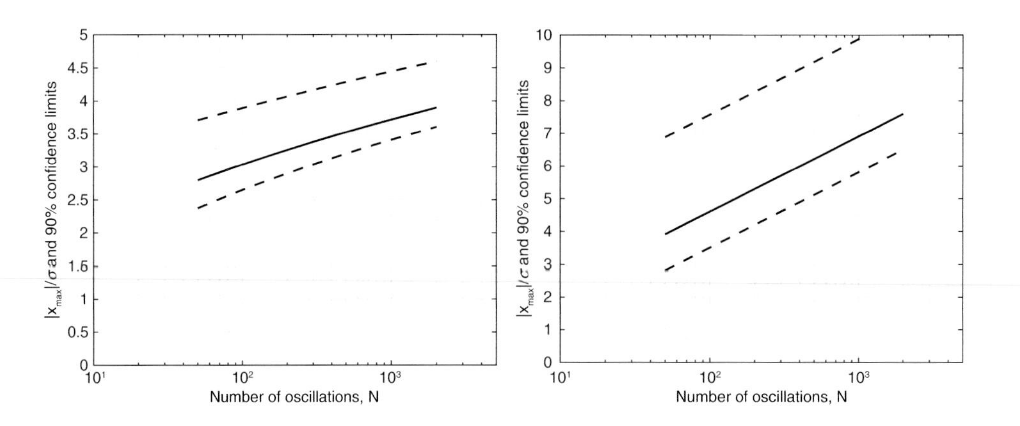

Figure 2.35 Most probable largest value in a narrow-banded process with Rayleigh-distributed (left) and exponential-distributed (right) amplitudes as a function of number of oscillations. Solid line: most probable value. Dashed lines: lower and upper limits for the 90th-percentile interval. Zero mean value is assumed.

periods. To present the long-term statistics of the key wave parameters, a "scatter diagram" is frequently used, as illustrated in Table 2.4. The data are based upon approximately 18 years of measured data, altogether 52 365 wave records. The measured significant wave heights are grouped in bins of 0.5 m and the peak periods are grouped in bins of 1 s. The numbers given in the scatter diagram are the number of observations within each $H_s - T_p$ bin combination. The marginal and cumulative probability with respect to H_s and T_p are given along the edges of the table. For these specific data the yearly average H_s is obtained as 2.77 m and the yearly average T_p is 9.52 s. It is also observed that the most probable value (the mode) of the significant wave height and the peak period are both lower than the corresponding average values. This illustrates the skewness in the probability distributions. The scatter diagram in Table 2.4 includes all wave directions. For design purposes, such diagrams are frequently made for several wave directions, typically 12 wave sectors, each of 30 deg.

In the design of marine structures, extreme value estimates are needed. This may be either the extreme wave height independent of spectral peak period or the extreme wave height given a spectral peak period.

For the long-term distribution of the significant wave height, a three-parameter Weibull distribution may be used. The cumulative probability using the three-parameter Weibull distribution may be written in the form:

$$P_{H_s}(h) = 1 - \exp\left[-\left(\frac{h-\gamma}{\alpha}\right)^{\beta}\right]. \quad [2.79]$$

$\alpha(>0)$ is the scale parameter, used to normalize the variable; $\gamma(\leq h)$ is the location parameter, used to define a lower threshold for the variable; and $\beta(>0)$ is the shape parameter, defining the shape of the distribution. Consider periods of duration of 3 h. If we want to estimate the significant wave height that is exceeded once every N^{th} year, the probability level we are seeking, the target probability, is given by:

$$P_N = \left(\frac{24}{3}\cdot 365.25\cdot N\right)^{-1} = N_i^{-1}. \quad [2.80]$$

Here, the first factor is the number of 3 h data points per day and 365.25 is the average number of days per year. Using the three-parameter Weibull distribution, the significant wave height that is expected to on average be exceeded once every N^{th} year is now given by:

$$H_{sN} = \gamma + \alpha\left(\ln(N_i)\right)^{1/\beta}. \quad [2.81]$$

In estimating the parameters in the Weibull distribution, all the available data are used. However, it is frequently observed that the tail of the distribution, representing the highest waves, deviates slightly from the distribution obtained using the complete dataset. Therefore, in estimating the extreme value for a long period of time, e.g., 50 years, two other methods are frequently employed: the annual maximum (AM) method and the peak over threshold (POT) method.

In the AM method, only the largest significant wave height measured every year is used in the statistical distribution. The data are fitted to a Gumbel distribution in the form:

$$P_{H_{se}}(h_e) = \exp\left(-\exp\left(-\frac{h_e - \mu}{\sigma}\right)\right). \quad [2.82]$$

Here, h_e is the distribution of the yearly extreme significant wave height; μ is the location parameter; and σ is the scale parameter. The significant wave height expected to be exceeded once every Nth year is now given by:

$$H_{sN} = \mu - \sigma\left[\ln\left(-\ln\left(1 - \frac{1}{N}\right)\right)\right]. \quad [2.83]$$

To obtain a reliable extreme value estimate by the AM method, data for many years are needed.

The POT method represents an alternative. Here, all measured significant wave heights above a certain threshold are used to establish the extreme value distribution. Thus, many values per year may enter the distribution without accounting for all the small H_s . A basic assumption in this approach is that all H_s entering the distribution are independent, i.e., they do not belong to the same storm event. A key issue related to the POT method is the choice of threshold. The choice of an excessively high threshold leads to few data entering into the distribution, causing greater uncertainty in the estimates than for the AM method. Choosing a threshold that is too low leads to a lot of low wave heights entering the distribution, introducing a possible bias to the extremes, in the same way as for the Weibull approach.

Details related to the use of the methods are found in, for example, Orimolade et al. (2016) and Naess and Moan (2013).

In Figures 2.36 and 2.37, the significant wave heights in the northern North Sea from measurements over a period of about 18 years are presented. The measurements are taken once every third hour. The stars are the distribution based upon binned data. 100 bins are used. The solid line is the fitted three-parameter Weibull distribution. The logarithmic scaling in Figure 2.37 is introduced to obtain a linear relation between the two axes for large H_s values. By this scaling it is also possible to read out the wave heights corresponding to low probability of exceedance.

Long-Term Extreme Value

*Find an estimate on the 3 h significant wave height expected to be exceeded once every N^{th} year. The aim is thus to find a wave height with probability of exceedance equal to 1/(N*8760/3). The significant wave height that is expected to be exceeded once during a period of 50 years corresponds to cumulative probability level of H_s equal to 1-1/146000. On the Weibull scale this becomes* $\ln(-\ln(1/146000)) = 2.4758$. *Using [2.81] and the distribution fitted to the data in Figure 2.36, the corresponding 50-year extreme 3 h significant wave height becomes:*

$$H_{s50} = 0.6234 + 2.370 \cdot \left(\ln(146000)\right)^{1/1.425} = 14.09m. \qquad [2.84]$$

Similar results for 1–100-year extreme values are shown in Table 2.3.

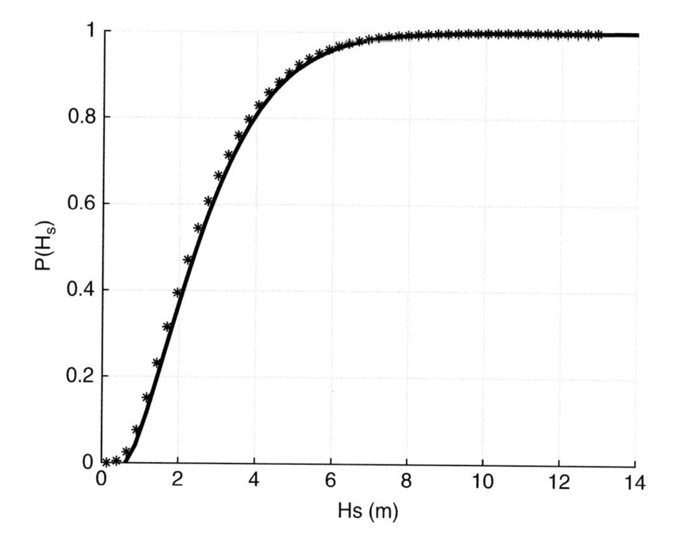

Figure 2.36 The cumulative probability of significant wave height. The stars are binned measured data over a period of approximately 18 years from the northern North Sea. The solid line is a fitted three-parameter Weibull distribution with $\alpha = 2.370, \beta = 1.425$ and $\gamma = 0.6234$.

Table 2.3 *Example of expected 3 h extreme significant wave height versus return period using the Weibull distribution obtained by fitting the data in Figure 2.36*

Return Period, N (years)	Probability Level $\ln(-\ln(1/2920 \cdot N))$	3 h extreme H_s (m)
1	2.0769	10.80
10	2.3304	12.78
50	2.4758	14.09
100	2.5325	14.64

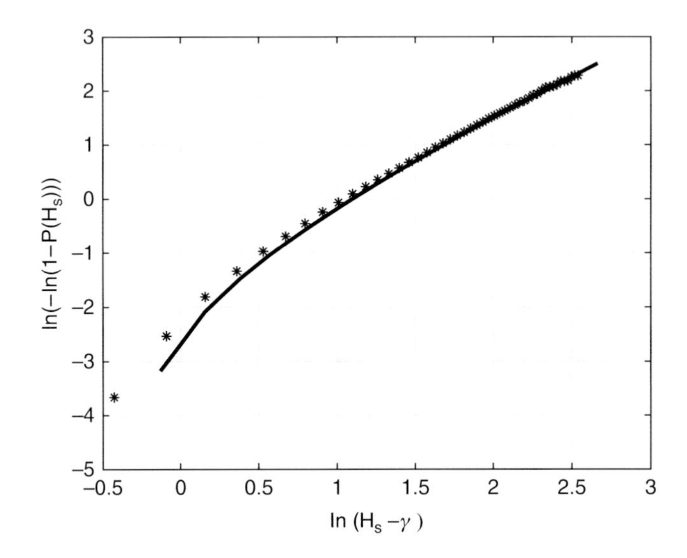

Figure 2.37 As Figure 2.36, but plotted on a Weibull scale.

Exercises Chapter 2

1. Which main assumptions are used to derive the logarithmic vertical profile for the mean wind velocity?
2. In engineering applications an exponential wind profile is frequently used rather than a logarithmic profile. Assume the wind speed at 10 m above sea level is known. Compute the wind profile by the logarithmic and the exponential approach in the range 1–250 m above sea level. Discuss how the two profiles deviate. How will the profiles match if you force the velocities to be equal at a higher level than at the reference level of 10 m?

Table 2.4 *Scatter diagram. Example based upon approximately 18 years of measured data from the North Sea.*

	T_p (s)																				
Hs (m)	<4	4–5	5–6	6–7	7–8	8–9	9–10	10–11	11–12	12–13	13–14	14–15	15–16	16–17	17–18	18–19	19–20	>=20	Marg sum	Cum sum	Marg prob	Cum prob
< 0.5	25	14	18	14	24	17	8	9	5	0	2	3	1	1	6	2	0	7	156	156	0.0030	0.0030
0.5–1	79	235	408	539	496	488	426	225	111	46	16	7	6	3	2	0	2	17	3106	3262	0.0593	0.0623
1–1.5	71	466	896	1223	1227	1219	911	626	470	252	108	70	36	12	13	2	2	49	7653	10915	0.1461	0.2084
1.5–2	3	134	768	1342	1443	1288	1143	854	618	404	205	129	44	21	10	3	4	21	8434	19349	0.1611	0.3695
2–2.5	0	22	292	1013	1328	1359	1056	921	669	431	188	186	56	27	17	5	1	16	7587	26936	0.1449	0.5144
2.5–3	1	2	79	491	1037	1276	1031	795	678	479	208	193	81	35	13	3	2	8	6412	33348	0.1224	0.6368
3–3.5	0	1	16	189	701	1038	896	729	532	386	235	197	97	40	15	4	1	5	5082	38430	0.0970	0.7339
3.5–4	0	0	3	57	318	814	767	562	447	371	193	168	89	50	17	3	0	3	3862	42292	0.0738	0.8076
4–4.5	1	0	1	11	130	568	731	500	341	306	146	142	67	29	25	4	3	2	3007	45299	0.0574	0.8651
4.5–5	0	0	0	3	44	329	533	444	337	248	122	95	40	31	20	2	4	1	2253	47552	0.0430	0.9081
5–5.5	0	0	0	1	15	112	339	396	280	195	121	98	31	26	16	5	2	0	1637	49189	0.0313	0.9393
5.5–6	0	0	0	1	4	46	193	244	224	138	73	70	27	15	15	3	0	1	1054	50243	0.0201	0.9595
6–6.5	0	0	0	0	0	18	91	177	156	125	65	51	20	15	7	1	2	2	730	50973	0.0139	0.9734
6.5–7	0	0	0	0	1	1	34	108	137	106	46	40	14	5	1	2	0	0	495	51468	0.0095	0.9829
7–7.5	0	0	0	0	0	1	11	53	93	74	50	30	14	7	8	0	1	0	342	51810	0.0065	0.9894
7.5–8	0	0	0	0	0	0	1	29	61	63	27	19	6	3	1	1	0	1	212	52022	0.0040	0.9934
8–8.5	0	0	0	0	0	1	0	14	31	43	24	13	6	5	1	0	0	0	138	52160	0.0026	0.9961
8.5–9	0	0	0	0	0	0	0	8	19	27	17	8	7	1	0	0	0	0	87	52247	0.0017	0.9977
9–9.5	0	0	0	0	0	0	0	2	5	18	13	9	2	1	1	0	0	0	51	52298	0.0010	0.9987
9.5–10	0	0	0	0	0	0	0	1	2	7	7	4	3	3	0	0	0	0	27	52325	0.0005	0.9992
10–10.5	0	0	0	0	0	0	0	0	3	3	7	4	4	1	0	0	0	0	22	52347	0.0004	0.9997
10.5–11	0	0	0	0	0	0	0	0	0	3	1	2	0	0	0	0	0	0	6	52353	0.0001	0.9998
11–11.5	0	0	0	0	0	0	0	0	0	4	2	1	0	0	0	0	0	0	7	52360	0.0001	0.9999
11.5–12	0	0	0	0	0	0	0	0	0	1	0	0	0	0	0	0	0	0	1	52361	0.0000	0.9999
12–12.5	0	0	0	0	0	0	0	0	0	1	0	0	1	0	0	0	0	0	2	52363	0.0000	1.0000
12.5–13	0	0	0	0	0	0	0	0	0	0	0	0	0	0	1	0	0	0	1	52364	0.0000	1.0000
13–13.5	0	0	0	0	0	0	0	0	0	0	1	0	0	0	0	0	0	0	1	52365	0.0000	1.0000
13.5–14	0	0	0	0	0	0	0	0	0	0	0	0	0	0	0	0	0	0	0	52365	0.0000	1.0000
14–14.5	0	0	0	0	0	0	0	0	0	0	0	0	0	0	0	0	0	0	0	52365	0.0000	1.0000
14.5–15	0	0	0	0	0	0	0	0	0	0	0	0	0	0	0	0	0	0	0	52365	0.0000	1.0000
>=15	0	0	0	0	0	0	0	0	0	0	0	0	0	0	0	0	0	0	0	52365	0.0000	1.0000
Marg sum	180	874	2481	4884	6768	8575	8171	6697	5219	3731	1877	1539	652	331	189	40	24	133				
Cum sum	180	1054	3535	8419	15187	23762	31933	38630	43849	47580	49457	50996	51648	51979	52168	52208	52232	52365				
Marg prob	0.0034	0.0167	0.0474	0.0933	0.1292	0.1638	0.1560	0.1279	0.0997	0.0712	0.0358	0.0294	0.0125	0.0063	0.0036	0.0008	0.0005	0.0025				
Cum prob	0.0034	0.0201	0.0675	0.1608	0.2900	0.4538	0.6098	0.7377	0.8374	0.9086	0.9445	0.9739	0.9863	0.9926	0.9962	0.9970	0.9975	1.0000				

3. [2.11] gives the flux Richardson number. A gradient Richardson number may be written as $R_g = \frac{g\bar{\theta}\frac{\partial\bar{\theta}}{\partial z}}{\left(\frac{\partial\bar{u}}{\partial z}\right)^2}$. Show that $R_g/R = K_m/K_h$, where K_m and K_h are the eddy diffusivity for momentum and heat respectively.
4. Turbulence is frequently described as a stationary, Gaussian, homogeneous and ergodic random process. Explain what is meant by these characteristics.
5. Use the wind time history in the data file Wind_vel.txt to compute the distribution of wind speeds 100 m above sea level using: a) all-year data; and b) seasonal data (winter (Dec–Feb); spring (Mar–May); summer (Jun–Aug); autumn (Sep–Nov). Discuss the differences.
   a. Plot the distribution of the mean wind speed on an all-year basis as well as a seasonal basis. Fit the distributions above to Weibull distributions. (You may assume $\gamma = 0$.)
   b. Use the same data as above and compute the yearly as well as seasonal mean, median, maximum and standard deviation of the wind speed 100 m above sea level. Can you see any trends over the investigated period?
   c. Choose four (or more) wind situations picked midday on a day in January, April, July and October. Investigate the wind profile by estimating the shear exponent*. Discuss your findings.

** = The shear exponent may be estimated using the wind velocity at two vertical levels. A better estimate is obtained using all levels you have data for. In the latter case you may take the logarithmic to the exponential relation and make a linear fit using a least-square fitting technique, i.e., you may do as follows:*

Assume the mean velocities U_i are given at the vertical levels z_i. We want to find the exponent α that fits the exponential profile $U(z) = U(z_r)\left(\frac{z}{z_r}\right)^\alpha$. Taking the logarithmic of this equation, we obtain a linear equation with two unknown quantities α and $\log\left(U(z_r)\right)$:

$$\log\left(U(z)\right) = \alpha\log\left(\frac{z}{z_r}\right) + \log\left(U(z_r)\right).$$

The two parameters can be found by fitting the equation in a least-square sense to the "observed" data. Use $z_r = 10$ m. Instructions on how to fit data to a linear equation may be found here: https://mathworld.wolfram.com/LeastSquaresFitting.html (accessed July 2023; compare the above equation to equation 3 at the website).

6. Use the wind time history in Wind_vel.txt and Wind_dir.txt and sort the data with respect to wind direction. Use four direction intervals: 315 deg–45 deg

(northerlies); 45 deg–135 deg (easterlies); 135 deg–225 deg (southerlies); and 225 deg–315 deg (westerlies).

a. Compute the distribution of wind speeds 100 m above sea level for each direction. Are there any differences?

b. Can you explain the reason for the differences? (The data are from a site located far offshore in the North Sea (about 5°0" east, 56°5" north).

7. The file WindTimeSeries.txt gives a 40-min-long record of wind velocity at five vertical levels at a coastal site. The data are sampled at an interval of 1.1719 s. The file is organized in five columns, one for each height. The heights, z, in meters are given at the first line of the file. The wind speeds are given in m/s.

   a. Calculate the mean wind speed, standard deviation and turbulence intensities at the five heights.

   b. Investigate how the observed mean wind profile compares to logarithmic and power law formulations.

   c. Calculate the vertical coherence for the wind speeds using the 10 m level and the three levels above. Plot the coherences in a common plot and discuss the result.

8. Use the wave time history in Wave_data.txt to compute the distribution of significant wave height, H_s, using: a) all-year data; and b) seasonal data (winter (Dec–Feb); spring (Mar–May); summer (Jun–Aug); autumn (Sep–Nov).

   a. Discuss the differences. The data are from the same location as the wind data above.

   b. Fit the distributions above to two-parameter Weibull distributions. (Use $\gamma = 0$.) Discuss the Weibull parameters found.

   c. From the distributions of H_s above, find how large fraction of the time H_s is below certain limits (e.g., 1, 2 and 3 m) during the various seasons.

9. Make a 30-min time series of wave elevation by summation of a (large) number of regular wave components. Pick the amplitudes so that the time series represents a realistic wave spectrum. Use, for example, $H_s = 5\,m$, $T_p = 10\,s$ and $\gamma = 3.3$ in the Jonswap spectrum. Assume infinite water depth.

   a. Make a histogram of the wave heights in the time series and compare it to a Rayleigh distribution.

   b. Repeat the generation of the time series and check how the maximum amplitudes compare with what should be expected from a Rayleigh distribution.

c. Compute the wave elevation spectrum and check how it compare to the input you used.
d. What does the horizontal velocity and acceleration time series look like at $z = 0\,m$ and $z = -10\,m$?
e. Compute the spectra of the velocity and acceleration time series.
f. Can you compute the spectra of the velocity and acceleration time series directly from the wave elevation spectrum?

3
Wind Energy and Wind Loads

This chapter outlines the principles for extracting wind energy, with a focus on horizontal-axis wind turbines (HAWT) as these are the most common offshore wind turbines. How much energy is it possible to extract from such a turbine? This question is addressed by classical, simplified momentum considerations. It is also shown that when energy is extracted, consequential wind forces, or thrust forces, are acting on the turbine. The structure supporting the wind turbine must be designed to withstand these forces.

Having established an upper limit for the available power, the next question to address is how it can be achieved. Lifting surfaces as used for airplanes have shown themselves to also be the most efficient principle to extract power from wind turbines. The basic principles for two-dimensional (2D) lifting surfaces are thus addressed. The 2D lifting surface, or aerofoil, together with the momentum theory is the starting point for the most common way of computing the power output and forces on a HAWT, through the beam element momentum (BEM) method.

Wind turbines have increased in size over the years, with turbines with diameters of greater than 200 m now becoming common offshore. This introduces new challenges to the design of the turbines; understanding the time-dependent loads in a turbulent wind field and three-dimensional (3D) flow effects is essential. To understand the basics of the time-dependent load effects, the 2D aerofoil is considered, while both time-dependent loads and 3D effects are efficiently described by using vortex methods. Vortex methods also constitute an intermediate step between the simple BEM method and time-consuming computational fluid dynamics (CFD) methods.

The power characteristic of a modern HAWT is discussed in some detail and compared to idealized turbines.

In principle, as much energy as possible should be extracted from a wind turbine. However, at high wind speeds, the available power is too large for the electrical generator. Therefore, efficient control systems are required, both to maximize the power extraction at low wind speeds and to limit the power extraction at high wind speeds.

3.1 The Betz Theory

Consider a flow of air in x-direction with constant speed U and a virtual stream tube with axis in the flow direction. The air is flowing in and out through the end surfaces and no flow is crossing the cylindrical surface. The cross-sectional area is A (see Figure 3.1). For a body with mass m the kinetic energy may be expressed as $E = (1/2)mU^2$. The mass crossing A during a time interval dt is $m = \rho AU\,dt$. The power in the air crossing the surface A is the time derivative of the energy and obtained as:

$$P = \frac{dE}{dt} = \frac{1}{2}\frac{dm}{dt}U^2 = \frac{1}{2}\rho AU^3. \qquad [3.1]$$

The power in the air flow is thus proportional to the third power of the wind speed.

How much of this power may be extracted? By considering the cross-section A, the following observations are made. If all the kinetic energy is extracted, the flow will have to stop, and the air will thus "pile" up in front of A. This is not possible. Similarly, if the wind speed is the same on both sides of the tube, the kinetic energy is the same at both sides of the tube, and no power is extracted. To find an answer to the question about maximum possible power extraction, some simplifications are made. The flow is assumed to be ideal and incompressible; thus, the density of the air is the same in the whole flow volume considered and no viscous losses are present. The flow is considered steady and unidirectional, thus any radial flow components are ignored. Assume now a virtual "actuator disc" is inserted into the stream tube at location A_1 (see Figure 3.2). The actuator disc is assumed to extract power from the flow. As power is extracted by the disc, the flow speed will be reduced, which again means that a force will act on the disc. The fluid volume between a cross-section far upstream, at A_0, and a cross-section far downstream, at A_2, is

Figure 3.1 Illustration of an airflow of speed U through a cylinder with cross-section A.

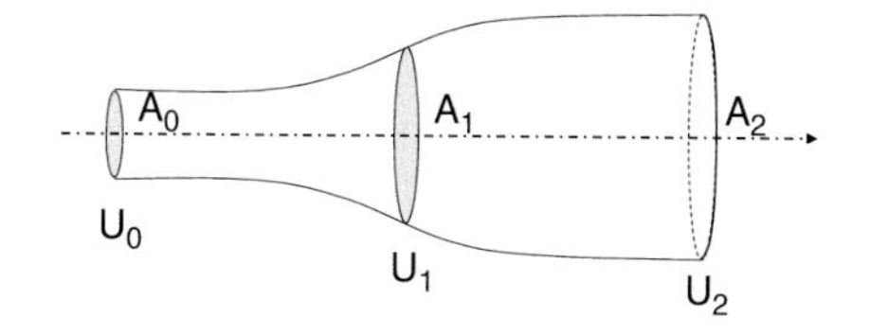

Figure 3.2 The stream tube with an actuator disc located at A_1.

considered. The streamlines limiting the volume in radial direction are such that the mass flow through all cross-sections is equal. To study the flow, three basic conservation relations are considered: the conservation of mass, the conservation of momentum and the conservation of energy.

As there is no flow through the sides of the stream tube, the conservation of mass implies that the mass flux through all cross-sections of the stream tube is equal:

$$\frac{dm}{dt} = U_0 \rho A_0 = U_1 \rho A_1 = U_2 \rho A_2. \qquad [3.2]$$

Here, the density is equal at all cross-sections as the flow is assumed to be incompressible.

The conservation of momentum in the stream tube implies that the change of momentum in the fluid volume considered must equal the forces acting on the volume. In the present 1D case the change in momentum is given from the difference in momentum between the flow into the stream tube and out of the stream tube:

$$T = \frac{dM_x}{dt} = \frac{dm}{dt}(U_0 - U_2) = \rho A_1 U_1 (U_0 - U_2). \qquad [3.3]$$

Here, T is the thrust acting from the fluid on the actuator disc. This is assumed to be the only axial force acting on the fluid volume. (One may add a correction to the thrust by accounting for the axial component of the pressure acting on the expanding part of the stream tube. This is, however, not done in the classical derivation of the Betz theory.) M_x is the momentum of the fluid in the $x-$ direction.

The conservation of energy is expressed by the Bernoulli equation. As a force is supposed to act on the actuator disc, there will be a pressure jump at this location. The Bernoulli equation must thus be applied separately upstream and downstream of the actuator disc:

$$\begin{aligned} p_0 + \frac{1}{2}\rho U_0^2 &= p_{1+} + \frac{1}{2}\rho U_1^2 \\ p_{1-} + \frac{1}{2}\rho U_1^2 &= p_2 + \frac{1}{2}\rho U_2^2. \end{aligned} \qquad [3.4]$$

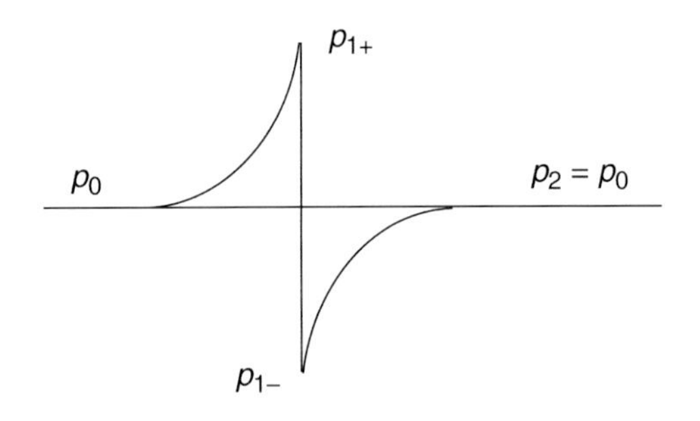

Figure 3.3 Pressure jump over the actuator disc.

Here, p_{1+} and p_{1-} are the pressures at the upstream and downstream side of the actuator disc respectively. Far away from the disc, the pressure is equal to the pressure in the ambient flow, i.e., $p_2 = p_0$. The development of the pressure in the stream tube is illustrated in Figure 3.3. Integrating the pressure jump over the actuator disc gives the force acting upon the disc:

$$T = A_1(p_{1+} - p_{1-}) = \frac{1}{2}\rho A_1 \left(U_0^2 - U_2^2\right). \quad [3.5]$$

The force acting upon the actuator disc is now obtained both by considering the conservation of momentum and the conservation of energy. By equating these results, we obtain:

$$\begin{aligned} T &= \frac{1}{2}\rho A_1 \left(U_0^2 - U_2^2\right) = \rho A_1 U_1 (U_0 - U_2) \\ &\text{i.e.} \\ U_1 &= \frac{1}{2}(U_0 + U_2). \end{aligned} \quad [3.6]$$

It is thus obtained that the velocity at the actuator disc is the average between the speed far upstream and far downstream. An axial induction factor, a is introduced. a expresses the velocity at the actuator disc relative to the upstream velocity:

$$U_1 = (1 - a)U_0. \quad [3.7]$$

Introducing this into the expression for the thrust force, we may write:

$$\begin{aligned} T &= \rho A_1 U_1 (U_0 - U_2) = 2\rho A_1 U_0^2 a(1 - a) = \frac{1}{2}\rho A_1 U_0^2 C_T \\ &\text{with} \\ C_T &= 4a(1 - a). \end{aligned} \quad [3.8]$$

C_T is denoted as the thrust coefficient. The power exerted upon the actuator disc is given by the force times the velocity of the fluid crossing the disc area, i.e.:

$$P = TU_1 = 2\rho A_1 U_0^3 a(1-a)^2 = \frac{1}{2}\rho A_1 U_0^3 C_P. \qquad [3.9]$$

It is observed that the power extracted from the fluid is the total power in the incident flow multiplied by a power coefficient, C_P. The power coefficient is given from the induction factor. The maximum possible power extraction is found by considering dP/da. $dP/da = 0$ is obtained for $a = 1/3$. This corresponds to a maximum value of P. Inserting this value for the induction factor in the expressions for the thrust and the power, the following values are obtained for the power and thrust coefficients:

$$\begin{aligned} C_{P_{\max}} &= C_{P,a=1/3} = \frac{16}{27} \simeq 0.59 \\ C_{T,a=1/3} &= \frac{8}{9} \end{aligned} \qquad [3.10]$$

This maximum value for the power coefficient is called the "Betz limit," after Betz (1926). It should be noted that this limit is derived without any assumptions related to how the power extraction is performed. It is assumed that the power extraction reduces the axial velocity only and no energy is used to set the fluid into rotation. The issue of rotation is discussed below. It is also observed that the slow-down of the fluid takes place both in front of and behind the actuator disc. Actually, according to [3.6], half of the speed reduction has taken place as the fluid passes the actuator disc. The reduction of the wind speed upstream of the actuator disc is frequently denoted as an induction effect.

In Figure 3.4, the power and thrust coefficients are plotted as functions of the axial induction factor. The ratio between the upstream and downstream fluid velocities is also included. It is observed that the downstream velocity becomes negative if $a > 0.5$. Thus, for $a > 0.5$ the flow is reversed, and the initial assumption of a stream tube is violated. The present assumptions and derivation are thus not valid as the induction factor approaches 0.5 and beyond.

3.2 Including the Effect of Wake Rotation

In deriving the Betz limit, it was assumed that only the flow in the $x-$ direction was of importance. However, for a HAWT the power is extracted from the wind by generating torque in the rotor. This torque must, considering the conversation of momentum, be balanced by a rotational motion in the fluid. A classical way of showing the effect of the rotation on power production is by considering an annular area through the rotor plane, as illustrated in Figure 3.5. Within the annulus, continuity of the flow in axial direction requires:

$$2\pi u_{x0} r_0 dr_0 = 2\pi u_{x1} r_1 dr_1 = 2\pi u_{x2} r_2 dr_2. \qquad [3.11]$$

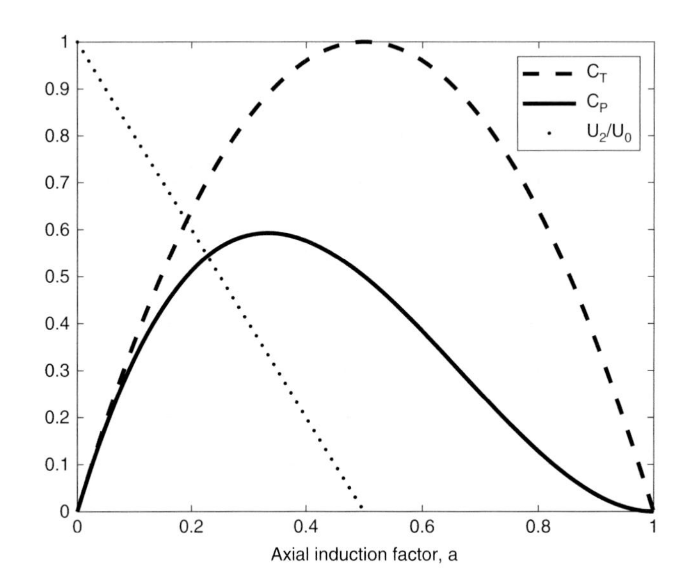

Figure 3.4 Power and thrust coefficient and far downstream velocity as a function of the axial induction factor.

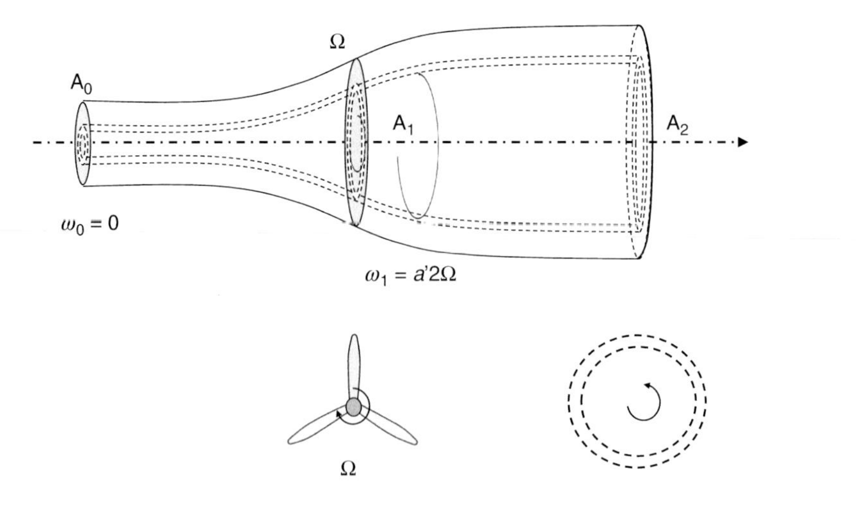

Figure 3.5 Illustration of stream tube and the annulus used in analyzing the effect of rotation.

Here, the indices 0, 1 and 2 as before refer to the three sections considered: far upstream, at the turbine and far downstream. r is the radius of the annulus, dr is the thickness and u_x is the axial velocity component in the annulus.[1]

Let u_r and u_θ denote the radial and azimuthal velocity components in the annulus respectively. $\omega = u_\theta / r$ is then the angular velocity of the fluid in the annulus considered. In this classical approach, it is assumed that the azimuthal velocity jumps from zero to $u_{\theta 1+}$ at the rotor disk (Burton et al., 2011). Continuity of momentum of the rotational fluid motion, the angular momentum, just behind the rotor and far downstream, requires that $u_{\theta 1+} r_1 = u_{\theta 2} r_2$, or $\omega_1 r_1^2 = \omega_2 r_2^2$, as there are no rotational forces acting on the fluid between these two sections and no exchange of momentum between the annuli are assumed. The torque acting on the rotor section from r to $r + dr$ must equal the change in angular momentum in the annulus of fluid considered. The angular momentum of a mass m of fluid in the annulus is given by:

$$L = m\omega r^2. \quad [3.12]$$

Consider the change in angular momentum from a far-upstream section to a section far downstream. The torque on the rotor is given by the time derivative of the angular momentum. At the upstream section, the azimuthal velocity is zero, and the torque may thus be written:

$$dQ = \frac{dL}{dt} = \dot{m} \cdot \omega_2 r_2^2 = 2\pi r_1 dr_1 \rho u_{x1} \omega_2 r_2^2 = 2\pi \rho u_{x1} \omega_1 r_1^3 dr_1. \quad [3.13]$$

$\dot{m}$ is the mass of fluid crossing the section per unit time. Using the Bernoulli equation from plane 0 to 1- and from 1+ to 2, we obtain:

$$C_1 = p_0 + \frac{1}{2}\rho u_{x0}^2 = p_{1-} + \frac{1}{2}\rho\left(u_{x1}^2 + u_{r1}^2\right)$$
$$C_2 = p_{1+} + \frac{1}{2}\rho\left(u_{x1}^2 + u_{r1}^2 + (\omega_1 r_1)^2\right) = p_2 + \frac{1}{2}\rho\left(u_{x2}^2 + (\omega_2 r_2)^2\right). \quad [3.14]$$

As can be observed from these equations, the radial velocity component u_r is included at the rotor plane, while this component is assumed to be zero in the far-upstream and far-downstream sections. Further, it is assumed, from continuity arguments, that the axial and radial velocity components are equal on both sides of the rotor plane, while there is a jump in the rotational velocity. In the Betz theory it was assumed that $p_0 = p_2$. Due to the rotational velocity component, this is not

[1] It is assumed that the annulus extends from r to $r + dr$ and as $dr \ll r$, the radius to the middle of the annulus can be set to r.

the case in the present approach. From [3.14] the following expression for the pressure difference between the upstream and downstream plane is obtained:

$$p_0 - p_2 = p_{1-} - p_{1+} + \frac{1}{2}\rho\left(u_{x2}^2 - u_{x0}^2 + (\omega_2 r_2)^2 - (\omega_1 r_1)^2\right). \qquad [3.15]$$

$\Delta p = p_{1-} - p_{1+}$ is the pressure jump over the rotor disc in the annular section considered. Consider now the pressure jump over the rotor disc by using a coordinate system following the rotation of the rotor (see, e.g., Manwell, McGowan and Rogers, 2009; Burton et al., 2011; Kramm et al., 2016). Just in front of the rotor the fluid is assumed to have zero rotation, while just behind the rotor plane the fluid has angular velocity ω_1. As seen from the rotor, the angular velocities of the fluid are Ω and $\Omega + \omega_1$ respectively. Ω is the angular velocity of the rotor and the plus sign appears because the rotation in the fluid is opposite to the rotor rotation according to the conservation of angular momentum. The Bernoulli equation relative to the rotating blades can then be written as:

$$p_{1-} + \frac{1}{2}\rho\left(u_{x1}^2 + u_{r1}^2 + \Omega^2 r_1^2\right) = p_{1+} + \frac{1}{2}\rho\left(u_{x1}^2 + u_{r1}^2 + (\Omega + \omega_1)^2 r_1^2\right)$$

$$\Delta p = \rho\left(\Omega + \frac{1}{2}\omega_1\right)\omega_1 r_1^2 = \rho\left(\Omega + \frac{1}{2}\omega_1\right)\omega_2 r_2^2. \qquad [3.16]$$

To replace $\omega_1 r_1^2$ by $\omega_2 r_2^2$ the conservation of angular momentum is invoked. Observing that $(\omega_2 r_2)^2 - (\omega_1 r_1)^2$ may be written as $\omega_1 r_1^2(\omega_2 - \omega_1)$, the result from [3.16] may be inserted into [3.15] to obtain:

$$p_2 = p_0 + \frac{1}{2}\rho\left(u_{x0}^2 - u_{x2}^2\right) - \rho\omega_2 r_2^2\left(\Omega + \frac{1}{2}\omega_2\right). \qquad [3.17]$$

The pressure far downstream is thus related to the rotational velocity in the fluid as well as the turbine rotor velocity. Within the above framework, an exact solution for the thrust and power can be obtained if it is assumed that the circulation is confined to the axis of the wake. In that case we obtain $\omega_1 r_1^2 = \omega_2 r_2^2$ (as already has been assumed above) and independent of the radius. It can then also be deduced that the axial velocity in the wake is independent of the radius, similar to in the axial actuator disc approach (for details, see Kramm et al., 2016).

The classical approach used is now to equate the thrust obtained from [3.16] to the thrust from the axial momentum approach of [3.8]. A relation between the axial and angular induction is thereby obtained, i.e.:

$$dT = 2\pi r_1 dr_1 \rho u_{x1}(u_{x0} - u_{x2}) = 2\pi r_1 dr_1 \rho \left(\Omega + \frac{1}{2}\omega_1\right)\omega_1 r_1^2$$
$$u_{x0}^2 2a(1-a) = (\Omega r_1)^2 2a'(1+a'). \qquad [3.18]$$

Here, the axial induction factor $a = 1 - u_{x1}/u_{x0} = 1 - u_{x1}/U_0$ and the angular induction factor $a' = \omega_1/2\Omega$ have been introduced. Further, introducing the local speed ratio defined by $\lambda_r = \Omega r_1/U_0$, the following equation relating the angular and axial induction factors is obtained:

$$\lambda_r^2 = \left[\frac{a(1-a)}{a'(1+a')}\right]. \qquad [3.19]$$

Using the induction factors in the expression for the torque in [3.13], we obtain:

$$dQ = 2\pi r_1 dr_1 \rho U_0 (1-a) a' 2\Omega r_1^2, \qquad [3.20]$$

and for the power:

$$dP = \Omega dQ = 2\pi r_1 dr_1 \rho U_0 (1-a) a' 2\Omega^2 r_1^2. \qquad [3.21]$$

The total power is obtained by integrating over the rotor disc:

$$P = 4\pi\rho U_0 \Omega^2 \int_0^R r_1^3 (1-a) a' dr_1. \qquad [3.22]$$

Here, it is assumed that the induction factors may vary over the rotor radius. On dimensionless form this reads:

$$C_P = \frac{P}{\frac{1}{2}\rho U_0^3 \pi R^2} = \frac{8}{\lambda^2}\int_0^\lambda \lambda_r^3 (1-a) a' d\lambda_r. \qquad [3.23]$$

Here, $\lambda = \Omega R/U_0$ is the tip speed ratio. In order to integrate this equation, a' is expressed in terms of a and λ_r using [3.19]. The following relation is obtained, using the condition that $a' > 0$:

$$a' = \frac{1}{2}\left[-1 + \sqrt{1 + \frac{4}{\lambda_r^2} a(1-a)}\right]. \qquad [3.24]$$

We are now searching for the relation between a' and a that maximizes the power coefficient within each annulus, i.e., we want to maximize the term $f = (1 - a)a'$ in [3.23]. This implies finding the combination of a and a' that satisfies:

$$\frac{df}{da} = (1 - a)\frac{da'}{da} - a' = 0. \quad [3.25]$$

From [3.24] the following is obtained:

$$\frac{da'}{da} = \frac{1 - 2a}{\lambda_r^2(1 + 2a')}. \quad [3.26]$$

Now, using [3.26] and [3.19] in [3.25], the following relation is found for optimum performance:

$$a' = \frac{3a - 1}{1 - 4a}. \quad [3.27]$$

The results from this approach are displayed in Figure 3.6. Here the optimum angular induction as obtained from [3.27] and the corresponding speed ratio as obtained from [3.19] are plotted to the left. To the right, the optimum induction factors are plotted versus the speed ratio. It is observed that physical results are obtained in the interval $1/4 < a < 1/3$ only. $a = 1/3$ corresponds to the Betz limit, and thus zero angular induction. This corresponds to the speed ratio going toward infinity. As the axial induction factor is reduced toward 0.25, the optimum angular induction tends to infinity and corresponds to zero speed ratio (see Figure 3.6 (left)). From Figure 3.6

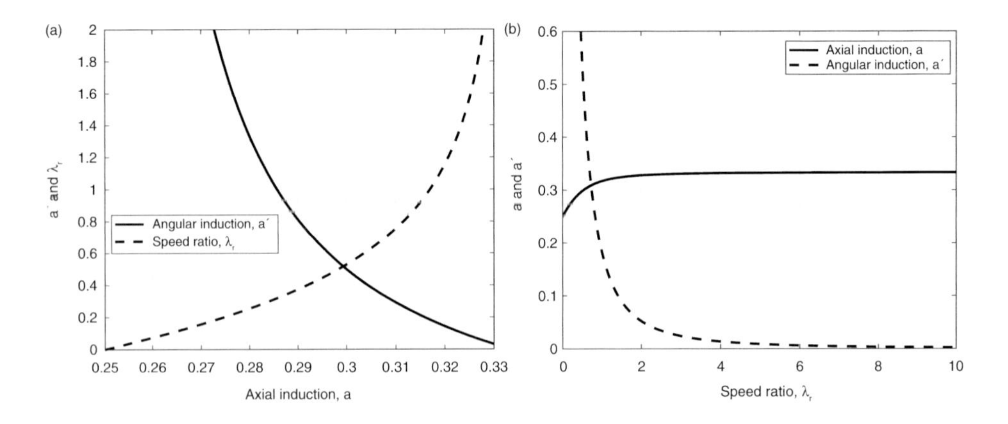

Figure 3.6 Left: optimum angular induction factor and corresponding speed ratio, $\lambda_r = \Omega r_1/U_0$, as a function of the axial induction factor, a. Right: optimum axial and angular induction factors as a function of the local speed ratio.

(right), it is observed that for speed ratios above 2, the optimum axial induction factor is very close to the Betz limit of 1/3, and that the angular induction becomes very small. This is justifying the frequently used assumption of zero angular induction at speed ratios above approximately 6.

In conjunction with [3.17] relating the upstream and downstream pressures, it was mentioned that if the circulation is confined to the axis of the wake, $\omega_2 r_2^2$ is independent of radius. This assumption has as a consequence that the axial induction factor also is equal for all radii. If the tip speed ratio is above approximately 6, a local speed ratio below 2 will be present for the inner third of the rotor only, i.e., an optimum axial induction factor of 1/3 for all annular sections is a reasonable assumption.

Using the above optimum relation between the axial and angular induction factors for each radii, the maximum power coefficient for the full rotor may be computed from [3.23]. This is called the "Glauert limit" for the power coefficient, after Glauert (1935). In Figure 3.7 the maximum power coefficient according to the Betz limit and the Glauert limit is plotted as a function of the tip speed ratio. For tip speed ratios above 6, the difference is less than 2%. The Glauert limit may be considered to be the limit for a rotor with an infinite number of blades and no viscous losses. A rotor with a finite number of blades has a lower theoretical maximum power. The fewer the blades, the lower the maximum. The maximum obtained depends upon the assumed circulation distribution along the radius of the blades; for a further discussion, see Kramm et al. (2016).

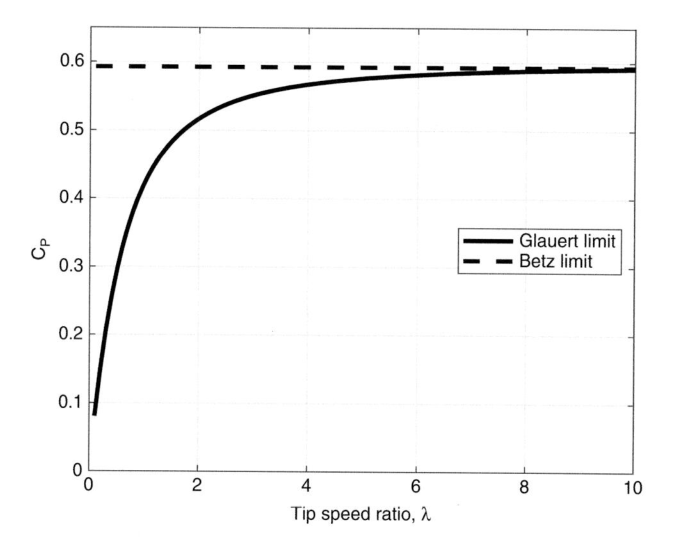

Figure 3.7 Maximum power coefficient according to the Betz and Glauert limits.

3.3 Two-Dimensional Lifting Surfaces

Above, it was assumed that the power is extracted from the air by a device rotating about a horizontal axis. To generate the necessary torque to drive the rotor, the principle of lifting surfaces is used. This is the same principle as used for airplane wings. In this section some of the basic concepts for two-dimensional (2D) lifting surfaces are described. For example, in the beam element momentum (BEM) method for computing the loads on a wind turbine, the assumption of a local 2D flow over the rotor blades is used. The BEM method is discussed in more detail in Section 3.4. For more details about 2D lifting surface theory, the reader should visit the special literature on aero- or hydrodynamics, e.g., Newman (1977) and Bramwell, Done and Balmford (2001).

A 2D aerofoil has a layout as illustrated in Figure 3.8. The straight line from the *leading edge* (front) to the *trailing edge* (tail) is denoted the *chord*. The line from the leading edge to the trailing edge representing the mean line between the upper surface and the lower surface is denoted as the *camber line*. For a symmetric aerofoil this line coincides with the chord. For most aerofoils, the front of the foil is rounded to secure a smooth flow around the foil. The trailing edge, however, is sharp. This is essential to the working principle of an aerofoil. Figure 3.10 illustrates the flow direction towards the aerofoil and the corresponding lift and drag forces. In the 2D steady flow case the lift force is perpendicular to the incident flow and the drag force is zero under ideal flow conditions.

Figure 3.9 illustrates the streamlines of an ideal flow, without viscous effects, around an aerofoil. The incident flow hits the foil close to the leading edge and splits into two flows, above and beneath the foil. These two flows meet at the trailing edge, which becomes a common separation point. According to the Kutta–Joukowski theorem, the joined flow should leave the trailing edge in a smooth manner. The force acting on the aerofoil can be derived from a local perspective or a global perspective. In the local perspective, it is observed that the flow at the upper side of the foil must move a longer distance and thus faster than the flow on the lower side. The pressure on the upper side is thus lower than on the lower side and a net upward force is created. In the global perspective, we observe that the foil gives the flow a small change in direction; a small

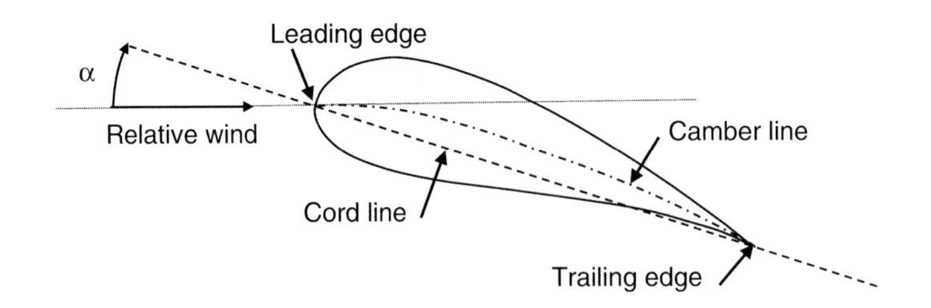

Figure 3.8 Aerofoil nomenclature. α is the angle of attack.

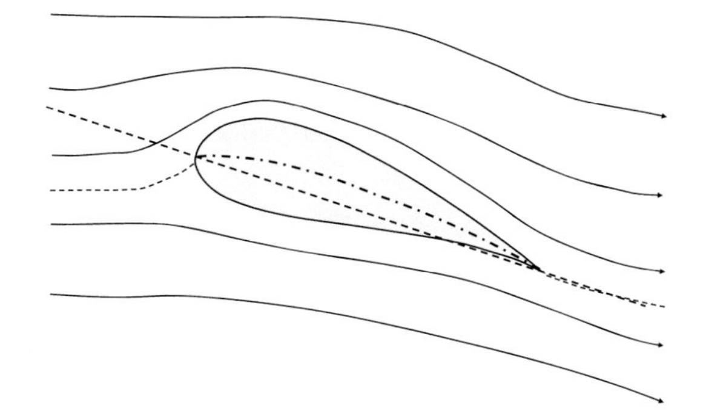

Figure 3.9 Illustration of streamlines around an aerofoil satisfying the Kutta–Joukowski condition at the trailing edge. The dashed streamline marks the division between flow over and beneath the aerofoil.

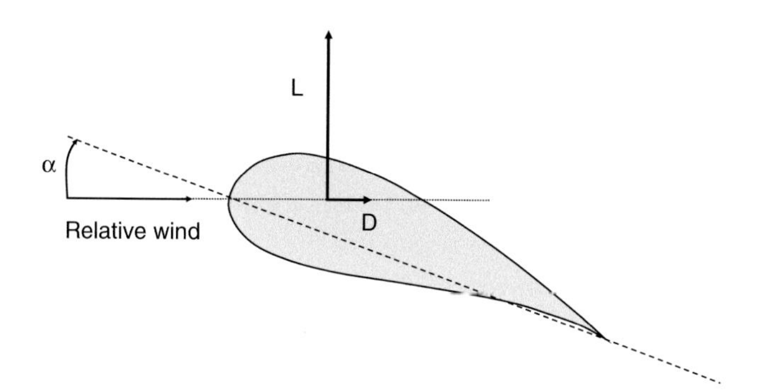

Figure 3.10 Wind direction toward the aerofoil and corresponding lift, L, and drag force, D.

vertical velocity component is added to the flow behind the foil. This means that a vertical momentum is introduced to the flow, which must be balanced by an opposite-directed vertical force on the foil.

If it were not for the sharp trailing edge forcing the upper and lower streams to join here, the ideal flow theory (potential theory) would predict a separation point at the upper side of the foil with zero force in vertical and horizontal directions in a stationary flow as a result, according to d'Alembert's paradox.

If the stream was not forced to separate at the trailing edge, the flow potential could be modeled by a source distribution along the camber line. The integral of the source distribution would be zero. Thus, the source distribution causes a local disturbance of the incident flow only. Far away from the foil, the disturbance vanishes. However, without the Kutta-Joukowski condition, the flow at the trailing edge becomes unphysical. This is illustrated in Figure 3.13 and discussed further in the following.

3.3.1 Lift by Vortex Theory

So far, the lift force has been explained by considering the flow and the effect of the trailing edge; the Kutta-Joukowski condition requires the flows from the two sides of the foil to join at the trailing edge and leave the foil smoothly.

The lift force can, however, also be derived in a more theoretical way. Assume a 2D incompressible and ideal flow that can be described by a velocity potential. Consider a closed curve S in the fluid. The circulation in the flow, Γ is obtained by considering the line integral of the tangential fluid velocity along this curve:

$$\Gamma = \oint_S U_s ds. \qquad [3.28]$$

Here, U_s is the velocity tangential to the curve S. We use a coordinate system with x in the direction of the free flow and y positive upward, then a positive rotation of a vortex, using the "right-hand rule" will induce a negative x-velocity above and a positive x-velocity below the vortex, as indicated in Figure 3.11. In a potential flow, $\Gamma = 0$ everywhere except when a singularity of vortex type is located inside the closed curve S. If a point vortex of strength Γ is present in the fluid, the tangential velocity at a radius r induced by the vortex becomes as follows, according to the Biot–Savart law:

$$U_s = \frac{\Gamma}{2\pi r}. \qquad [3.29]$$

Consider now a circular cylinder with radius r inserted into a steady incident flow with velocity U_0, as illustrated in Figure 3.11. According to potential theory, the tangential velocity component of the incident flow along the cylinder surface is given by:

$$U_{sc} = -2U_0 \sin \beta. \qquad [3.30]$$

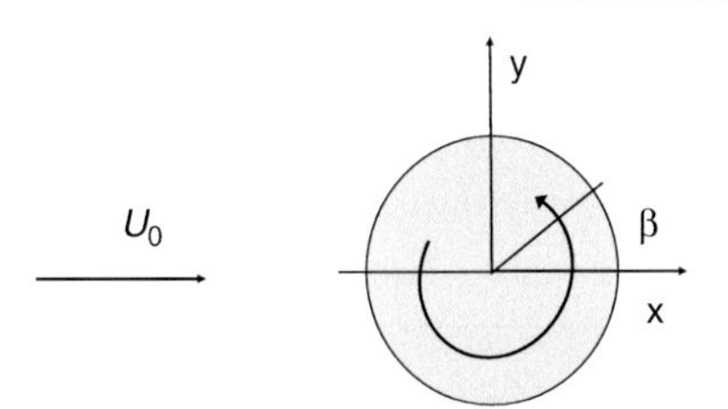

Figure 3.11 A circular cylinder in an incident flow with velocity U_0. A vortex with positive circulation is indicated.

Here, a positive tangential velocity has the same direction as the direction of a positive vortex. This result can be derived by modeling the cylinder by a dipole in the incident flow. If we insert a vortex of strength Γ in the center of the cylinder, the corresponding induced velocities are given from [3.29]. The total velocity at the cylinder surface is tangential and given by the sum of the two contributions. The difference in pressure at the surface and the far field is thus obtained as:

$$\Delta p = \frac{\rho}{2}\left[U_0^2 - (U_{sc} + U_s)^2\right]. \qquad [3.31]$$

Consider now the pressure difference between two points at the cylinder surface, one located at (R, β), the other at $(R, -\beta)$:

$$\begin{aligned} \Delta p' &= \Delta p(R,\beta) - \Delta p(R,-\beta) \\ &= 4\rho U_0 U_s \sin\beta. \end{aligned} \qquad [3.32]$$

The vertical force, the lift, becomes thus:

$$\begin{aligned} L = -\int_0^\pi \Delta p' R \sin\beta \, d\beta = -4\rho U_0 U_s R \int_0^\pi \sin^2\beta \, d\beta = -4\rho U_0 U_s R \frac{\pi}{2} \\ = -\rho U_0 \Gamma. \end{aligned} \qquad [3.33]$$

The lift force is thus given from the product of the free flow velocity and the vortex strength. This result is valid more generally, and is applied in vortex methods, to compute the lift force on aerofoils. The result for the cylinder may also be used directly by mapping the cylinder into the shape of an aerofoil by conformal mapping; see, e.g., Benson (1996). By the above approach, it can be shown that in a steady, 2D flow, the force in the direction of the free flow, the drag force, is zero. However, this is not the case in time-dependent flow, 3D flow or when the effect of viscosity is included.

3.3.2 Two-Dimensional Aerofoils

For a general-shaped 2D aerofoil in ideal flow, the flow around the foil and the corresponding lift force can be computed by distributing vortices and sources along the camber line of the foil. Frequently, linearized foil theory is used for this purpose. In linearized foil theory (see, e.g., Newman, 1977; Faltinsen, 2005), it is assumed that the induced velocities due to the presence of the foil are small compared to the incident flow velocity. This implies also that the thickness to chord length ratio must be small. Following the derivation in Faltinsen (2005), we may denote the upper and lower surface of the foil by $y_u(x)$ and $y_l(x)$ respectively (see Figure 3.12). The velocity

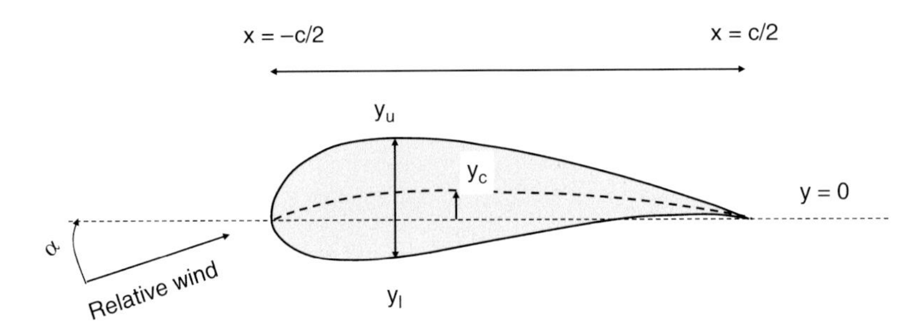

Figure 3.12 2D aerofoil considered.

potential due to the presence of the foil is denoted $\phi(x,y)$. Consider now a steady fluid flow in an infinite fluid. The boundary condition for the potential is given from the zero normal flow condition, i.e.:

$$\frac{\partial\phi}{\partial n} = \begin{cases} -U_0 n_1 & \text{at } y = y_u \\ +U_0 n_1 & \text{at } y = y_l \end{cases}. \qquad [3.34]$$

(n_1, n_2) are the x- and y-components of the outward-directed body surface normal. We may write:

$$\frac{\partial\phi}{\partial n} = n_1 \frac{\partial\phi}{\partial x} + n_2 \frac{\partial\phi}{\partial y}. \qquad [3.35]$$

Except for close to the leading edge (and at the trailing edge), the x-component of the surface normal is much smaller than the y-component, justifying the approximation $n_1 \simeq \pm\,\partial y_{u/l}/\partial x$ on the upper and lower foil surface respectively. Again, under the assumption of $|n_1| \ll |n_2|$, we have that $n_1 \simeq \pm\,\partial y_{u/l}/\partial x$ on the upper and lower surface. From [3.34] the following is thus obtained:

$$\frac{\partial\phi}{\partial y} \simeq \begin{cases} U_0 \dfrac{\partial y_u}{\partial x} & \text{at } y = y_u \\ U_0 \dfrac{\partial y_l}{\partial x} & \text{at } y = y_l \end{cases} \qquad [3.36]$$

The linearization now implies that the chord is transformed to a straight line along the x-axis. The foil extends from $x = -c/2$ to $x = c/2$. Further, by a Tailor expansion of $\partial\phi/\partial y$ about $y = 0$, the boundary condition in [3.36] is satisfied on $y = 0_{+/-}$ rather than on $y = y_{u/l}$.

As the camber line is the midline between the lower and upper surface, it is defined by:

$$y_c(x) = \frac{1}{2}\left[y_u(x) + y_l(x)\right]. \qquad [3.37]$$

The thickness of the foil is similarly given by $t(x) = y_u(x) - y_l(x)$. Within the approximations given, the linearized boundary conditions in [3.36] may thus be written:

$$\frac{\partial \phi}{\partial y} = \begin{cases} U_0 \left[\dfrac{\partial y_c}{\partial x} + \dfrac{1}{2} \dfrac{\partial t}{\partial x} \right] & \text{at } y = 0_+ \\ U_0 \left[\dfrac{\partial y_c}{\partial x} - \dfrac{1}{2} \dfrac{\partial t}{\partial x} \right] & \text{at } y = 0_-. \end{cases} \qquad [3.38]$$

The potential describing the flow around the foil is split into a symmetric and an antisymmetric part, ϕ_s and ϕ_a, satisfying the following boundary conditions at the foil, between $x = -c/2$ and $x = c/2$:

$$\begin{aligned} \frac{\partial \phi_s}{\partial y} &= \pm U_0 \frac{1}{2} \frac{\partial t}{\partial x} \quad \text{on } y = 0_{+/-} \\ \frac{\partial \phi_a}{\partial y} &= U_0 \frac{\partial y_c}{\partial x} \quad \text{on } y = 0. \end{aligned} \qquad [3.39]$$

The symmetric part of the potential represents a source distribution along the camber line and accounts for the thickness distribution of the foil. The antisymmetric part accounts for the slope of the camber line. This part of the solution can be represented by a vortex distribution.

From [3.39] it is observed that the source and vortex distribution along the camber line is obtained from the local thickness gradient and local camber line gradient only. This is due to the linearization, involving that the boundary conditions are satisfied on $y = 0$. I.e., a source at a certain $x-$ position contributes to vertical velocities on the foil at this $x-$ location only. Similarly, a vortex at a certain $x-$ position contributes to $x-$ velocities only at that position on the foil. However, the vortex distribution induces $y-$ components of the velocity all along the foil. In the more general case (see the description of the boundary element method in Section 3.4), one has to consider that any source and vortex will induce velocities at any part of the body surface, thus involving boundary-integral methods to solve for the source and vortex strength.

Integrating the source distribution along the camber line, the result is zero. Thus, the source distribution causes a local disturbance of the incident flow only. Far away from the foil, the disturbance vanishes. The Kutta–Joukowski condition at the trailing edge is not satisfied by the source distribution and the flow will not leave the trailing edge in a smooth manner.

Adding the antisymmetric part of the solution, the vortex distribution along the camber line is obtained. The Kutta–Joukowski condition is used to obtain a unique

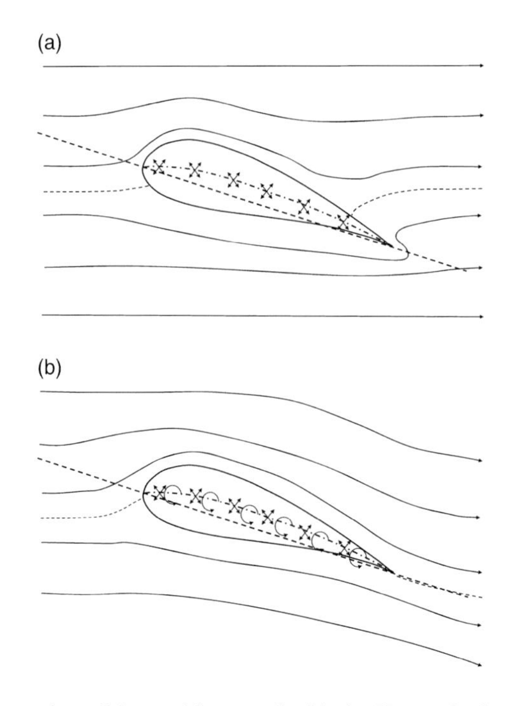

Figure 3.13 Illustration of flow without and with the Kutta–Joukowski condition at the trailing edge. Top: flow around an ideal foil without circulation. Bottom: circulation included to obtain a smooth flow at the trailing edge. The dashed lines indicate separation lines between flow above and below the foil. The straight arrows indicate a sink source distribution along the camber line. The circular arrows indicate the distribution of vortices.

solution. The requirement imposed is that the flow shall leave the trailing edge tangential to the chord line at the training edge. The solution involves singular integral equations (see Newman, 1977; Faltinsen, 2005). The impact of the vortex distribution on the flow is illustrated in Figure 3.13 (bottom). It should also be noted that the vortex distribution causes the flow, which initially is in the positive x-direction, to obtain a small velocity component in the negative y-direction behind the foil. This change of momentum in the flow has its counterpart in a force in positive y-direction on the foil: the lift force.

It may be considered strange that a physical lift force can be modeled by introducing a distribution of vortices that represents mathematical singularities. It is, however, important to note that the circulation set up by this vortex distribution exists physically. The boundary layer around the foil creates a circulation so that the Kutta–Joukowski condition is fulfilled. Faltinsen (2005) discusses this issue in more detail.

The relation between the velocity potential and a circulation distribution along the line $x = [-c/2, c/2], y = 0$ can be written as:

$$\phi_a(x,y) = \frac{1}{2\pi}\int_{-c/2}^{c/2} \gamma(\xi)\arctan\left(\frac{y}{x-\xi}\right)d\xi. \qquad [3.40]$$

Faltinsen (2005) shows that the vortex distribution in the linearized case can be written by a Fourier series given as:

$$\gamma(\xi) = 2U_0\left[a_0\frac{1+\cos\theta}{\sin\theta} + \sum_{n=1}^{\infty} a_n \sin n\theta\right], \qquad [3.41]$$

where $\theta = \arccos\left(-2\frac{\xi}{c}\right)$ and where the coefficients are related to the slope of the camber line by:

$$a_n = \frac{k_n}{\pi}\int_0^{\pi} \frac{\partial y_c}{\partial x} \cos n\chi d\chi, \quad k_n = \begin{cases} 1 & \text{for } n = 0 \\ -2 & \text{for } n > 0 \end{cases}. \qquad [3.42]$$

a_0 is thus an expression for the angle of the chord line, or angle of attack, while a_1 is a weighted average camber of the foil. As for the example above, with the cylinder, the lift force on the foil is given from the total circulation around the foil, i.e.:

$$L = -\rho U_0 \Gamma = -\rho U_0 \int_{-c/2}^{c/2} \gamma(\xi)d\xi. \qquad [3.43]$$

From the definition of θ and the Fourier coefficients above, it is found that only the first two coefficients in the series for the vortex distribution contribute to the integral and the lift force may be written as:

$$L = -\rho U_0^2 \pi c(a_0 + 0.5a_1). \qquad [3.44]$$

Similarly, the moment about $x = 0$ is obtained as:

$$\begin{aligned} M &= -\rho U_0 \int_{-c/2}^{c/2} \gamma(\xi)\xi d\xi \\ &= \frac{\rho U_0^2}{4}\pi c^2(a_0 + 0.5a_2). \end{aligned} \qquad [3.45]$$

A "point of attack" of the lift force can thus be obtained as $x_L = M/L = -\frac{c}{4}\left(\frac{a_0+0.5a_2}{a_0+0.5a_1}\right)$.

Flat Plate

A flat plate has a constant slope, $\partial y_c/\partial x = -\alpha$, where α is the angle of attack. In this case all $a_n = 0$ except for $a_0 = -\alpha$, see [3.42]. The lift force and moment relative to the mid-point of the foil are then obtained as:

$$L = \rho U_0^2 c\pi\alpha$$
$$M = -0.25\rho U_0^2 c^2 \pi\alpha. \quad [3.46]$$

The point of zero moment, x_p, is obtained as:

$$x_p = \frac{M}{L} = -\frac{c}{4}. \quad [3.47]$$

That is, the lift force on a 2D flat plate acts $0.25c$ behind the leading edge.

Figure 3.14 shows the vortex distribution along the flat plate according to [3.41]. Note that the distribution tends to minus infinity at the leading edge and tends to zero at the trailing edge.

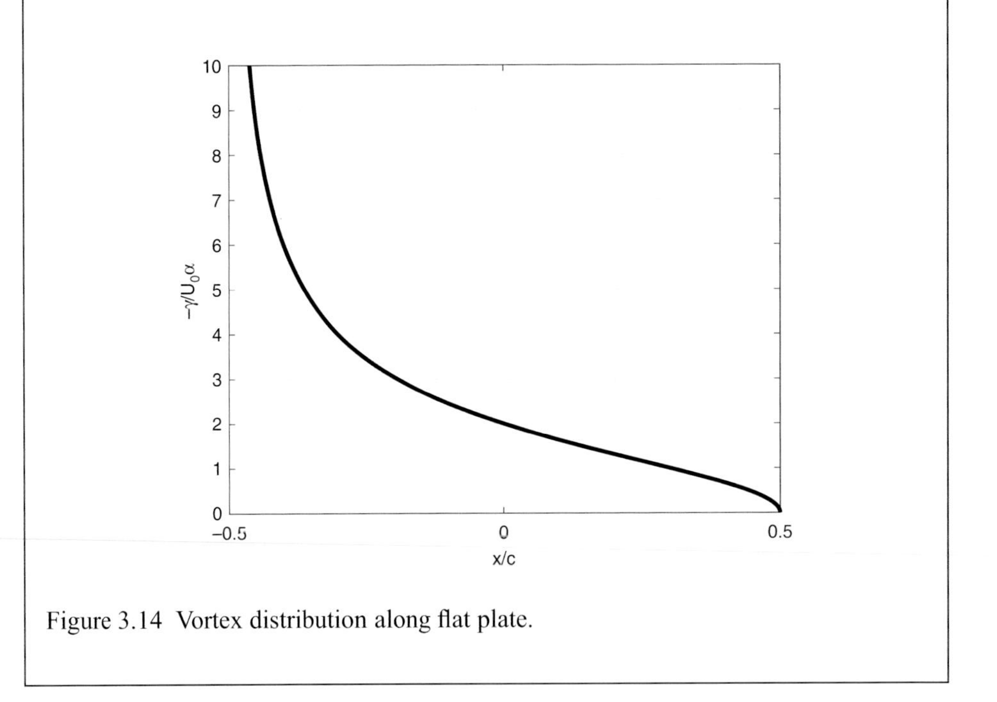

Figure 3.14 Vortex distribution along flat plate.

Faltinsen (2005) shows that for a foil with parabolic shape where the camber line is given according to:

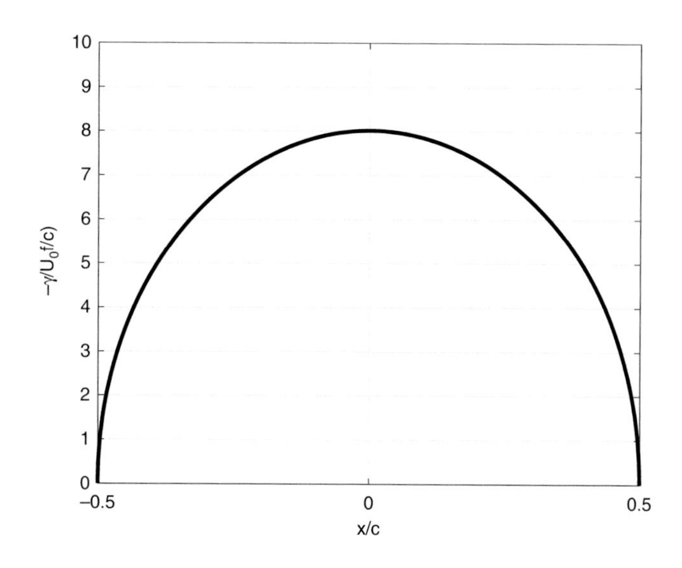

Figure 3.15 Vortex distribution along a parabolic foil with camber f/c and zero angle of attack.

$$y_c = -\alpha x + f\left(1 - \left(\frac{2x}{c}\right)^2\right), \quad [3.48]$$

the lift, within the framework of linear theory, is obtained as:

$$L = \rho U_0^2 c\pi\left(\alpha + \frac{2f}{c}\right). \quad [3.49]$$

The moment is as for the flat plate. From [3.49] it is observed that in this case the zero lift force is obtained for a negative angle of attack. The vortex distribution over the parabolic foil at zero angle of attack is shown in Figure 3.15.

The above linear, ideal flow solution of the flow around a 2D foil is useful in understanding the basic principles of how an aerofoil works. The approach also provides reasonable results for small angles of attack for real foils when the flow separation satisfies the Kutta–Joukowski condition. In the real case, separation may take place at the suction side forward of the leading edge and the lift coefficient may be very sensitive to the Reynolds number of the flow, i.e., the viscous effects become important.

In Figure 3.16, typical lift and drag coefficients for a real aerofoil with camber are shown. The lift coefficient at zero angle of attack is thus different from zero. The dotted line shows the lift coefficient corresponding to $C_L = 2\pi\alpha + 0.6$, with α in radians. The addition of 0.6 is to ease the comparison with the typical lift curve. The slope of the lift curve for real aerofoils is normally close to this value for small angles of attack.

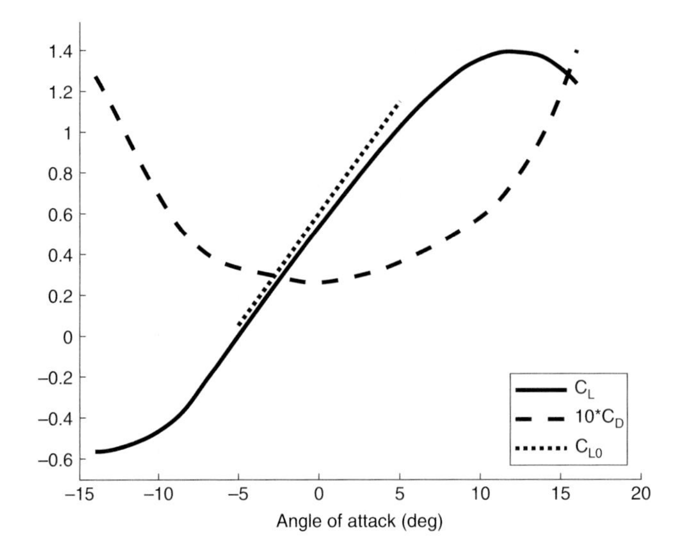

Figure 3.16 Typical lift and drag coefficients for aerofoil sections used in rotor blade design. C_{L0} corresponds to the theoretical lift of a 2D flat plate with an addition of 0.6 to ease the comparison with the typical lift curve.

For greater angles of attack, flow separation from the suction side starts, and the lift curve flattens out and may even decay. This phenomenon is called stalling. Separation effects are very sensitive to the Reynolds number. For small angles of attack, the drag forces are very small, with a drag coefficient typically less than 0.05. As flow separation starts, a significant increase in drag will occur. The drag coefficient is even more sensitive to the Reynolds number and surface roughness than the lift coefficient. In the design of aerofoils for wind rotor blades, low drag is an important criterion.

In a real operational case, the angle of attack will vary over time. If, in a dynamic case, the angle of attack is moving from the attached flow region into the stall region, there may be some delay in the stalling process, meaning that the lift in the dynamic case accepts some higher angle of attack than in the stationary case. On the other hand, if separation has first taken place, and the angle of attack is reduced, there may be a significant delay compared to the stationary case before the attached flow again is established. This phenomenon is called dynamic stall and is usually implemented in computational tools for wind turbine dynamics.

3.4 The Blade Element Momentum Method

The blade element momentum (BEM) method is a well-established and commonly used method for computing the forces acting on a blade as well as the complete rotor of a wind turbine. The method may be considered as an extension of the

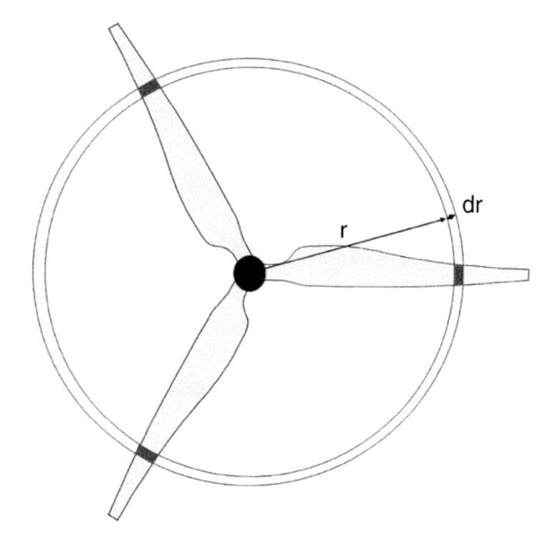

Figure 3.17 A three-bladed horizontal-axis wind turbine. Each blade is assumed to be composed of several 2D aerofoil sections. One such section is illustrated. The annulus described by the rotating section is used in the momentum considerations.

actuator disc approach used in deriving the Betz limit. In the following, only horizontal-axis wind turbines (HAWTs) are discussed. However, the method is also used for vertical-axis turbines. The starting point is the 2D lift and drag coefficients for an aerofoil, as discussed in the previous section. In most cases these coefficients are established from experiments, as both a wide range of angles of attack and reliable estimates on the drag forces are required. The wind turbine blade is assumed to be composed of a large number of 2D sections, as illustrated in Figure 3.17. When the blade rotates, each blade section describes an annulus in the rotor plane (see Figure 3.17). This annulus is used in the momentum considerations. At the outset, the BEM method rests upon three main assumptions. First, the flow over each of the blade sections can be assumed to be 2D and independent of the flow over the neighboring sections. Second, the momentum equations apply within each annulus for solving for the loads on the blade segment. Third, the flow is stationary.

To account for deviations from these assumptions, several corrections are usually introduced, e.g.: correcting for tip and hub vortices (violating the 2D flow assumption); correcting for nonstationary flow conditions; correcting for large induction factors; and correcting for the incident flow not being perpendicular to the rotor plane. Details of the BEM theory as well as practical examples of implementation are given by, e.g., Moriarty and Hansen (2005) and Hansen (2015). The following is a short summary of the approach.

Consider an aerofoil section as displayed in Figure 3.18. As the geometry of the neighboring sections are similar as the one considered, the flow over the section is assumed to be 2D. The inflow velocity experienced by the foil is composed of the

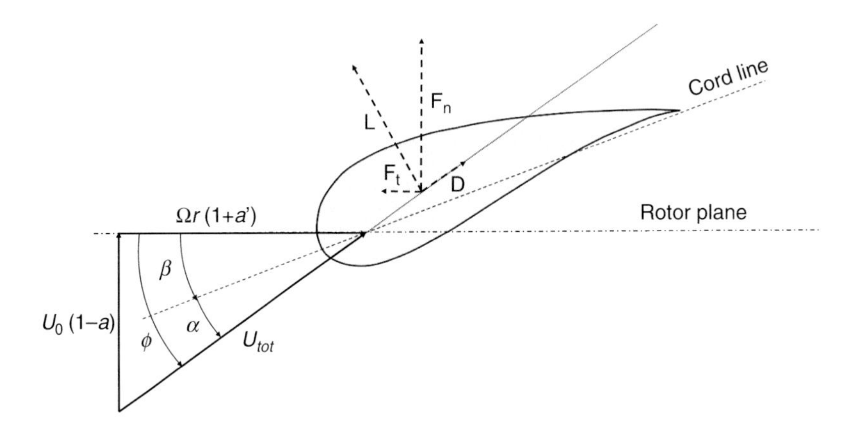

Figure 3.18 Section of rotor blade with incident flow velocities and forces acting on it.

wind velocity, perpendicular to the rotor plane, and the velocity due to the rotation of the blade in the rotor plane. It is observed that the incident, free wind velocity is modified due to the axial induction, i.e., the axial velocity as seen by the aerofoil is $U_0(1-a)$. Similarly, the air velocity into the foil due to the rotation of the foil is given by $\Omega r(1+a')$. Here, Ω is the angular velocity of the rotor, r is the radius of the section considered and a' is the angular induction factor, as discussed in Section 3.2. In many implementations (see, e.g., Moriarty and Hansen, 2005), additional velocity components accounting for the local elastic deformation velocities of the blade are added to the above two velocity components. The angle of attack is defined as the angle between the incident flow and the chord line and is denoted α. The angle β is the angle between the chord line and the rotor plane, or the local pitch of the 2D section. This is given as the sum of the twist of the blade section, θ, and the blade pitch angle, β_b. For most state-of-the-art offshore wind turbines, the blade pitch angle is varied during operation to control the power output. The twist, however, is specific to the design of the blade and is introduced to obtain the wanted angle of attack at the various blade sections. The angle between the flow and the rotor plane is ϕ. This is given from the two velocity components:

$$\phi = \arctan\left(\frac{U_0(1-a)}{\Omega r(1+a')}\right) = \arctan\left(\frac{1-a}{\lambda_r(1+a')}\right). \qquad [3.50]$$

Here, $\lambda_r = \Omega r/U_0$ is the local speed ratio. In Table 3.1 the sectional blade twist for the NREL 5 MW reference turbine is given (Jonkman et al., 2009). Note that the twist angle is large close to the hub and decays to almost zero at the tip. From the table it is also observed that the geometry of the blade sections differs along the

Table 3.1 *Sectional properties for the turbine blades of the NREL 5 MW reference turbine (Jonkman et al., 2009). The rotor radius is 63 m. Key data for the turbine are given in Table 3.2. Data for the various aerofoils used are also given in Jonkman et al. (2009).*

Node (–)	Radius (m)	Twist Angle θ (°)	Length of Section (m)	Chord Length (m)	Aerofoil Type (–)
1	2.8667	13.308	2.7333	3.542	Cylinder1
2	5.6000	13.308	2.7333	3.854	Cylinder1
3	8.3333	13.308	2.7333	4.167	Cylinder2
4	11.7500	13.308	4.1000	4.557	DU40_A17
5	15.8500	11.480	4.1000	4.652	DU35_A17
6	19.9500	10.162	4.1000	4.458	DU35_A17
7	24.0500	9.011	4.1000	4.249	DU30_A17
8	28.1500	7.795	4.1000	4.007	DU25_A17
9	32.2500	6.544	4.1000	3.748	DU25_A17
10	36.3500	5.361	4.1000	3.502	DU21_A17
11	40.4500	4.188	4.1000	3.256	DU21_A17
12	44.5500	3.125	4.1000	3.010	NACA64_A17
13	48.6500	2.319	4.1000	2.764	NACA64_A17
14	52.7500	1.526	4.1000	2.518	NACA64_A17
15	56.1667	0.863	2.7333	2.313	NACA64_A17
16	58.9000	0.370	2.7333	2.086	NACA64_A17
17	61.6333	0.106	2.7333	1.419	NACA64_A17

blade. The 2D lift and drag forces, acting perpendicular and in line with the incident flow direction, are given as L and D in Figure 3.18. Decomposing these forces in the axial and in-plane directions, the axial and tangential forces are obtained as:

$$\begin{aligned} F_n &= L\,\cos\phi + D\,\sin\phi \\ F_t &= L\,\sin\phi - D\,\cos\phi \end{aligned}. \quad [3.51]$$

Introducing the lift and drag coefficients and setting the width of the aerofoil section to dr, the contributions of the section to the thrust and torque are obtained as:

$$\begin{aligned} dT &= N_B F_n = 0.5 N_B \rho U_{tot}^2 (C_L \cos\phi + C_D \sin\phi) c\, dr \\ dQ &= N_B F_t r = 0.5 N_B \rho U_{tot}^2 (C_L \sin\phi - C_D \cos\phi) cr\, dr \end{aligned}. \quad [3.52]$$

Here, N_B is the number of blades and c is the length of the chord. Here, a homogeneous incident flow is assumed. For large rotors, the incident flow varies as the blade rotates, both due to the wind shear and turbulence.

The thrust and torque may also be obtained by applying the momentum theory to the annulus as described by the rotating section. According to [3.8] and [3.20] we may thus write:

$$dT = 4\pi\rho U_0^2(1-a)a\ rdr$$
$$dQ = 4\pi\rho U_0\Omega r(1-a)a'\ r^2dr. \qquad [3.53]$$

The two sets of equations describe the thrust and torque forces on the section as obtained from a "near-field" and a "far-field" perspective. The equations may be used to solve for the sectional forces by iteration. Note that the induction factors may vary from section to section. Before solving the equations, state-of-the-art computer programs introduce several corrections to the 2D approach, as mentioned above, in the text following Figure 3.17 (see, e.g., Moriarty and Hansen, 2005).

A tip-loss model is used to account for the induced velocities due to the tip vortices. This effect is discussed in more detail below in relation to the 3D vortex model. In the BEM implementation, a frequently used model is given by Moriarty and Hansen (2005). The sectional thrust and torque as found by the momentum equation [3.53] are corrected by the factor F, given by:

$$F = \frac{2}{\pi}\arccos\left[\exp\left(-\frac{N_B}{2}\frac{R-r}{r\sin\phi}\right)\right]. \qquad [3.54]$$

The value of this correction factor is displayed for three different inflow angles in Figure 3.19. The inflow angle is here assumed to be constant along the blade. The theoretical background for this correction factor was developed by Prandtl. He considered the effect of a helical vortex shed from the tip of each blade. It is observed that the correction depends upon the ratio $N_B/\sin\phi$. As the number of blades increases or the inflow angle is reduced, the strength of the tip vortex is reduced, and the correction is thus reduced.

As a strong free vortex also will be shed from the hub, a similar correction as for the tip may be used for the hub correction, replacing $R-r$ with $r-R_{hub}$ in [3.54].

As discussed in Section 3.1, the Betz theory is not valid for axial induction factors above 0.5; even for values below this, the accuracy ceases. This condition occurs for a rotor operating at high rotational speed in a low-incident wind speed. The so-called "Glauert correction" may be introduced to correct the thrust force as obtained by the Betz theory at axial induction factors above 0.4. According to the Betz theory, the thrust coefficient has its maximum value of 1.0 for an axial induction factor of 0.5 (see Figure 3.4.).Using the Glauert correction, the thrust coefficient increases almost linearly from a value of 0.96 at $a = 0.4$ to 2.0 at $a = 1$. The Glauert correction was developed for a complete axial actuator disc but is also applied on each annulus in the BEM approach. The Glauert correction is also related to the tip-loss factor. For details, see Moriarty and Hansen (2005).

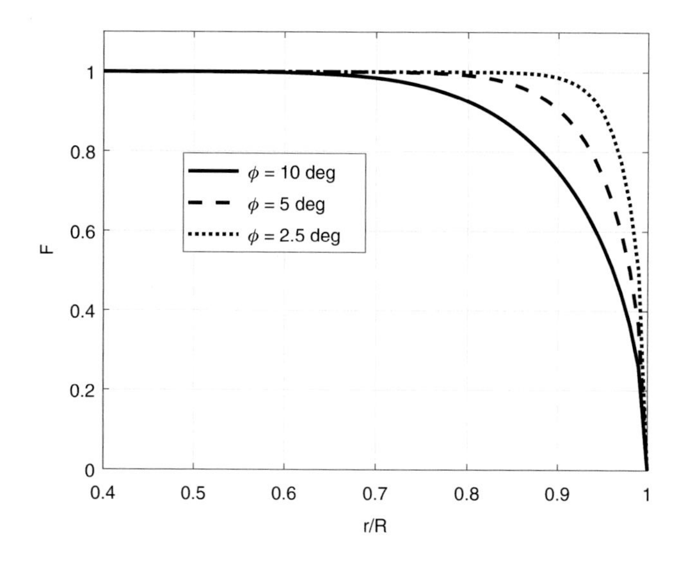

Figure 3.19 Tip-loss correction factor according to [3.54] for a three-bladed turbine and three different inflow angles.

If the incident flow is not perpendicular to the rotor plane, a skewed wake correction should be introduced.

The relations needed to compute the loads on the blades are now established. The following procedure may be applied to solve for the thrust and power of the rotor.

1) Start with one section of a blade and make an initial estimate on the axial and angular induction factors a and a'. A convenient choice may be zero for both.
2) Compute the inflow angle, ϕ, from [3.50] and the local angle of attack, $\alpha = \phi - \beta$. The pitch angle, β which is the sum of the local twist angle and blade pitch angle, is input.
3) Find the lift coefficient, C_L and C_D, from look-up tables. These tables may be based upon experimental data or 2D numerical analysis.
4) Compute thrust and torque from [3.52].
5) Introduce tip-loss and hub-loss corrections to the computed thrust.
6) New estimates on the axial and angular induction factors a and a' are computed by invoking [3.53]. (If after the tip- and hub-loss corrections the thrust coefficient is above 0.96, the Glauert correction must be used in computing a.)
7) If the values of a and a' obtained in Step 6 differ from the input values with more than a set tolerance, estimate new values and repeat from Step 2; otherwise, go to Step 8.
8) Repeat the above procedure for all sections.
9) Compute final lift and torque distribution for the rotor.

As mentioned above, the BEM method is widely used in industry due to its computational efficiency. One should, however, note that the implementations may differ as various methods for corrections may be applied. Further, as each annual section is considered independently, the flow conditions between neighboring sections should not differ much. Differences between neighboring sections will cause radial flow components not accounted for in the theory. The theory assumes stationary flow. That implies dynamic effects due to wind shear, turbulence and structural vibrations are not directly handled by the method. Various corrections are introduced to mitigate these shortcomings; see, e.g., Hansen et al. (2006) and Leishman and Martin (2002).

3.5 Drag-Based Devices

In the above derivations, the lift force on aerofoils is utilized to extract power from the wind. Devices utilizing drag forces are frequently proposed for extracting wind power. However, these are far less efficient than lift-based devices. This is illustrated in the following.

Consider a cup anemometer, usually applied for measuring wind speed. However, assuming it is used for power extraction, the following simplified considerations may be done. The layout is illustrated in Figure 3.20. It is assumed that the radius of the cups, R_c, is much smaller than the rotor radius, R. The force on the cups is governed by drag, i.e., the drag force on each cup is given by:

$$F = \frac{1}{2}\rho C_D A u_{rel}|u_{rel}|, \qquad [3.55]$$

where u_{rel} is the relative velocity between the air and the cup in the tangential direction of the rotor, $u_{rel} = (U\cos\theta - \omega R)$. The angular velocity, ω, is assumed constant. $\theta = 0$ corresponds to the cup in the top position. The drag coefficient will

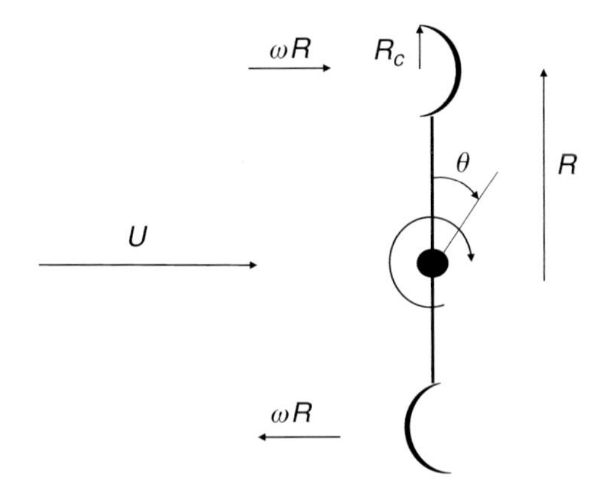

Figure 3.20 Illustration of a cup anemometer with two cups. The projected area of each cup is denoted as A.

vary with the direction of the relative velocity. For simplicity it is assumed that $C_D = C_{D1}$ for $u_{rel} > 0$ and $C_D = C_{D0}$ for $u_{rel} < 0$, $C_{D1} > C_{D0}$. C_{D1} is the drag coefficient for flow into the cup.

The drag force acts in the direction of the rotation if $u_{rel} > 0$. This is the case for $\omega R < U$ and when the cup is in the interval $-\theta_0 < \theta < \theta_0$, $\theta_0 = \arccos(\omega R/U)$. During this part of the revolution, the wind force drives the rotation. If $\omega R > U$, u_{rel} is negative during the complete rotation and no power is extracted.

The instantaneous power extracted from one cup is given by the torque multiplied by the angular velocity:

$$P = FR\omega. \qquad [3.56]$$

To find the average power production, [3.56] is integrated over one complete revolution, taking into account the change in drag coefficient as the sign of the relative velocity changes; i.e., the following expression for the average power from one cup is obtained:

$$\overline{P} = \overline{FR\omega} = \frac{R\omega}{\pi}\int_0^{\pi} F(\theta)d\theta = \frac{R\omega}{\pi}\frac{1}{2}\rho A\left[C_{D1}\int_0^{\theta_0} u_{rel}^2 d\theta + C_{D0}\int_{\theta_0}^{\pi} u_{rel}|u_{rel}|d\theta\right]. \qquad [3.57]$$

Here, the integration is performed at the interval $0-\pi$ as the result is equal for the second part of the revolution. The integration of u_{rel}^2 involves the following integrals:

$$U^2\int_0^{\theta_0}\cos^2\theta d\theta = U^2\left[\frac{\theta_0}{2} + \frac{1}{4}\sin(2\theta_0)\right]$$

$$2U\omega R\int_0^{\theta_0}\cos\theta d\theta = 2U\omega R\sin\theta_0$$

$$(\omega R)^2\int_0^{\theta_0} d\theta = (\omega R)^2\theta_0$$

$$U^2\int_{\theta_0}^{\pi}\cos^2\theta d\theta = U^2\left[\frac{\pi-\theta_0}{2} - \frac{1}{4}\sin(2\theta_0)\right]$$

$$2U\omega R\int_{\theta_0}^{\pi}\cos\theta d\theta = -2U\omega R\sin\theta_0$$

$$(\omega R)^2\int_{\theta_0}^{\pi} d\theta = (\omega R)^2(\pi-\theta_0). \qquad [3.58]$$

The average power from each cup can thus be written as:

$$\overline{P} = \frac{R\omega}{2\pi}\rho A \left\{ \begin{array}{c} C_{D1}\left[U^2\left(\frac{\theta_0}{2} + \frac{1}{4}\sin(2\theta_0)\right) - 2U\omega R\sin\theta_0 + (\omega R)^2\theta_0\right] \\ -C_{D0}\left[U^2\left(\frac{\pi-\theta_0}{2} - \frac{1}{4}\sin(2\theta_0)\right) + 2U\omega R\sin\theta_0 + (\omega R)^2(\pi-\theta_0)\right] \end{array} \right\}$$

$$= \frac{1}{2}\rho A C_{D1} U^3 \frac{\lambda}{\pi} \left\{ \begin{array}{c} \left[\left(\frac{\theta_0}{2} + \frac{1}{4}\sin(2\theta_0)\right) - 2\lambda\sin\theta_0 + \lambda^2\theta_0\right] \\ -\alpha\left[\left(\frac{\pi-\theta_0}{2} - \frac{1}{4}\sin(2\theta_0)\right) + 2\lambda\sin\theta_0 + \lambda^2(\pi-\theta_0)\right] \end{array} \right\}. \qquad [3.59]$$

Here, the "cup speed ratio," $\lambda = \omega R/U$, and the "drag ratio," $\alpha = C_{D0}/C_{D1}$, are introduced. For N cups without any interaction, the power is N multiplied by this value.

In Figure 3.21, the angle for which the drag force changes sign is shown as a function of the cup speed ratio. It is observed that for a cup speed ratio of zero, the shift in sign takes place at $\theta_0 = \pi/2$. This angle decreases as λ increases. As λ tends toward 1, $\theta_0 \rightarrow 0$. For $\lambda > 1$ there is no change in sign, i.e., the relative velocity is acting toward the "back side" of the cup all the time.

In Figure 3.22, the average power from each cup is plotted as a function of the cup speed ratio for five different drag coefficient ratios. The power is

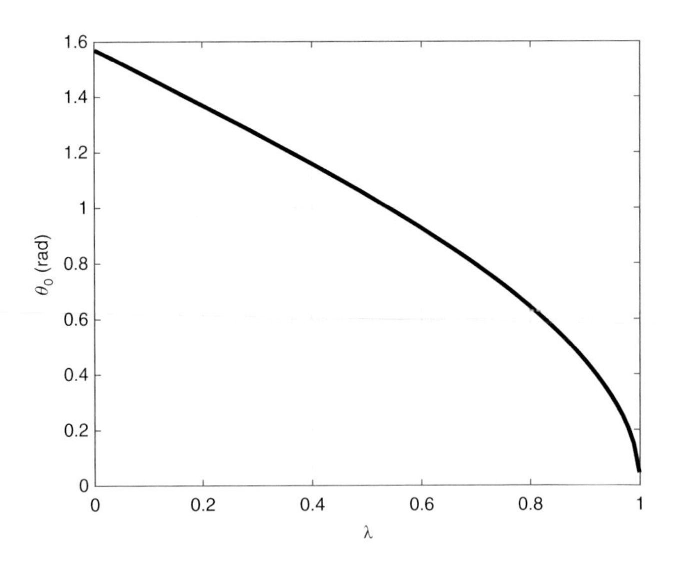

Figure 3.21 The angle θ_0 for which the force on the cup changes sign as a function of the cup speed ratio.

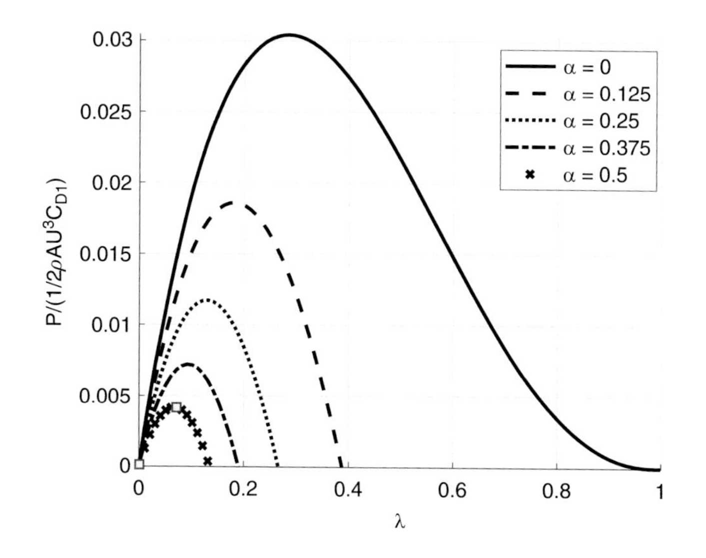

Figure 3.22 Average power produced by each cup as a function of the cup speed ratio for various drag ratios.

scaled by $1/2\rho AU^3C_{D1}$. If $C_{D1} \simeq 1$, $C_{D0} \ll 1$ and the rotor has four cups, it is observed that the system will at most extract about 0.1 times the power available at a cross-sectional area equal to the projected area of one cup. More realistic values of the drag coefficients are $C_{D1} = 1.33$ and $C_{D0} = 0.34$, i.e., $\alpha \simeq 0.25$. With these figures, a maximum power coefficient, defined as above, for a four-cup system becomes about 0.05. This is very low compared to the Betz limit. As the drag coefficient ratio increases from zero, the maximum power is considerably reduced.

It is observed that the maximum power is obtained for a rotational speed that is about half the zero-power speed. The zero-power speed is the speed of an anemometer with zero rotational resistance. According to Paschen and Laurat (2014), typical cup speed ratios for standard anemometers with conical cups are 0.25–0.3, depending upon radius ratio and drag coefficients. Conical cups have typical drag coefficients of 1.4 and 0.5, corresponding to $\alpha = 0.36$. From Figure 3.22, this should correspond to a zero-moment cup speed ratio of about 0.2, i.e., somewhat lower than reported by Paschen and Laurat (2014). This may partly be explained by the assumption of small cup radius relative to the rotor radius used in the present derivation, as well as the assumption that the driving force depends upon the square of the tangential velocity only.

3.6 Unsteady Effects

Unsteady effects appear on wind turbines for various reasons. The turbines operate in an atmospheric boundary layer. This causes the mean wind inflow velocity to a blade section to vary during one revolution. In addition, the wind field has turbulent velocity components, causing fluctuations at a wide range of frequencies. This effect may be enhanced if the turbine is operating in the wake of an upwind turbine. Due to the control system of the wind turbine, rotational speed as well as blade pitch angle may vary. Further, dynamic response of the wind turbine, due to both wind and wave loading, may cause elastic vibrations in the structure as well as rigid body motions for floating wind turbines.

The following section discusses the effect of transient and harmonic variation in the inflow condition, or motions of an aerofoil. The discussion is based upon classical results for 2D foils.

3.6.1 Step Changes: Two-Dimensional Aerofoils

Following Leishman and Martin (2002), four different unsteady effects for a 2D aerofoil are considered.

Change in angle of attack
Vertical motion of the aerofoil
Time-varying incident flow
Variation in flow velocity normal to the incident flow, due to turbulence

The four cases are illustrated in Figure 3.23. For a blade on a HAWT, the vertical motion corresponds to variation in the wind speed or motion of the nacelle in axial direction. Analytical solutions exist for these conditions under ideal flow conditions, i.e., attached flow over the foil with the Kutta–Joukowski condition fulfilled.

Characteristic for all these conditions is that the change in lift force is delayed relative to the motion of the aerofoil or change in flow condition. The reason for this is found by considering the lift force as expressed via the circulation; see [3.43]. Consider an aerofoil sitting in a steady flow at an angle of attack α. The corresponding circulation around the foil, the "bound vortex," is $-\Gamma$. Then, suddenly, the angle of attack is increased by a value $d\alpha$. To fulfil the boundary conditions at this new angle of attack, the bound vortex gets an addition $-d\Gamma$. As the circulation in the system must be conserved, a vortex of opposite strength will be shed from trailing edge and advected downstream (see Figure 3.24). This shed vortex will induce a downward velocity, v_{ind}, on the foil, causing a reduction in the angle of attack. The modified incident velocity is denoted $U_{0\mathrm{mod}}$ in Figure 3.24. As the shed vortex moves downstream, the induction effect on the incident velocity diminishes, and

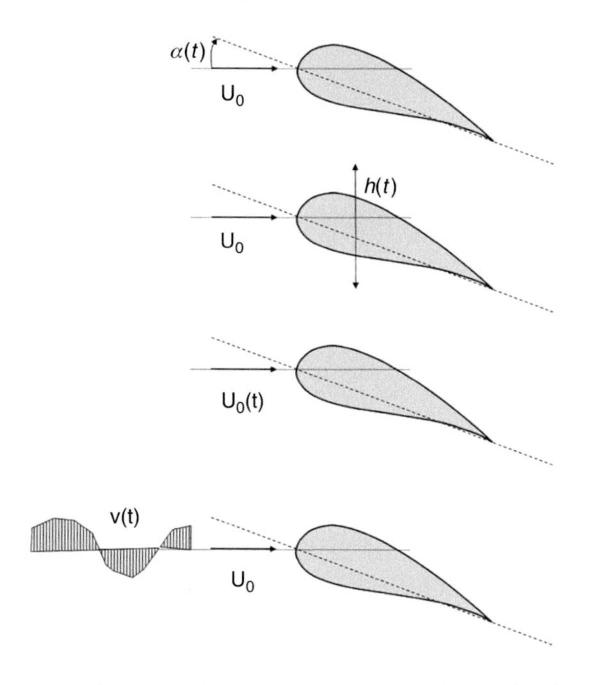

Figure 3.23 Two modes of motion and two modes of variation for the incident wind field, from top: variation in angle of attack; pure vertical motion; variation in incident wind speed and variation in vertical wind speed due to, e.g., turbulence.

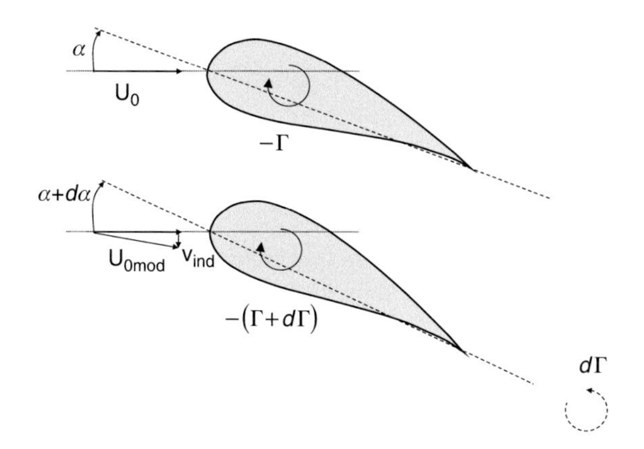

Figure 3.24 The shed vortex due to an indicial change in angle of attack. The negative value of the bound vortices is due to the direction of the circulation.

the angle of attack and the corresponding lift force converge toward the steady-state value corresponding to the angle of attack $\alpha + d\alpha$.

Wagner (1925) solved the problem of an indicial[2] change in the angle of attack for a flat plate (see, e.g., Leishman and Martin, 2002; Newman, 1977). This problem can be shown to be the same as an indicial change in the incident velocity at a foil at a constant angle of attack. The two problems correspond to case one and three in Figure 3.23 (numbered from the top). Even if the angle of attack or the incident velocity perform a step change (is instantaneous), the lift force needs some time to adjust to the new stationary value. The analytical solution to the change in lift force is denoted the Wagner solution and may be approximated (Leishman and Martin, 2002) as:$\simeq$):

$$\frac{\Delta C_L(s)}{\Delta C_L(s \to \infty)} \simeq 1 - 0.165e^{-0.091s} - 0.335e^{-0.6s}. \tag{3.60}$$

Here, $s = Ut/c$ is the number of chord lengths by which the flow has moved since the indicial change. $\Delta C_L(s)/\Delta C_L(s \to \infty)$ is the change in lift coefficient relative to the difference between the new and the previous stationary value. The function is shown in Figure 3.25. It is observed that half of the change in lift takes place immediately as the change in angle of attack takes place, but to obtain 90% of the change corresponding to the new steady-state value, the flow must move approximately six chord lengths.

Consider the indicial version of the fourth problem in Figure 3.23, i.e., a sharp gust approaching the foil, or the foil moving into a shadow where the vertical velocity component suddenly changes (e.g., moving into the tower shadow for a downwind HAWT). In this case the change in lift force is described by the Küssner function (see, e.g., Leishman and Martin, 2002; Newman, 1977). An approximation of the Küssner function is given by:

$$\frac{\Delta C_L(s)}{\Delta C_L(s \to \infty)} \simeq 1 - 0.5\left(e^{-0.26s} + e^{-2.0s}\right). \tag{3.61}$$

The function is shown in Figure 3.25. In this case the change in lift starts from zero. This can be explained by the fact that for $0 < s \ll 1$, only a small fraction of the aerofoil, close to the leading edge, experiences the gust. For $s > 1$, the gust is acting on the complete foil and more than half the steady-state lift force is obtained.

In the vertical motion case in Figure 3.23, there is no change in the lift force after a change in vertical position. However, if the foil is performing a harmonic motion in this mode, vertical forces will act on the foil. This issue will be addressed below.

[2] By «an indicial change» is meant that the value changes abruptly from one steady value to another, i.e. changed by a step function.

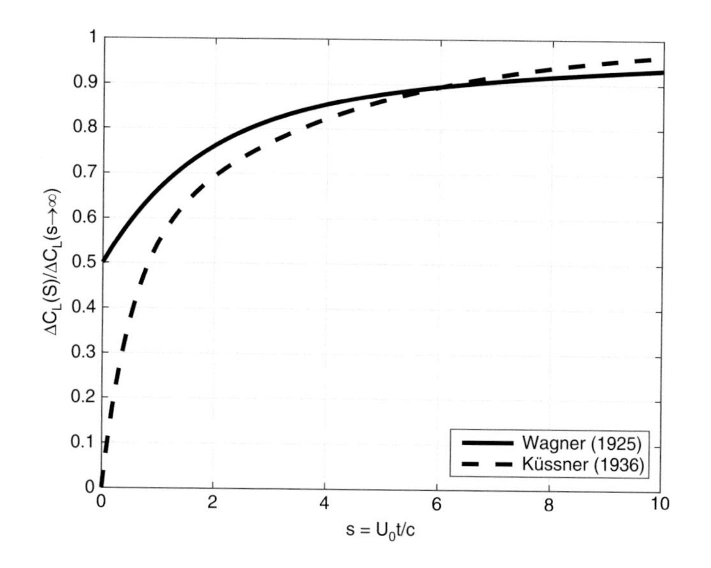

Figure 3.25 Change in lift coefficient due to an indicial change in angle of attack (Wagner) or a vertical gust moving with the incident wind speed (Küssner). Approximate expressions according to [3.60] and [3.61].

3.6.2 Harmonic Oscillations: Two-Dimensional Aerofoils

Consider a flat 2D foil in steady forward velocity and performing harmonic oscillations in heave, $h(t)$, perpendicular to incident fluid velocity, and a pitch motion, $\alpha(t)$, around the mid-point of the foil. The heave and pitch motions may be written as:

$$h(t) = \mathrm{Re}\{h_0 e^{i\omega t}\}$$

$$\alpha(t) = \mathrm{Re}\{\alpha_0 e^{i\omega t}\}. \quad [3.62]$$

Here, the index "0" denotes the complex amplitude. This is to accommodate for a phase angle between the two motions. t denotes time and ω denotes the angular frequency of oscillation. The dynamic forces obtained under these assumptions were derived by Theodorsen (1935). Faltinsen (2005) writes the lift force on the 2D foil as:

$$L = -\frac{\rho\pi}{4}c^2\left(\ddot{h} - U\dot{\alpha}\right) - \rho\pi c U C\left(k_f\right)\left[\dot{h} - U\alpha - \frac{c\dot{\alpha}}{4}\right]. \quad [3.63]$$

Here, the dots refer to time derivatives, c is the chord length of the foil and U is the incident steady fluid velocity (the index "0" is omitted for convenience); $C(k_f)$ is the so-called Theodorsen function (see next paragraph). $k_f = \omega c/2U$ is the reduced (nondimensional) frequency of oscillation. Inspecting the various terms in [3.63], the following is observed. The first part of the equation represents an inertia effect; $-\frac{\rho\pi}{4}c^2\ddot{h}$ corresponds to the added mass force in heave motion of a 2D plate with chord length c in an infinite fluid volume; $\frac{\rho\pi}{4}c^2U\dot{\alpha}$ is a coupling between the forward velocity and rotational velocity, contributing with a positive heave force when the pitch velocity is positive. This first part of the equation is frequently denoted as the "noncirculatory"[3] part.

The second, "circulatory" part of the equation contains terms proportional to the Theodorsen function, which is complex and can be written as:

$$C(k_f) = F(k_f) + iG(k_f) = \frac{H_1^{(2)}(k_f)}{H_1^{(2)}(k_f) + iH_0^{(2)}(k_f)}. \quad [3.64]$$

Here, $H_n^{(2)}(k_f) = J_n(k_f) - iY_n(k_f)$ is the Hankel function of second kind and order n, and J_n and Y_n are the Bessel functions of first and second kind and order n. The real and imaginary parts of $C(k_f)$ are plotted in Figure 3.26. It is observed that for low and high frequencies the imaginary part of $C(k_f)$ tends to zero, while the real part of $C(k_f)$ tends to 1 and 0.5 in the two limits respectively, i.e., in both these limits $C(k_f)$ becomes real.

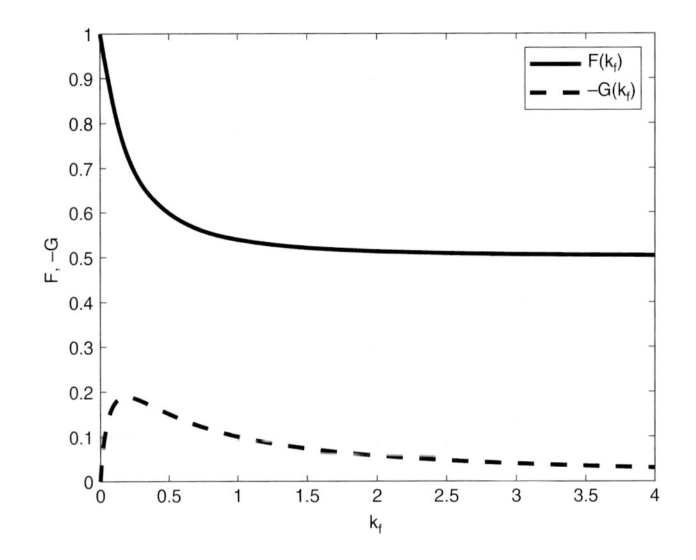

Figure 3.26 Real (F) and imaginary (G) part of the Theodorsen function as a function of the reduced frequency of oscillation, k_f.

[3] The term noncirculatory is used to denote the terms that do not involve vorticity. The term circulatory involves vorticity to satisfy the Kutta–Joukowski condition.

Reverting to [3.63] and assuming $C(k_f)$ to be real and positive, it is observed that the term proportional to $\dot{h}$ represents a positive damping force (heave force opposing the heave velocity), while the term proportional to α represents a quasi-static lift force:

$$\rho\pi c U C(k_f) U\alpha = \begin{cases} 2\pi\alpha\frac{1}{2}\rho c U^2 \text{ for } k_f \to 0 \\ \pi\alpha\frac{1}{2}\rho c U^2 \text{ for } k_f \to \infty \end{cases}. \qquad [3.65]$$

Thus, in the low-frequency limit ($k_f = 0$), the lift coefficient for a 2D plate in steady flow, 2π is approached, while in the high-frequency limit the lift coefficient is obtained as π.

The last term in [3.63] represents a negative lift force due to a pitch velocity, i.e., it adds to the frequency-independent lift force due to a pitch velocity in the first part of the equation.

3.6.3 Inertia and Damping Effects

As illustrated above, heaving and pitching of an aerofoil generate inertia and damping forces. In the following, the results of the Theodorsen theory are used to estimate the damping of the tower's fore aft motion due to the frequency-dependent forces on the rotor. A HAWT is considered. For simplicity, a homogeneous inflow to the rotor is assumed. Further, it is assumed that the tower motion causes a pure axial motion of the rotor, and that no pitching of the rotor blades takes place, i.e., $\dot{\alpha} = 0$. As will be discussed in detail in Section 3.8, the operational mode for a HAWT differs below and above rated wind speed. The rated wind speed is the wind speed at which the turbine obtains rated (full) power output. No pitching of the blades is the normal mode of operation below rated wind speed, and may also be assumed for rapid oscillations at wind speeds above rated.

Consider a section of length dr at a location r at the rotor blade. With the above assumptions, the following expression is obtained for the dynamic lift force on the section:

$$dL_{Dyn} = -\frac{\rho\pi}{4}c^2\left[\ddot{h} + 4\frac{U}{c}C(k_f)\dot{h}\right]dr. \qquad [3.66]$$

$\dot{h}$ is according to the previous derivations the velocity of the blade section normal to the total incident flow, i.e., perpendicular to U_{tot} in Figure 3.18. For a case with steady wind and constant rotor velocity, the induction effects will be fully developed, i.e., the incident velocity in [3.66] is given from:

$$U = \left[\left(U_0(1-a) \right)^2 + \left(\Omega r(1+a') \right)^2 \right]^{1/2}, \qquad [3.67]$$

and the angle relative to the wind is given by [3.50]. For simplicity it is assumed that ϕ is small, so $\cos\phi \simeq 1$. Harmonic motion of the blade section is considered, i.e., $\dot{h} = i\omega h = i\omega h_0 e^{i\omega t}$. [3.66] may then be rewritten as

$$dL_{Dyn} = -\frac{\rho\pi}{4} c^2 \left[-\omega^2 + \frac{4U}{c} C(k_f) i\omega \right] dr h_0 e^{i\omega t}$$

or

$$\frac{dL_{Dyn}}{\pi\rho U^2 h_0 dr} = -\left[-k_f^2 + 2iC(k_f) k_f \right] e^{i\omega t}. \qquad [3.68]$$

$dL_{Dyn}/dr = L_{D2D}$ is the dynamic lift force per unit length perpendicular to the total incident velocity at the foil section. For small and large values of k_f, the following asymptotic results are obtained using $C(k_f = 0) = 1.0$ and $C(k_f \to \infty) = 0.5$ (see discussion in Section 3.6.2 and Appendix B):

$$\frac{L_{D2D}}{\pi\rho U^2 h_0} \approx \begin{cases} -i2k_f e^{i\omega t} & \text{for } k_f \to 0 \\ k_f^2 e^{i\omega t} & \text{for } k_f \to \infty \end{cases}. \qquad [3.69]$$

The real part of the dynamic lift force corresponds to an inertia effect, while the imaginary part corresponds to a damping effect. In the low-frequency limit the dynamic lift is thus dominated by the damping force, while the inertia effect dominates at high frequencies.

From Figure 3.27, it is observed that the imaginary part dominates and that the asymptotic value is a reasonable approximation for $k_f < 0.1$. The real part dominates for high frequencies and the high-frequency asymptotic value for the real part approximates the full expression well even for $k_f < 1$. Note also that for $k_f < 0.35$ the real term is negative, corresponding to a negative added mass, as shown in the following paragraphs.

Defining the added mass force as the force in phase with and proportional to the acceleration, and the damping effect as the force in phase with the velocity, the dynamic lift force may be written as:

$$\begin{aligned} L_{D2D} &= L_I + L_B \\ &= -[\ddot{h}A + \dot{h}B] \\ &= -[-\omega^2 A + i\omega B] h_0 e^{i\omega t}. \end{aligned} \qquad [3.70]$$

Combining [3.68] and [3.70], A and B are obtained as:

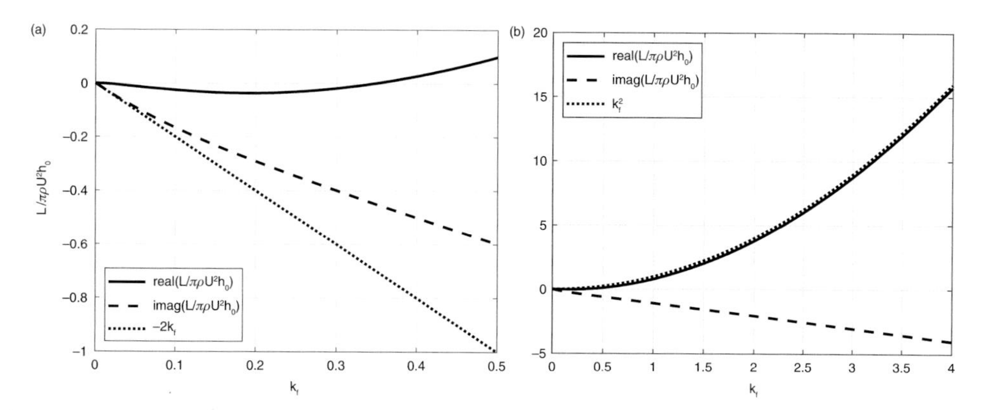

Figure 3.27 Real and imaginary part of the lift force. Solid and dashed lines: full expressions [3.68]; dotted lines: low- and high-frequency asymptotic values according to [3.69].

$$A = \frac{\rho\pi}{4}c^2\left[1+\frac{4U}{\omega c}G\left(k_f\right)\right] = \frac{\rho\pi}{4}c^2\left[1+\frac{2}{k_f}G\left(k_f\right)\right] \simeq \begin{cases} \dfrac{\rho\pi}{4}c^2\left[1+\dfrac{2\ln\left(k_f\right)}{1+\pi k_f}\right] & \text{for } k_f \to 0 \\ \dfrac{\rho\pi}{4}c^2 \text{ for } k_f \to \infty \end{cases}. \tag{3.71}$$

$$B = \rho\pi c U F\left(k_f\right) \simeq \begin{cases} \rho\pi c U \dfrac{1+\dfrac{\pi}{2}k_f}{1+\pi k_f} & \text{for } k_f \to 0 \\ 0.5\rho\pi c U \text{ for } k_f \to \infty \end{cases}. \tag{3.72}$$

The approximate expressions correspond to low- and high-frequency asymptotic values of the Theodorsen function.[4] Further details are given in Appendix B. The full expressions for added mass and damping values are shown in Figure 3.28, together with the above asymptotic expressions.

It is observed that the damping is positive for all frequencies and that the zero-frequency asymptotic value is twice the high-frequency asymptotic value. The added mass tends to minus infinity as the frequency of oscillation tends to zero. Even if the added mass tends to minus infinity, the inertia force tends to zero, as can be observed from Figure 3.27. An added mass tending to minus infinity may seem unphysical; an alternative interpretation is obtained by denoting $-\omega^2 A$ an aerodynamic stiffness, i.e., the aerodynamic stiffness may be written as:

[4] A formal series expansion has not been performed to obtain low-frequency asymptotic value of added mass. By inspection of the asymptotic behavior of the Theodorsen function shown in [B.4] in the Appendix B, the logarithmic term is retained, as this term is of leading order at low frequencies.

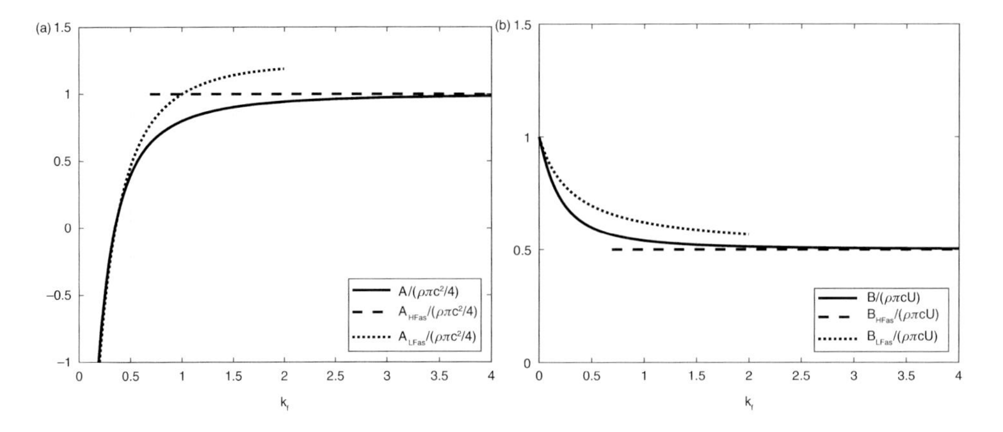

Figure 3.28 Added mass (left) and damping coefficients (right) for a heaving foil. Solid lines: full expressions; dashed lines: high-frequency asymptotic values; dotted lines: low-frequency asymptotic values.

$$
\begin{aligned}
K = -\omega^2 A &= -\left(\frac{2U}{c}k\right)^2 \frac{\rho\pi c^2}{4}\left[1+\frac{2}{k_f}G(k_f)\right]. \\
&= -\rho\pi U^2\left[k_f^2\left(1+\frac{2}{k_f}G(k_f)\right)\right].
\end{aligned}
\quad [3.73]
$$

The low-frequency asymptotic behavior of the aerodynamic stiffness can be written as:

$$
K_{low} = -\rho\pi U^2\left[k_f^2\left(1+\frac{2\ln(k_f)}{1+k\pi}\right)\right] \quad \text{as } k_f \to 0. \quad [3.74]
$$

The results from the full and asymptotic expressions are displayed in Figure 3.29. As observed from the right-hand part of the figure, the stiffness is positive in a limited range for $k_f < 0.35$. I.e., for very low frequencies there is a force in phase with the displacement tending to restrict the oscillations. For higher frequencies this force acts in opposite phase of the displacement and behaves thus as a positive inertia. The use of a stiffness definition rather than inertia demonstrates clearly how the force tends smoothly to zero as the frequency tends to zero.

3.6.4 Implementation on a Wind Turbine Rotor

The above results will be used to study the inertia and damping effects of a wind turbine rotor during operation. For simplicity, assume that the rotor is operating in a steady and homogeneous wind field. Each segment of the rotor blade is thus

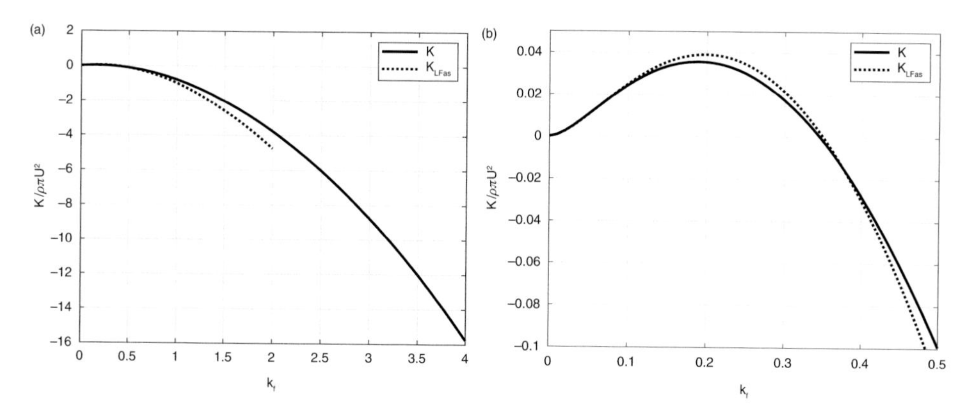

Figure 3.29 Aerodynamic stiffness defined according to [3.73] and [3.74]. Solid line: full expression; dotted line: low-frequency asymptotic expression; right: range of positive stiffness.

operating at a "steady forward velocity," U, as given by [3.67]. If the dynamic axial motions of the rotor are assumed fast, the corresponding steady-state-induced velocities are not established. No induction factors are thus to be applied on these velocities. "Fast" in the context may be related to the reduced frequency, $k_f = \omega c/2U$. The reduced frequency may, under the assumption of a rotational velocity much larger than the wind velocity and small rotational induction, approximately be written as:

$$k_f = \frac{\omega c}{2U} \simeq \frac{\omega c}{2\Omega r(1+a')} \simeq \frac{\omega c}{2\Omega r} \, . \tag{3.75}$$

Using the low frequency, the asymptotic values of added mass and damping, [3.71] and [3.72], the two-dimensional added mass and the damping force at the section at radius r may be approximated by:

$$\begin{aligned} F_A &= \ddot{h}A = -\omega^2 hA \simeq -\rho\pi(\Omega r)^2 hk_f^2\left[1+\frac{2\ln(k_f)}{1+\pi k_f}\right] \\ F_B &= \dot{h}B = i\omega hB \simeq i\rho\pi(\Omega r)^2 h\left[\frac{4+2\pi k_f}{1+k_f}\right] \end{aligned} \, . \tag{3.76}$$

Here, h is the amplitude of the motion perpendicular to the incident flow at radius r. The damping force dominates the total force. The damping is positive as it opposes the motion velocity. For $k_f < 0.35$ the added mass is negative, resulting in a small but negative inertia force.

The 2D forces derived above are acting perpendicular to the local incident flow velocity. Decomposing the axial rotor motion into a local "heave" mode on the blade section, the local 2D sectional heave velocity is obtained as:

$$\dot{h}(r) = \dot{\eta}_x \cos(\phi(r)). \tag{3.77}$$

Here, $\dot{\eta}_x$ is the velocity of the rotor in axial direction, positive in the downwind direction, and ϕ is the local inflow angle relative to the rotor plane (see Figure 3.18). Note that no induction factors are applied to this velocity as we are considering high-frequency oscillations for which the induction effects are not yet developed. The axial motion of the rotor also has a component in the direction of incident flow. The effect of this motion is presently ignored. This can be justified as follows. The motion of the rotor in axial direction will have a velocity component along the total inflow vector direction of $\Delta U = -\dot{\eta}_x \sin\phi$. The inflow angle is given by $\phi = \sin^{-1}(U_0(1-a)/U) \simeq U_0(1-a)/U$ for small inflow angles. As U is changed to $U + \Delta U$, there will be a corresponding minor change in the angle of attack corresponding to approximately $\Delta\phi \simeq U_0(1-a)\Delta U/U^2$. As this change in inflow direction may be assumed to be $\ll 1$, the effect of this change is disregarded.

The dynamic lift force at a small radial segment of length Δr may now be written as:

$$\Delta L(r) \simeq [-\omega^2 A(r) + i\omega B(r)]\eta_x e^{i\omega t} \Delta r \cos(\phi(r)). \tag{3.78}$$

The axial-directed force on the complete rotor is obtained as:

$$F_{Dx} = N_b \sum \Delta L(r) \cos(\phi(r)). \tag{3.79}$$

Here, N_b is the number of blades. The summation is taken over all the sections along one rotor blade. The sectional added mass and damping are given in [3.71] and [3.72].

Referring the sectional added mass and damping to the axial motion rather than to the local sectional heave motion, the following approximate results for added mass and damping for the entire rotor in axial motion are obtained:

$$\begin{aligned} A_x &= N_b \sum A(r) \cos^2(\phi_r) \Delta r \\ &= N_b \frac{\rho\pi}{4} \sum \left[c_r^2 \left(1 + \frac{2}{k_{fr}} G(k_{fr}) \right) \cos^2 \phi_r \Delta r \right] \\ &\simeq N_b \frac{\rho\pi}{4} \sum \left[c_r^2 \left(1 + \frac{2 \ln k_{fr}}{1 + \pi k_{fr}} \right) \cos^2 \phi_r \Delta r \right]. \end{aligned} \tag{3.80}$$

$$
\begin{aligned}
B_x &= N_b \sum B(r) \cos^2\phi_r \Delta r \\
&= N_b \sum \rho\pi c_r U_r F(k_{fr}) \cos^2\phi_r \Delta r \\
&\simeq N_b \rho\pi \sum c_r U_r \frac{1 + \dfrac{\pi k_{fr}}{2}}{1 + \pi k_{fr}} \cos^2\phi_r \Delta r \ .
\end{aligned} \tag{3.81}
$$

The index r refers to the value at the actual radius. Alternatively to the added mass, the aerodynamic stiffness is obtained from $K_x = -\omega^2 A_x$, which becomes positive for low frequencies and finite as the frequency tends to zero.

Salzmann and van der Tempel (2005) estimate the damping by considering the gradient of the lift coefficient with respect to the angle of attack, $dC_L/d\alpha$, for each section of the rotor. They obtain the damping estimate as:

$$
B_x = \frac{N_b \rho \Omega}{2} \int_{R_h}^{R} \frac{dC_L}{d\alpha} b_r c_r r dr. \qquad [3.82]
$$

Here, R_h is the hub radius. The factor b_r is a correction factor to account for the effect of variation in rotor speed. As a first estimate, the gradient of the lift coefficient may be assumed to be 2π. To determine the slope of the lift coefficient and the proper correction factor, more detailed analysis must be performed (see Liu et al., 2017), where alternative methods for estimating the damping are also discussed. One of them implies using the slope of the thrust force versus incident wind speed rather than the slope of the lift force.

As can be observed from the results in the example "Estimated Aerodynamic Mass and Damping of a 5 MW Turbine" below, the largest damping and the largest absolute value of the aerodynamic mass is obtained close to the rated wind speed. This is to be expected as both the forces and their gradients are large in this region. In the present approach it is assumed that the control system does not react on motions in the frequency range considered. If the motion frequency is slower and the controller may react by either changing the rotational speed or the blade pitch angle, much larger dynamic forces may be experienced.

Souza and Bachynski (2019) show that for floating wind turbines with natural period in pitch in the range of 25–40 s and surge in the range of 80–140 s, the natural periods may change significantly depending upon wind speed. This change is explained partly by a change of mooring stiffness and partly by the controller action. Also, a significant change in the surge and pitch damping is observed, mainly explained by a phase shift in the turbine thrust introduced by the controller. For further discussion of the floater dynamics and controllers, see Chapter 7.

Estimated Aerodynamic Mass and Damping of a 5 MW Turbine

Consider the NREL 5 MW reference turbine (Jonkman et al., 2009). Key data for the turbine are given in Table 3.2 and for the rotor blade in Table 3.1. The tip speed at rated power is 80 m/s and the rotor radius $R = 63$ *m, corresponding to rotational speed* $\Omega = 1.27$ *rad/s. It is assumed that the motion frequency of interest is the frequency corresponding to the first bending mode of the tower. This mode causes an almost pure forward-aft, or axial, motion of the rotor. The natural frequency of this mode is estimated at* $\omega_0 = 2.01$ *rad/s. At a radius of 70% of the rotor radius, the chord of the blade is about 3.0 m; a characteristic reduced frequency becomes then* $k_f = \omega c/(2\Omega \cdot 0.7R) = 0.054$*. I.e., in an aerodynamic context for this 2D rotor section, this is a low-frequency phenomenon.*

For the blade pitch and torque controller system proposed by Jonkman et al. (2009), a low-pass filter with a cut-frequency of 1.57 rad/s is applied, implying that the blade pitch controller is filtering out responses at frequencies corresponding to the first bending mode of the tower. Thus, from the perspective of the wind turbine as a system, the elastic mode of motion may be considered to be a high-frequency phenomenon.

The modal mass for the first bending mode of the tower is estimated at $m_{el} = 4.0 \cdot 10^5$ *kg, and the critical damping thus becomes* $B_{cr} = 2m_{el}\omega_0$*. The aerodynamic added mass and damping are obtained as shown in Figure 3.30. It is observed that the aerodynamic mass is negative and reduces the modal mass in this case in the order of 1%. It will thus not have any significant impact on the tower dynamics. However, the damping is significant, with a magnitude of 3–5% of the critical damping,*

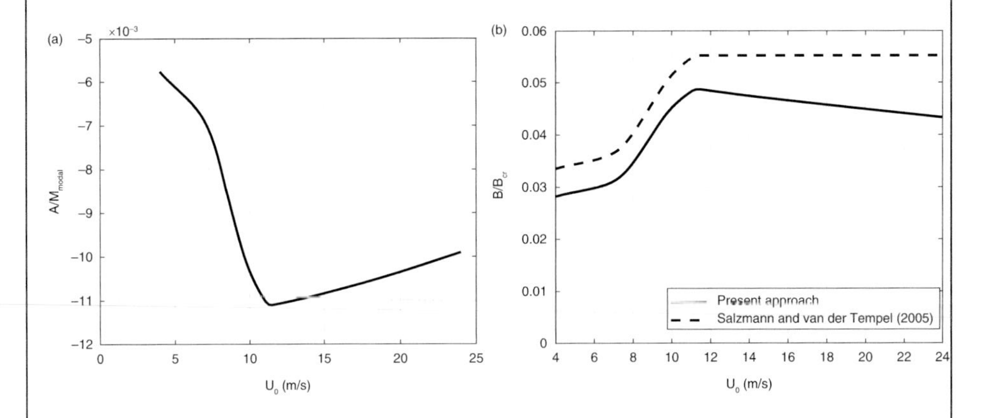

Figure 3.30 Left: aerodynamic added mass in axial motion for the NREL rotor assuming a modal mass of $4.0 \cdot 10^5$ kg and a natural frequency of 0.32Hz. Right: the corresponding relative damping, including the result using Salzmann and van der Tempel (2005) with correction factor of 1 and gradient of lift coefficient of 2π.

(cont.)

which will contribute to a significant damping of the first bending mode of the tower. It is also observed that the result for the damping is similar to what is obtained by using the method proposed by Salzmann and van der Tempel (2005) using a lift coefficient of 2π and a correction factor of 1. A somewhat lower correction factor is most likely realistic.

3.7 Vortex Methods

As has been shown in Section 3.3, the introduction of circulation, by a vortex distribution, may be used to satisfy the Kutta–Joukowski condition in the 2D steady flow case. In the case of non-stationary 2D flow, vortices will be shed from the trailing edge of the aerofoil, as illustrated in Figure 3.25. This results in time-varying forces on the foil. The vortex method may be extended to three dimensions, the simplest form of which is the "vortex line" approach. In this approach, the lift on an aerofoil with finite span is modeled by a single vortex line along the span, the bound vortex, as illustrated in Figure 3.31. Due to conservation of vorticity, the vortex line cannot end at the tips of the aerofoil or wing, but will continue as a "trailing vortex" following the fluid flow from each wing tip. The vortex line method may be extended to a vortex lattice method, solving for the vorticity distribution over the wing surface. These issues are discussed in some details in the following sections.

3.7.1 Velocity Induced by a Three-Dimensional Vortex Line

Consider a small segment of length dl of a vortex line. The segment is located at (x, y, z) and has a unit direction vector $\mathbf{s}$. The circulation is given by Γ (see Figure 3.32). The velocity induced at the point (x_1, y_1, z_1) by this segment is given from:

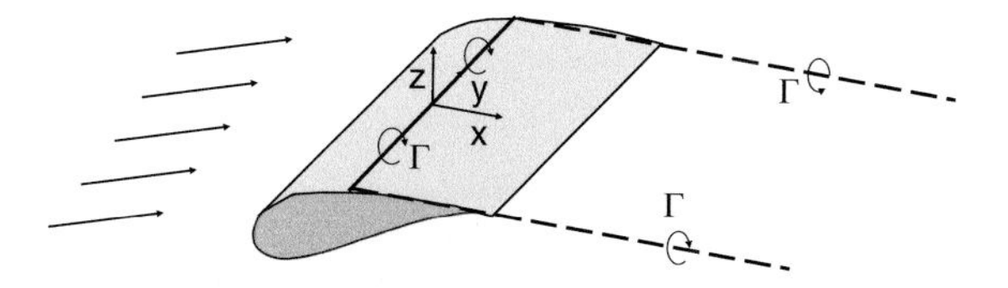

Figure 3.31 Illustration of vortex line representation of the lift on a wing with finite span. The solid line is the "bound" vortex; the dashed lines are the "trailing" vortices.

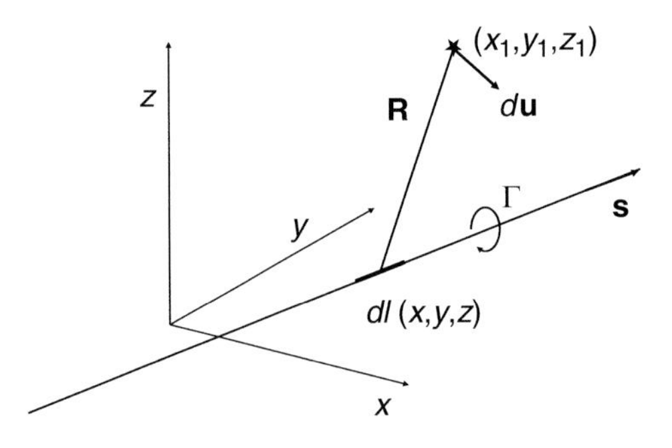

Figure 3.32 A vortex line with direction S and circulation Γ. The induced velocity in (x_1, y_1, z_1) by the short segment dl at (x, y, z) is $d\mathbf{u}(x_1, y_1, z_1)$. Note that with these 3D coordinates, a positive lift implies a positive vortex.

$$d\mathbf{u}(x_1, y_1, z_1) = \frac{\Gamma}{4\pi} \frac{\mathbf{s} \times \mathbf{R}}{R^3} dl, \qquad [3.83]$$

(see, e.g., Faltinsen, 2005). The vector between the vortex segment and the point considered is given by $\mathbf{R} = (x_1 - x)\mathbf{i} + (y_1 - y)\mathbf{j} + (z_1 - z)\mathbf{k}$, $R = |\mathbf{R}|$. As the circulation is constant along a vortex line, and the vortex has to form a closed line, the total induced velocity in (x_1, y_1, z_1) is given by:

$$\mathbf{u}(x_1, y_1, z_1) = \frac{\Gamma}{4\pi} \oint \frac{\mathbf{s} \times \mathbf{R}}{R^3} dl. \qquad [3.84]$$

Velocities Induced by a Straight Vortex Line

A) Consider an infinitely long vortex line coinciding with the x axis, $\mathbf{s} = \mathbf{i}$, $x = [-\infty, \infty].y = z = 0$. *We want to compute the velocity in* $(x_1, y_1, z_1) = (0, y_1, 0)$. *In this case* $\mathbf{s} \times \mathbf{R}$ *becomes* $\mathbf{i} \times (-x\mathbf{i} + y_1\mathbf{j}) = y_1\mathbf{k}$. *From [3.84] the velocity is obtained as:*

$$\mathbf{u}(0, y_1, 0) = \frac{\Gamma}{4\pi} \int_{-\infty}^{\infty} \frac{y_1\mathbf{k}}{\left(x^2 + y_1^2\right)^{3/2}} dx = \frac{\Gamma}{2\pi y_1}\mathbf{k}. \qquad [3.85]$$

This is the same result as for the 2D case; see [3.29].

B) Consider the trailing vortices in Figure 3.31. *Assume that these vortices may be approximated by a semi-infinite, straight horizontal vortex. The vertical induced velocity along the bound vortex,* $(x_1 = z_1 = 0)$, *due to the trailing vortex at* $y_{tip} = S/2$, *becomes* $w = \frac{\Gamma}{4\pi(y_1 - y_{tip})}$. *Here, S is the span of the wing. Accounting also for the trailing vortex at* $y_{tip} = -S/2$, *the total vertical velocity along the wingspan, the so-called "downwash," becomes:*

(cont.)

$$w_d(0, y_1, 0) = \frac{\Gamma}{4\pi}\left[\frac{1}{y_1 - S/2} - \frac{1}{y_1 + S/2}\right]. \quad [3.86]$$

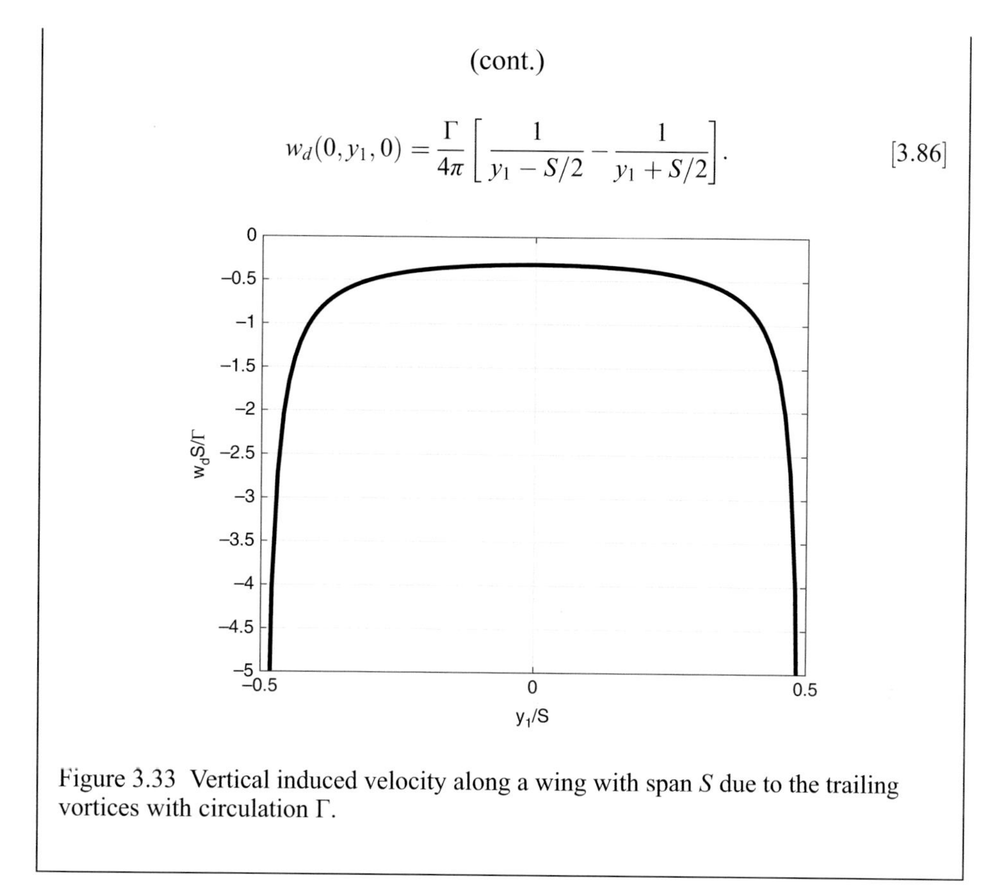

Figure 3.33 Vertical induced velocity along a wing with span S due to the trailing vortices with circulation Γ.

3.7.2 Variation in Vortex Strength along Wingspan

In Figure 3.31, a constant circulation is assumed along the wingspan. If the circulation strength of the bound vortices varies along the span, the difference in circulation will appear as trailing vortices. This is a consequence of the requirement that the circulation must be conserved. This is illustrated in Figure 3.34.

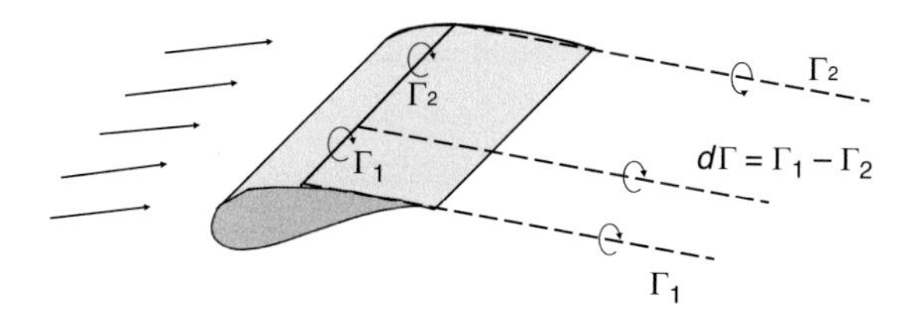

Figure 3.34 Illustration of trailing vortices due to variation in the strength of the bound vortex.

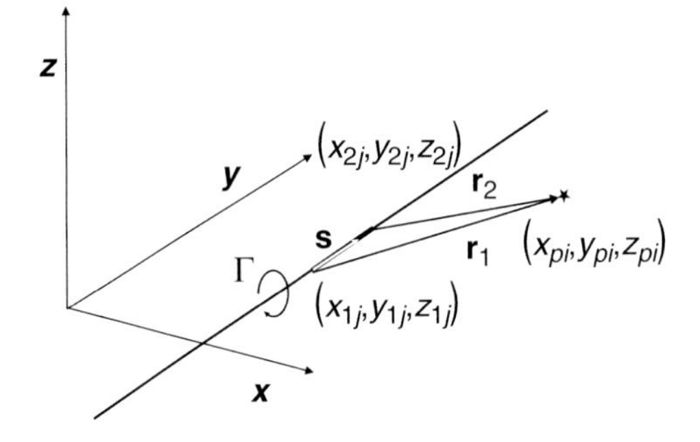

Figure 3.35 Illustration of the bond vortex segment $\mathbf{s}_j$, parallel to the $y-$ axis, together with the distance vectors from the ends of the segment to the collocation point of segment i.

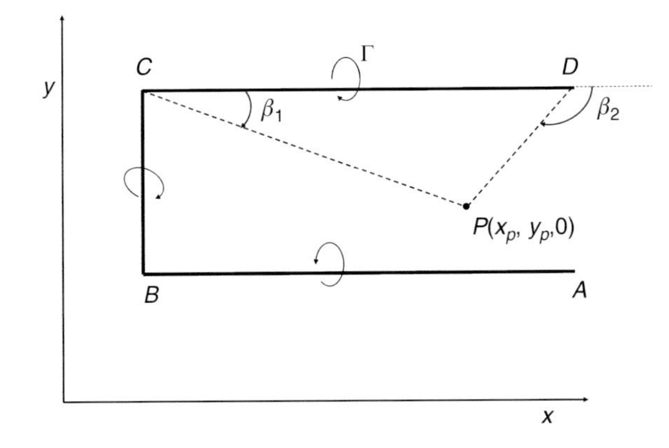

Figure 3.36 A horseshoe vortex located in the $z = 0$ plane.

From this principle, it is observed that an aerofoil may be modeled similarly to in the BEM method, using many 2D sections. The strength of the bound vortex may vary between the sections. The difference of the strength of the bound vortices in two neighboring sections results in a trailing vortex carrying the difference circulation. In the simplest version of the vortex line method, the track of each of the trailing vortices is assumed to be known a priori. The track is assumed to follow the mean flow, accounting for the downwash behind the wing. To solve for the strength of the bound vortices along each segment, a system of equations is set up where the number of unknowns is equal to the number of bound vortex segments. The total incident flow velocity and angle of attack is computed as a function of the strength of the vortices at each segment.

The vortex system illustrated in Figure 3.36 may be modeled as a system of "horseshoe" vortices. Each horseshoe is composed of the bound vortex on a given segment and the two semi-infinite trailing vortices. The bound vortex and the two trailing vortices have the same strength. The net strength of the trailing vortices leaving the border between two adjacent segments is the difference between the strength of the bound vortices in the two adjacent segments.

To solve for the vortex distribution along the span the following procedure may be applied. For simplicity a plan wing with small thickness and small angle of attack is assumed.

1) The span of the wing S is divided into N_S sections. The circulation Γ_i along each of the sections i is unknown.
2) The chord length at each segment is c_i. The bound vortex line is placed a distance $c_i/4$ from the leading edge. A collocation point is placed at the middle of each segment, a distance $3c_i/4$ from the leading edge. The collocation point is the point where the boundary condition of no flow perpendicular to the surface is satisfied. The choice of vortex location and the location of the collocation point are based upon results obtained for 2D sections; for details, see, e.g., Katz and Plotkin (2001).
3) The induced velocity in collocation point number i due to a horseshoe vortex in segment j is computed. For the bound vortex in segment j, the induced velocities, using [3.83], becomes:

$$\left(u_{bij}, v_{bij}, w_{bij}\right) = K(\mathbf{r}_1 \times \mathbf{r}_2)_{(x,y,z)}. \qquad [3.87]$$

Here, the constant K is given by:

$$K = \frac{\Gamma_j}{4\pi|\mathbf{r}_1 \times \mathbf{r}_2|^2}\left(\frac{\mathbf{s_j} \cdot \mathbf{r}_1}{r_1} - \frac{\mathbf{s_j} \cdot \mathbf{r}_2}{r_2}\right). \qquad [3.88]$$

The index (x, y, z) denotes the three directional components. $\mathbf{s}_j$ denotes the vector length of the bound vortex in segment j . $|\mathbf{s}_j| = s_j$; $\mathbf{r}_1$ and $\mathbf{r}_2$ are the distances from the ends of the vortex segment considered to the collocation point i (see Figure 3.35); $|\mathbf{r}_j| = r_j$. $K = \frac{\Gamma_j}{4\pi|\mathbf{r}_1 \times \mathbf{r}_2|^2}\left(\frac{\mathbf{s_j} \cdot \mathbf{r}_1}{r_1} - \frac{\mathbf{s_j} \cdot \mathbf{r}_2}{r_2}\right)$. The contributions from the two trailing vortices are obtained similarly; see the simplified case below.
4) Having computed the normal component of the induced velocity in all N_s collocation points due to a unit vortex strength in all bound and trailing vortices, a $N_s \times N_s$ influence matrix $\mathbf{A}_\Gamma$ is obtained. The unknown vortex strengths for each horseshoe vortex are found by solving the linear set of equations:

$$\mathbf{A}_\Gamma \Gamma = -\mathbf{U}_0 \cdot \mathbf{n}. \qquad [3.89]$$

Here, $\boldsymbol{\Gamma}$ is a vector containing the N_s vortex strengths, $\mathbf{U}_0$ is the incident velocity vector and $\mathbf{n}$ is the vector of unit surface normal at the collocation points.

Discussion of the solution method and further details are found in, e.g., Katz and Plotkin (2001).

The general expression for the induced velocity as given in [3.87] becomes very involved if expanded in the various components. However, if it is assumed that the bound vortex is parallel to the y-axis, the two trailing vortices are parallel to the x-axis and the complete horseshoe is located in the plane $z = 0$ (see Figure 3.36), the expressions become simpler.

Using [3.84] and considering a straight vortex line from C to D in Figure 3.36, the vertical velocity induced in the point P is obtained as:

$$w_{P(C-D)} = -\frac{\Gamma}{4\pi(y_C - y_P)}(\cos\beta_1 - \cos\beta_2). \qquad [3.90]$$

Here, $\cos\beta_{1C} = \frac{x_P - x_C}{\sqrt{(x_P-x_C)^2+(y_P-y_C)^2}}$ and $\cos\beta_{2D} \rightarrow -1$ as $x_D \rightarrow \infty$. Thus, for a semi-infinite vortex line the vertical velocity is obtained as:

$$w_{P(C-D)} = -\frac{\Gamma}{4\pi(y_c - y_P)}[\cos\beta_{1C} + 1]. \qquad [3.91]$$

For the bound vortex element, assumed to be parallel to the y-axis, the angles are similarly obtained as:

$\cos\beta_{1B} = \frac{y_P - y_B}{\sqrt{(x_P-x_B)^2+(y_P-y_B)^2}}$ and $\cos\beta_{2C} = \frac{y_P - y_C}{\sqrt{(x_P-x_C)^2+(y_P-y_C)^2}}$, and the induced velocity:

$$w_{P(B-C)} = -\frac{\Gamma}{4\pi(x_p - x_B)}[\cos\beta_{1B} - \cos\beta_{2C}]. \qquad [3.92]$$

For the trailing vortex extending from B to A the angles are obtained as $\cos\beta_{1A} \rightarrow 1$ as $x_A \rightarrow \infty$ and $\cos\beta_{2B} = -\frac{x_P - x_B}{\sqrt{(x_P-x_B)^2+(y_P-y_B)^2}}$, and the induced velocity is obtained as:

$$w_{P(A-B)} = -\frac{\Gamma}{4\pi(y_P - y_B)}[1 - \cos\beta_{2B}]. \qquad [3.93]$$

The total vertical velocity in P is obtained by summation of the three contributions. In Figure 3.37, an example of the induced vertical velocity, the "downwash" from such a horseshoe vortex is given. Note the significant downwash in the "interior" of the horseshoe and the singularities along the vortex lines. The downwash causes a modification of the effective inflow magnitude and direction. This also causes a change in the direction of the lift force. There will be a component of the lift force

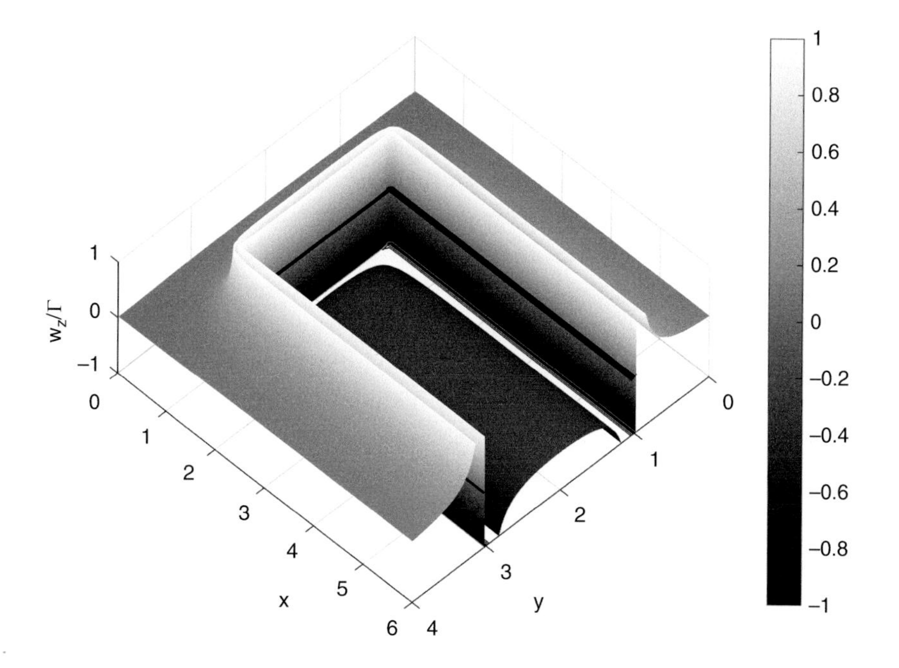

Figure 3.37 Induced vertical velocities from a horseshoe vortex with unit vortex strength and with $x_B = x_C = 1$ and $y_B = 1, y_C = 3$. $z = 0$; see Figure 3.36. Black line indicates the location of the vortex line.

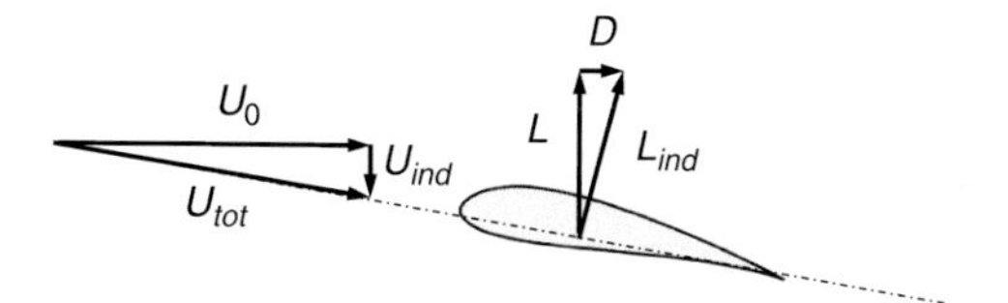

Figure 3.38 Illustration of the induced drag force on a section of an aerofoil. U_0 is the undisturbed incident velocity; U_{ind} is the induced velocity due to the vortices; L_{ind} is the lift force, decomposed into a lift force L perpendicular to the incident velocity and a drag force D in line with the incident velocity.

acting in the direction of the undisturbed incident velocity (see Figure 3.38). This component is called the induced drag force and is not related to viscous effects, but the trailing vortices and thus the finite length of the wingspan.

In the above simplistic considerations, it was assumed that the geometry of the trailing vortices is known a priori. An improved and more realistic solution is obtained by letting the geometry of the trailing vortices be part of the solution. The trailing vortices move with the fluid flow, but also induce velocities on each other, causing a complex flow pattern. The solution of such a flow problem is normally done by a time stepping procedure where the vorticity is developed over time (see, e.g., Katz and Plotkin, 2001). The numerical scheme for such solutions is not straightforward as the number of unknowns increases for each time step. Also, the geometry of the vortices develops over time, making it necessary to recompute the influence matrix for each time step. The vortices will tend to get close and "roll up," causing instabilities in the integration scheme. To improve the numerical stability, it may thus be necessary to combine nearby vortex elements into one vortex element and to introduce some dissipation of the vorticity. An example of an implementation of lifting line theory including vortex wakes is given in Marten et al. (2015).

Induced Velocities on a Rectangular Wing

Assume a rectangular wing with span to chord ratio $S/c = 10$. The wing profile has a 2D lift characteristic similar to a flat plate. Assuming 2D lift along the full span, the lift is thus obtained as $L = \rho U_0 \Gamma S = \rho U_0^2 c \pi \alpha S$. Using $c = 1$ m, $U_0 = 10$ m/s, $\alpha = 0.1$ rad and $\rho = 1.225$ kg/m^3 the estimated lift becomes 384.8 N.

Under this assumption all the bound circulation must be continued in one trailing vortex at each end of the span. The strength of the vortex becomes $\Gamma = 3.142$ m^2/s. For the sake of simplicity, assume that the trailing vortices are aligned with the flow direction, then the two trailing vortices will induce vertical velocities along the wingspan, as illustrated in Figure 3.33. The induced velocity will modify the local angle of attack along the span. The modified angle of attack becomes $\alpha_m = \alpha + w(y)/U_0$. Here, $w(y)$ is the vertical velocity along the span induced by the trailing vortices. As seen from Figure 3.33, large, induced velocities are present close to the wing tips. Using the modified angle of attack and using the above 2D expression for the lift, the lift on the wing is reduced to less than 100 N.

However, as the angle of attack is influenced by the trailing vortices, the strength of the bound vortex will vary along the span, causing trailing vortices to be shed along the full length of the span, not only from the wing tips. To estimate the distribution of the circulation along the span, the wing may be modeled by N horseshoe vortices with the bound vortex at $c/4$ and the collocation point at $3c/4$. N equations with N unknown vortex strengths are thus obtained. From the computed vortex strength, the distribution of the lift over the span can be computed. In Figure 3.39, the computed lift distribution is illustrated for three different span to chord ratios. In the case with $S/c = 10$ the estimated total lift accounting for 3D effects becomes 78% of the value obtained by the pure 2D

(cont.)

approach. From Figure 3.39, it is observed how the 3D lift distribution over the span approaches the 2D distribution as the S/c ratio increases.

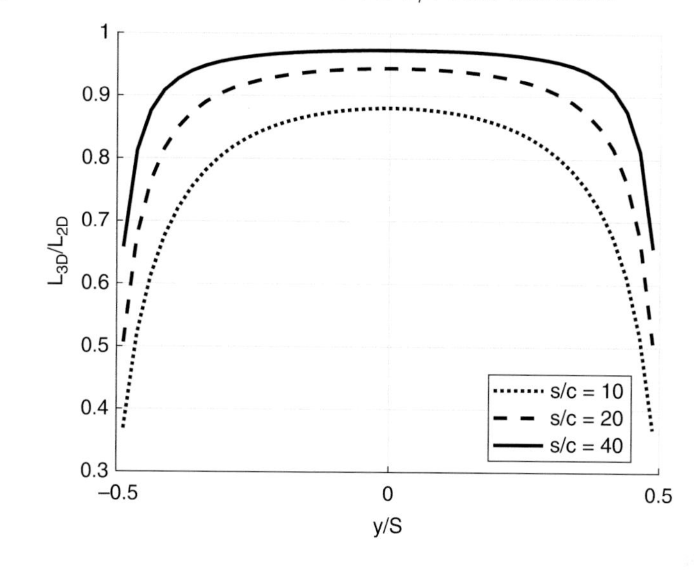

Figure 3.39 Lift distribution over a wingspan with various S/c ratios. Obtained using 40 horseshoe vortices. Bound vortices at c/4, collocation at 3c/4. Trailing vortices in the plane of the incident flow.

3.7.3 Transient Effects: The Start Vortex

As discussed in Section 3.6.1, a change in lift on an aerofoil is related to a corresponding change in circulation. Thus, as the bound vortex is changed by $\Delta\Gamma$, a corresponding vortex of strength $-\Delta\Gamma$ parallel to the bound vortex is shed from the trailing edge and advected with the flow downstream. If the aforementioned "horseshoe" model for the vortices is used in modeling transient phenomena, it implies that every horseshoe is completed by a shed vortex parallel to the bound vortex. Each horseshoe vortex is thus closed and fulfilling the requirement of continuity of the vorticity. Transient effects may thus be modeled by using ring vortices, i.e., quadrilateral vortices shed at every time step. The vortex strength of each of the sides is equal. Initially, the strength of the bound vortex is governed by the lift. During the interval of a time step, two trailing vortices are generated, moving with the free flow, and connected at the end by a shed vortex. At the next time step, the first ring vortex is advected downstream, and a new one generated.

During a stationary situation, the two ring vortices will have the same vortex strengths, causing the upstream part of the first ring vortex, the bound vortex during the first time step, to cancel the strength of the shed vortex of the second ring vortex. If the lift changes between the two time steps, there will be a difference vorticity shed and also appearing as a difference in strength along the trailing vorticities.

Start Vortex

A striking example of shed vortices is an airplane taking off from a runway. As the angle of attack of the wings is changed and lift is generated, the circulation around the wings has its counterpart in a shed vortex, the "start vortex" left on the runway. The bound vortex on the wings and the start vortex on the runway are in principle (except for dissipation) connected by the trailing vortices throughout the complete journey, as illustrated in Figure 3.40. As the start vortex may remain on the runway for a long time, it may represent a potential danger if a small airplane is coming in to land or taking off just after the takeoff of a larger airplane.

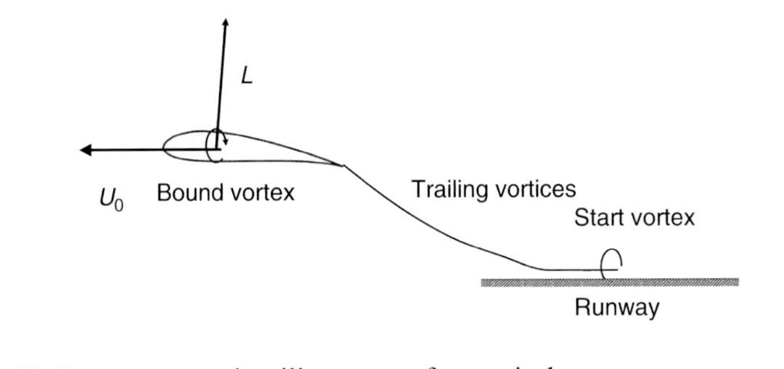

Figure 3.40 Start vortex and trailing vortex for an airplane.

3.7.4 The Vortex Lattice Method

In the vortex line method discussed in Section 3.7.1, the strength of the bound vortex is determined from the relation between the lift coefficient and the vortex strength. However, by distributing "horseshoe" vortices or vortex lattices over the surface represented by the camber line of the aerofoil and introducing a zero-through-flow condition as certain control points along the mean camber line, the strength of the vortices may be determined from the geometry of the wing. The principle is illustrated in Figure 3.41.

The span of the aerofoil is divided into a number of sections, and each section is divided into a number of panels. Within each panel, a bound vortex is placed parallel to the leading edge. Typically, the vortex is placed one

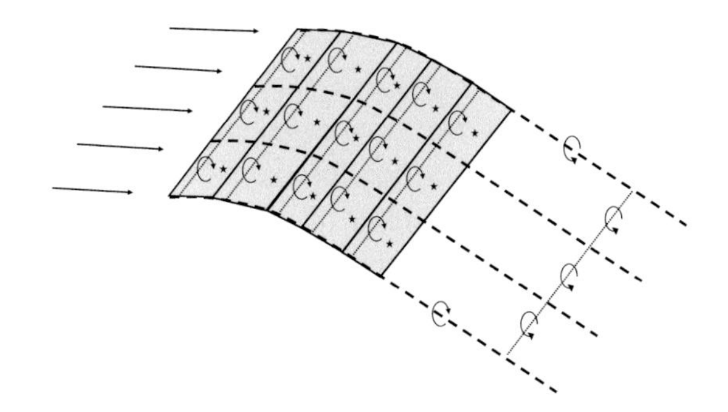

Figure 3.41 Illustration of the vortex lattice method. The panels distributed over the camber are indicated by solid lines. Dotted lines represent bound and shed vortices; dashed lines represent trailing vortices. The stars indicate collocation points.

fourth of the panel length from the forward edge. A collocation point is located at typically three-quarters of the panel length from the forward edge. Trailing vortices are located along the sides of the panels and leaving the aerofoil in a smooth manner at the trailing edge, satisfying the Kutta–Joukowski condition. A system of equations is formulated where the normal velocity at each collocation point is computed due to a unit strength of each horseshoe vortex. The strength of the vortices is computed from the condition of zero through-flow at each collocation point. As the track of the trailing vortices is part of the solution, an iterative solution is required as the down-wash will influence the track of the trailing vortices. The trailing vortices may initially be assumed to follow the undisturbed free flow and, by iteration, the correct track may be found. The common way to solve the problem is by solving it in time domain, starting with short or zero-length trailing vortices, including the shed vortices, and letting the flow and trailing vortices evolve downstream as the time progresses. This approach is similar to that for the ring vortices discussed in Section 3.7.3. Detailed description of the approach is found in, e.g., Katz and Plotkin (2001). Even if the principles involved are fairly straightforward, the implementation in robust computer codes involves several numerical issues. The shed and trailing vortices will change geometry over time, thus the influence matrix used to compute the velocities at the collocation points must be updated at every time step, causing a significant increase in computational effort as the number of time steps increases. Further, the roll-up of vortices may cause numerical problems by the high induced velocities close to a vortex. This may be handled by merging vortices, using a finite diameter of the vortex, introducing dissipation of vorticity etc. However, all such measures must be evaluated toward the convergence and accuracy of the solutions.

As an alternative to a vortex formulation of the problem, a dipole formulation is frequently used. The methods are in principle equivalent, but have different issues related to the numerical implementation. The dipole or the vortex distribution is used to handle the boundary condition given by the angle of attack and the mean camber of the aerofoil. To compute the effect of the thickness distribution, a sink-source distribution may be used. A sink-source distribution does not contribute to lift or drag, but to local modification of the flow pattern and pressure distribution over the aerofoil.

3.8 Characteristics of Horizontal-Axis Wind Turbines

One of the key characteristics of a horizontal-axis wind turbine (HAWT) is the power curve. The power curve relates the power production to the mean wind velocity at the nacelle height. An example on a power curve is given in Figure 3.42. Here, both the kinetic power in the wind, the maximum extractable power according to the Betz theory and the power extracted by an idealized HAWT are shown. At very low wind speeds there is no power extraction. The wind speed at which the power production starts is denoted the cut-in speed, typically in the range 3–5 m/s. For state-of-the-art variable-speed wind turbines, the rotor rotates with increasing wind speed as the wind speed increases beyond the cut-in speed. Normally the blade pitch is fixed in this "below-rated wind speed" range. Increasing the rotational speed with the wind speed makes it possible to work close to the optimum tip speed ratio and maximize

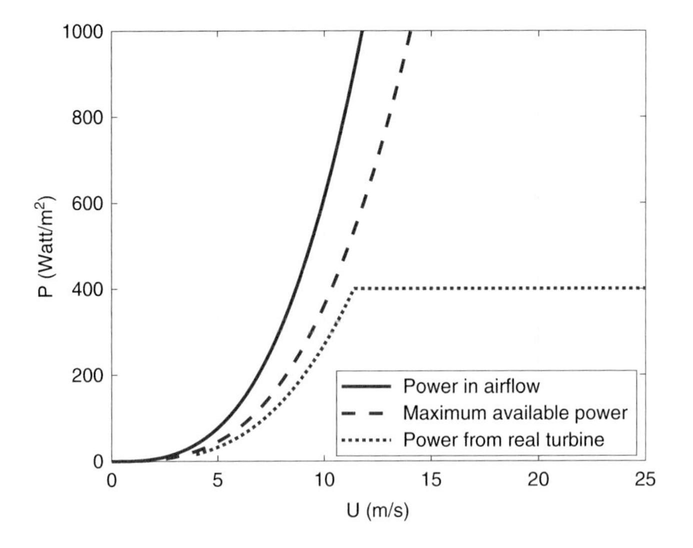

Figure 3.42 Total power in wind per square meter rotor area; available power according to the Betz theory; and power extracted from an idealized HAWT. Density of air used is 1.225 kg/m^3

the power extraction. In this range, the power curve will almost be proportional to the cube of the wind velocity. As the power reaches the rated (design) power, the power extraction is limited by the generator capacity, and all the available wind power cannot be utilized. The normal control strategy in this "over-rated wind speed" range is to keep a constant rotational speed of the rotor and increase the blade pitch angle to keep a constant turbine torque. For wind speeds close to rated wind speed, a smooth transition of the control regimes below and above the rated wind speed is used.

The power coefficient, thrust coefficient and the axial induction factors for a real wind turbine are shown in Figure 3.43. The characteristics of the NREL 5 MW reference turbine are used as an illustration (Jonkman et al., 2009). It is observed that the power coefficient and axial induction factor both are almost constant for below-rated wind velocities, except for at very low wind speeds, but drop just before rated wind speed is obtained. This smooth transition reduces the peak in the thrust force that otherwise would occur at rated wind speed. At wind speeds below 6 m/s the tip speed ratio of the turbine is not ideal, thus the power coefficient is reduced and the thrust coefficient increases.

In Figure 3.44, the thrust, power and axial induction factors for a virtual, ideal wind turbine running at the Betz limit below rated wind speed and at constant power above rated wind speed are shown. The key values 8/9, 16/27 and 1/3 are recognized for the below-rated state.

In Figure 3.45 (left), the power coefficient for the ideal and real turbine are compared. The most significant difference is observed close to rated wind speed. This is due to the smooth transition between the below-rated and above-rated control regimes. In Figure 3.45 (right), the corresponding relative thrust forces are compared. The forces are scaled with the thrust according to the Betz theory at

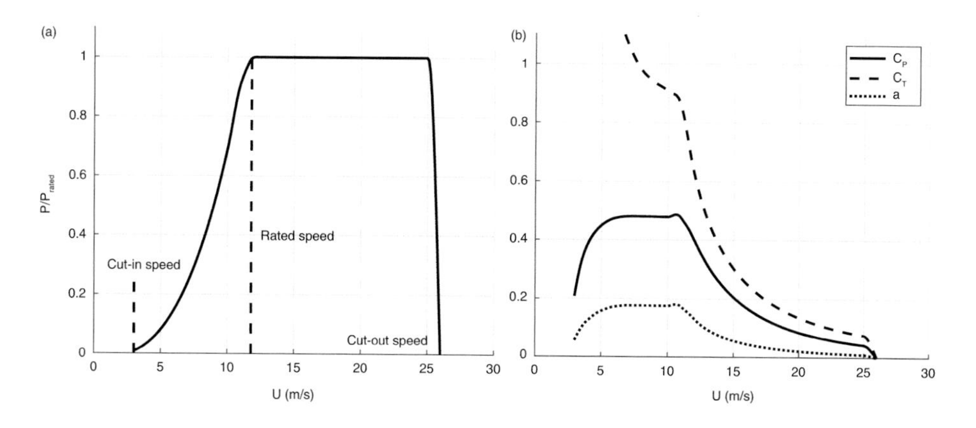

Figure 3.43 Left: power curve for a real HAWT. Right: corresponding thrust, power and axial induction factor.

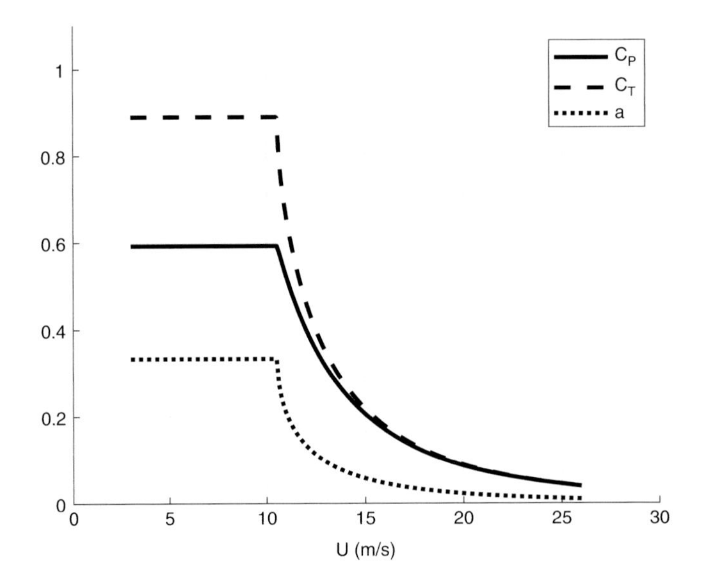

Figure 3.44 Illustration of thrust, power and axial induction factor for a virtual turbine running at the Betz limit below rated wind speed and at constant power above rated wind speed.

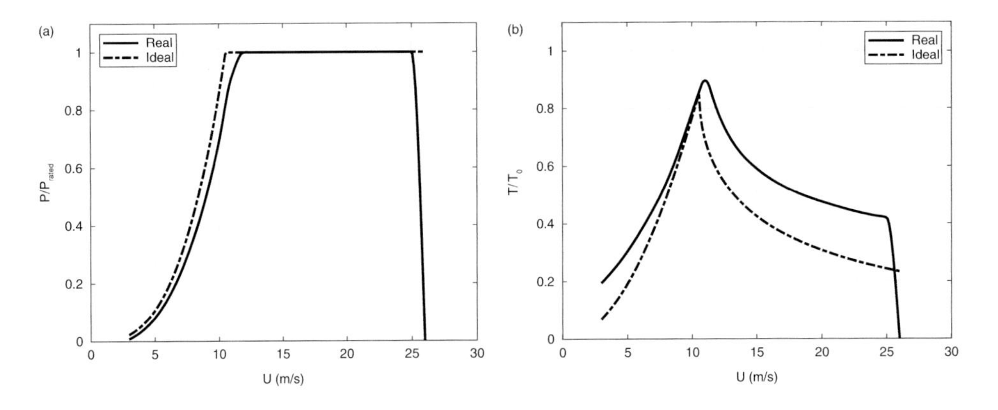

Figure 3.45 Left: comparison of the real power coefficient of the turbine in Figure 3.43 and the power curve based upon the Betz limit. Right: the thrust force for the ideal turbine and the real turbine. Both forces are scaled by $1/2\rho AU^2_{rated}$, where $U_{rated} = 11.4\ m/s$ is the rated wind speed for the real turbine.

rated wind speed, 11.4 m/s. It is observed that the real and ideal thrust forces are almost equal in the range where the turbine is designed for optimum performance (about 8–10 m/s); see the description of the reference turbine in Section 3.8.1. In the remaining wind speed ranges, the real turbine has higher thrust than the ideal turbine. This is because the ideal turbine achieves rated power at a lower wind speed than the real turbine.

3.8.1 Reference Turbines

Jonkman et al. (2009), Bak et al. (2013) and Gaertner et al. (2020) have developed and described a 5 MW, 10 MW and 15 MW HAWT turbine respectively. As the descriptions of these turbines are complete and well defined, they are frequently used as reference turbines for testing computational tools as well as being used as sample turbines for design and analyses of offshore support structures. Table 3.2 gives the main characteristics of the three turbines.

Figure 3.46 shows the results from steady state analysis of the NREL 5 MW turbine as reported by Jonkman et al. (2009) . The operational wind speeds, up to the cut-out wind speed at 25 m/s, are divided into five regions:

Table 3.2 *Comparison of the NREL 5 MW reference turbine (Jonkman et al., 2009), the DTU 10 MW reference turbine (Bak et al., 2013) and the 15 MW IEA reference turbine (Gaertner et al., 2020).*

	NREL 5 MW	DTU 10 MW	IEA 15 MW
Rating (output) (MW)	5	10	15
Rotor	Upwind three blades	Upwind three blades	Upwind three blades
Control	Variable-speed, collective-pitch	Variable-speed, collective-pitch	Variable-speed, collective-pitch
Drive train	High-speed, multiple-stage gearbox	Medium-speed, multiple-stage gearbox	Direct-drive
Rotor diameter (m)	126	178.3	240
Hub diameter (m)	3.0	5.6	7.94
Hub height (m)	90	119	150
Cut-in, rated, cut-out wind speeds (m/s)	3.0, 11.4, 25	4.0, 11.4, 25	3.0, 10.59, 25
Cut-in, rated rotor speed (rpm)	6.9, 12.1	6.0, 9.6	5.0, 7.56
Rated tip speed (m/s)	80	90	95
Design tip speed ratio	7.5	7.5	9.0
Overhang (m)	5.0	7.07	11.35
Shaft tilt, pre-cone (deg)	5.0, -2.5	5.0, -2.5	6.0, -4.0
Pre-bend (m)		3.0	4.0
Mass per blade (10^3 kg)	17.7	41	65
RNA mass (10^3 kg)	350	674	1017
Tower mass (10^3 kg)	347	628	860
Tower base diameter (m)	6.0	8.0	10.0

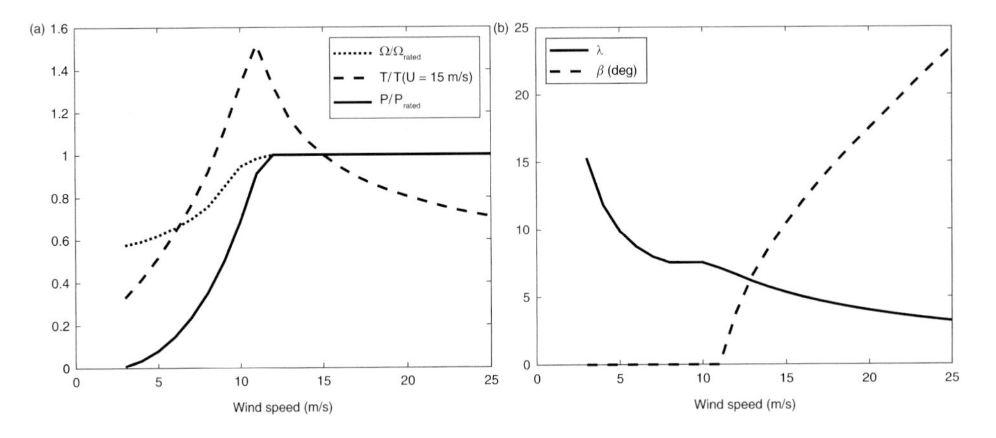

Figure 3.46 Results from steady-state aerodynamic analysis of the NREL 5 MW reference turbine, as reported by Jonkman et al. (2009). Left: rotor speed relative to rated speed (12.1 rpm), thrust relative to thrust at 15 m/s (520.5 kN) and rotor power relative to rated rotor power (5.297 MW). Right: tip speed ratio and blade pitch angle.

1, 1.5, 2, 2.5 and 3. Region 1 covers wind speeds below the cut-in wind speed at 3 m/s. Here, the controller sets the generator torque to zero. A slow rotation without power production may occur in this region. As the wind speed increases, the turbine moves into Region 1.5, where power production starts. This is a start-up region where the torque is set below the optimum value to speed up the turbine. This causes a tip speed ratio which is too high for optimum power production. As Region 2 is reached, the tip speed ratio has obtained the optimum value and the controller is set to optimize the power production, i.e., the torque is controlled so that the rotational speed is proportional to the wind speed. The blade pitch angle is kept constant (at zero) in this region. In Region 3, above the rated wind speed, the generator speed is kept constant, and the blade pitch angle is controlled to keep the torque constant as well. The tip speed ratio is gradually reduced from the optimum value in this region. Region 2.5 is a transitional zone between below and above the rated wind speed. Here, the rotational speed, and thus the tip speed ratio, is gradually decreased relative to the optimum value to secure a smooth transition to Region 3. In Region 2.5, the torque is thus increased above the optimum value, unless blade pitching is initiated.

From Figure 3.46, it is observed that the thrust force has a marked peak at rated wind speed. As wind speeds close to rated occur frequently, designers wish to avoid this peak load. One way of doing that is by so-called "peak-shaving." That implies that the controller in this region is operated to lower the thrust peak, which may be done by starting blade pitching at lower-than-rated wind speeds. The downside of this approach is lower power production around rated wind speed.

3.9 Control of Horizontal-Axis Wind Turbines with Variable Speed and Blade Pitch

As illustrated above for the NREL 5 MW wind turbine, the rotational speed and blade pitch are controlled to obtain the wanted behavior of the turbine. Variable-speed turbines with blade pitch control are the "standard" solution for multimegawatt HAWTs. For older and smaller turbines, fixed rotational speed and stall regulation exist. For further details about various control approaches, see, e.g., Burton et al. (2011). In this and the following sections, some of the basic control principles used for multimegawatt HAWTs are illustrated. Further details can be found in Burton et al. (2011), Jonkman et al. (2009) and Anaya-Lara et al. (2018). Additional control issues appear if the turbine is mounted on a floating support structure. This is discussed in Chapter 7.

The control system for a variable speed, multimegawatt turbine such as the 5 MW NREL turbine consists of two main components: a generator torque controller and a blade pitch controller that acts on all three blades collectively. The general principle is that the torque controller operates alone below rated wind speed and acts together with the blade pitch controller above rated wind speed. This is illustrated in Figure 3.46, which illustrates how the rotational speed and blade pitch angle vary below and above rated wind speed. Other, more special purpose control functions are needed to handle, e.g., normal and emergency shut-down, operation above shutdown wind speed and corresponding start-up procedures, handling of grid failure etc. These issues are not addressed here. The controller can also be used to add damping to mitigate, for example, resonant response of the tower bending mode. The main operational control functions are:

- control of generator torque to keep a desired (optimum) rotational speed. See, for example, how torque is varying with wind speed;
- control of blade pitch angle to control power output above rated wind speed;
- control of yaw angle to ensure the turbine heads into the mean wind direction.

The torque and blade pitch controllers need to be fast-acting, while the yaw control may be based upon an averaging of the wind direction over several minutes.

3.9.1 Simple Controllers

General, simple controllers may be composed of a proportional (P) component, an integral (I) component and a derivative (D) component. Thus, in a PID controller the output is controlled by one term proportional to the input; one term proportional to the time derivative of the input; and one term proportional to the integrated value over time of the input. In a PI controller, the derivative term is ignored. The input is normally a difference between an observed or measured value and a desired value, e.g., the measured torque and the optimum torque. In time domain the PID controller may be written as:

$$y(t) = K_d \frac{dx}{dt} + K_p x + K_i \int_0^t x \, dt. \qquad [3.94]$$

Here, x is the input signal and y the output from the controller. K_d, K_p and K_i are the gains for the derivative, proportional and integral terms respectively. To analyze the properties of controllers and the corresponding response of dynamic systems, Laplace transform is a useful mathematical tool. Some key properties of Laplace transform are given in Appendix C. Here, the response of a single-degree-of-freedom (SDOF) dynamic system is also derived using Laplace transform. For further discussion of Laplace transform, reference is made to mathematical textbooks.

Using Laplace transform, [3.94] may be written as:

$$Y(s) = \left[K_d s + K_p + \frac{K_i}{s} \right] X(s). \qquad [3.95]$$

Here, the capital X and Y denote the Laplace transform of x and y. The proportional term secures that the corrective signal increases proportional to the deviation from the wanted value, while the integral term secures that the average deviation over time between actual and wanted value is close to zero. The derivative term works on the speed of the deviation, i.e., the faster the deviation increases, the more correction is done. Using the derivative of x as input to the controller speeds up the action of the controller but may also cause too great a reaction to high-frequency input and even unstable response. A proper filtering of the input signal to the controller is important. For example, Jonkman et al. (2009) use the following simple discrete time recursion filter for the measured (or computed) generator speed:

$$\Omega_f(t_n) = (1 - \alpha)\Omega(t_n) + \alpha \Omega_f(t_{n-1}). \qquad [3.96]$$

Here, $\Omega(t_n)$ is the unfiltered generator speed at time t_n, $\Omega_f(t_{n-1})$ is the filtered generator speed at the previous time step, α is a low-pass filter coefficient, given by

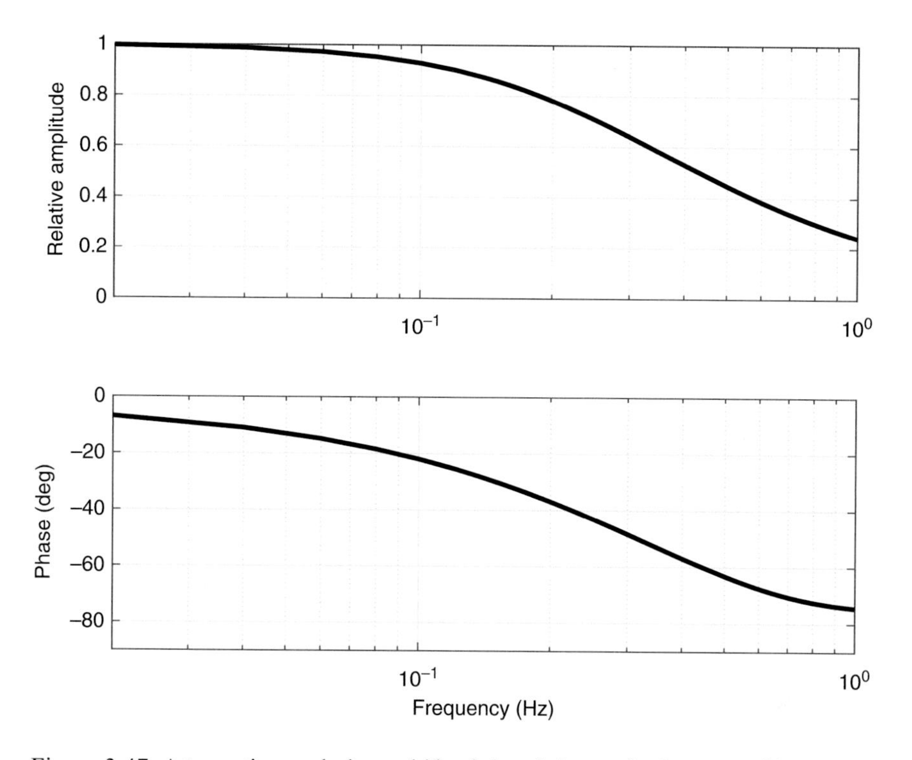

Figure 3.47 Attenuation and phase shift of signal due to the low-pass filter given by [3.96].

$\alpha = \exp(-2\pi\ \Delta t\, f_c)$, where $\Delta t = t_n - t_{n-1}$ is the discrete time step and f_c is the "corner frequency." The corner frequency is the frequency where the amplitude is reduced by 3 dB or to $1/2\sqrt{2}$ times the input amplitude. The filter characteristics using $f_c = 0.25Hz$ are shown in Figure 3.47. The value of f_c is by Jonkman et al. (2009), set to be well below the first edgewise natural frequency of the blade. This secures that the controller does not excite the blade vibrations. Several other low-pass filters could have been used, providing a sharper characteristic and less phase shift at low frequencies. However, the need for such filters is case-specific. In general, all measured signals should be filtered prior to use in a controller.

3.9.2 Control below Rated Wind Speed

To find the optimum tip speed ratio of the NREL 5 MW turbine, Jonkman et al. (2009) ran several numerical analyses of the wind turbine at below rated wind speed, at 8 m/s. From these analyses they found that the tip speed ratio giving the maximum power coefficient C_{Po} was $\lambda_o = 7.55$. This optimum is obtained at zero blade pitch angle. Figure 3.48 shows the power coefficient $C_P(\beta, \lambda)$ as a function of

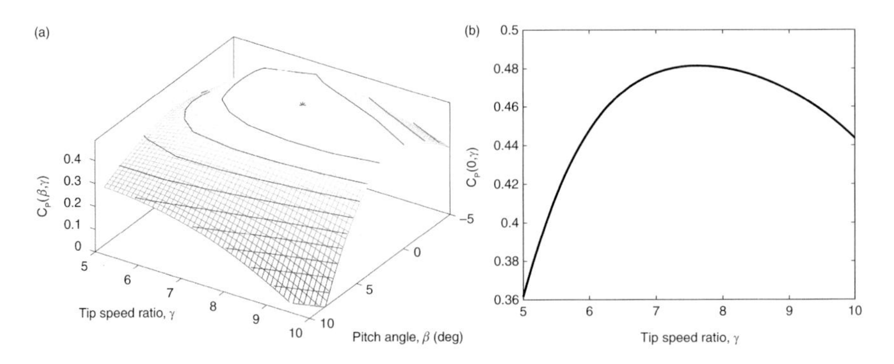

Figure 3.48 Left: power coefficient as a function of tip speed ratio and blade pitch angle. Data for the NREL 5 MW reference turbine. Maximum power coefficient $C_{Po} = 0.485$ is shown by star symbol. Right: power coefficient for zero blade pitch angle as a function of tip speed ratio. Data reproduced from Chaaban (2012).

blade pitch angle and tip speed ratio for the NREL 5 MW turbine. The data are reproduced from Chaaban (2012). It is observed that the function is fairly flat close to the optimum point, meaning that if the rotor speed or blade pitch angle are not exactly at the optimum value, the power coefficient is not reduced very much. Figure 3.48 (right) shows $C_P(0, \lambda)$. Here, the optimum tip speed ratio of $\lambda_o = 7.55$ is observed.

Below rated wind speed, in Region 2, one aims at running the turbine at constant tip speed ratio and zero blade pitch angle to obtain maximum power production. Thus, the rotational speed should increase proportionally to the wind speed. The aerodynamic power acting on the rotor is given from:

$$P_a = \frac{1}{2}\rho\pi R^2 C_p U^3 = \frac{1}{2}\rho\pi R^5 \frac{C_p(0,\lambda)}{\lambda^3}\Omega_r^3 = Q_r\Omega_r. \quad [3.97]$$

Here, the index r refers to rotor torque and angular frequency. $C_p(0, \lambda)$ is the power coefficient at zero blade pitch angle and a tip speed ratio λ. The power increases with the cube of the wind speed. A constant tip speed ratio thus implies that the rotor torque Q_r is proportional to U^2 or Ω_r^2. It is therefore reasonable to control the generator torque so that the generator torque q_g is proportional to the square of the rotational speed, i.e.:

$$q_g = K\omega_g^2. \quad [3.98]$$

Here, ω_g is the generator rotational speed. The gear ratio is the ratio between the generator and rotor speeds, $G = \omega_g/\Omega_r$. Ignoring losses in the gear system, the generator torque $q_g = Q_r/G$. Assume the factor K is set to:

$$K = \frac{1}{2}\rho C_{Po}\frac{\pi R^5}{\lambda_o^3 G^3}. \qquad [3.99]$$

C_{Po} is the optimum power coefficient obtained at the optimum tip speed ratio λ_o. The difference between the aerodynamic torque and the electrical (generator) torque will cause the rotor speed to change, i.e.:

$$I_d\frac{d\Omega_r}{dt} = Q_r - Gq_g - Q_l. \qquad [3.100]$$

Q_l is the torque due to losses in the gear transmission system. These losses are ignored in the following. I_d is the rotational inertia, accounting for the rotational inertia of the blades, drive train and generator as seen from the low-speed side of the system. The drivetrain is assumed to be infinitely stiff, so a simple SDOF system is obtained. Inserting for the aerodynamic torque [3.97] and generator torque [3.98] into [3.100], the dynamic equilibrium of the rotor speed is obtained as:

$$I_d\frac{d\Omega_r}{dt} = \frac{1}{2}\rho\pi R^5\Omega_r^2\left[\frac{C_P(0,\lambda)}{\lambda^3} - \frac{C_{Po}}{\lambda_o^3}\right]. \qquad [3.101]$$

Figure 3.49 shows the two terms in the brackets of [3.101]. It is observed that for $\gamma < \gamma_o$ the difference between the two terms is positive, causing the rotor speed to increase, while when $\gamma > \gamma_o$ the difference is negative, causing the rotor speed to slow down. The rotor speed will thus tend to the optimum speed.

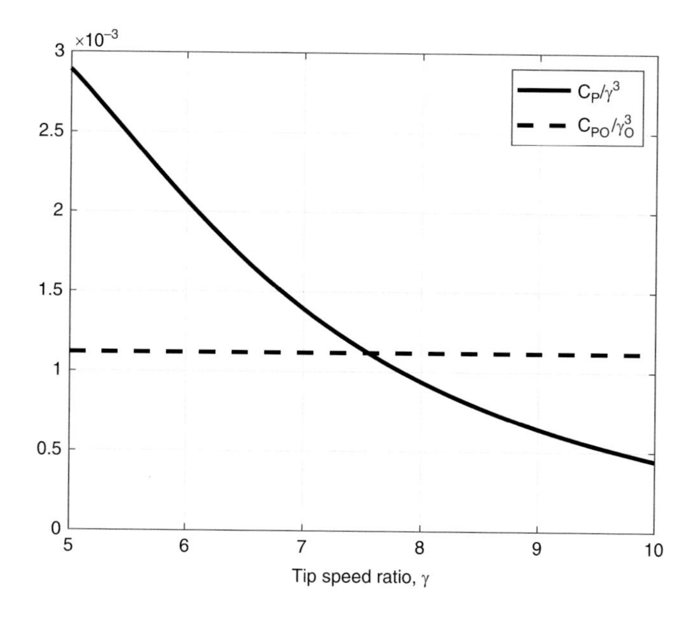

Figure 3.49 The two terms in the bracket of [3.101] as a function of tip speed ratio.

If this electrical load is lost due to, e.g., cable failure, the rotational speed will increase until the aerodynamic torque tends to zero. This may occur at unacceptably high rotational speeds. Emergency shut-down procedures are thus implemented to avoid damages in case of load failure. Various improvements of the above simple controller are discussed by Burton et al. (2011). Here, among others, improvements by considering the inertia of the rotor–gear–generator system are discussed. Such considerations may be needed to obtain a sufficient rapid response to variations in the wind velocity due to, for example, turbulence.

To improve the reaction to varying wind speeds, the generator torque may be modified by including a controller term, B, proportional to the generator acceleration, i.e., [3.98] is modified to:

$$q_g = \frac{1}{2}\rho C_P \frac{\pi R^5}{\lambda_o^3 G^3}\omega_g^2 - B\frac{d\omega_g}{dt}. \quad [3.102]$$

Thus, the following modified expression for the variation in the rotor speed is obtained.

$$\left(I_d - G^2 B\right)\frac{d\Omega_r}{dt} = \frac{1}{2}\rho\pi R^5 \Omega_r^2 \left[\frac{C_P(0,\lambda)}{\lambda^3} - \frac{C_{Po}}{\lambda_o^3}\right]. \quad [3.103]$$

With a positive value of the controller constant B, it is observed that the effective inertia of the system is reduced, making it possible for the rotor to more rapidly adapt to variations in the wind speed.

The inertia effect will contribute to a time delay between the actual and desired rotational speed. As discussed in Section 3.6, there is also a time delay between the change in wind speed and the corresponding change in lift force on the turbine blade. Thus, the rotational speed of the rotor will always be delayed relative to the optimum speed.

From the plot of the tip speed ratio in Figure 3.46 it is observed that the rotor is running faster than optimum in an interval of wind speeds just above "cut-in." This region is defined as a start-up region, between zero rotational speed and the optimum speed in Region 2.

Transient Response below Rated Wind Speed

Consider the NREL 5 MW reference turbine in a homogeneous wind field without shear and turbulence. The wind speed is constant at 6 m/s, then stepped up to 8 m/s and back to 6 m/s again. We assume that the turbine operates in Region 2, i.e., the controller is set to zero blade pitch and constant tip speed ratio. To simplify the case, transient

(cont.)

aerodynamic effects are ignored, and it is assumed that the aerodynamic power is given by the C_P-curve for zero blade pitch angle (Figure 3.48). The rotational moment of inertia as seen from the low-speed side is set to $40.45 \cdot 10^6 kgm^2$. *Solving [3.101], the results as displayed in* Figure 3.50 *are obtained. The sudden jump in the aerodynamic power is unphysical as the transient aerodynamic effects are ignored. Also, a real controller may have implemented limits to the torque rate, avoiding the jump in power observed in this example. It is clearly observed how the introduction of the modified generator torque (as in [3.103]) improves the rotor speed response.*

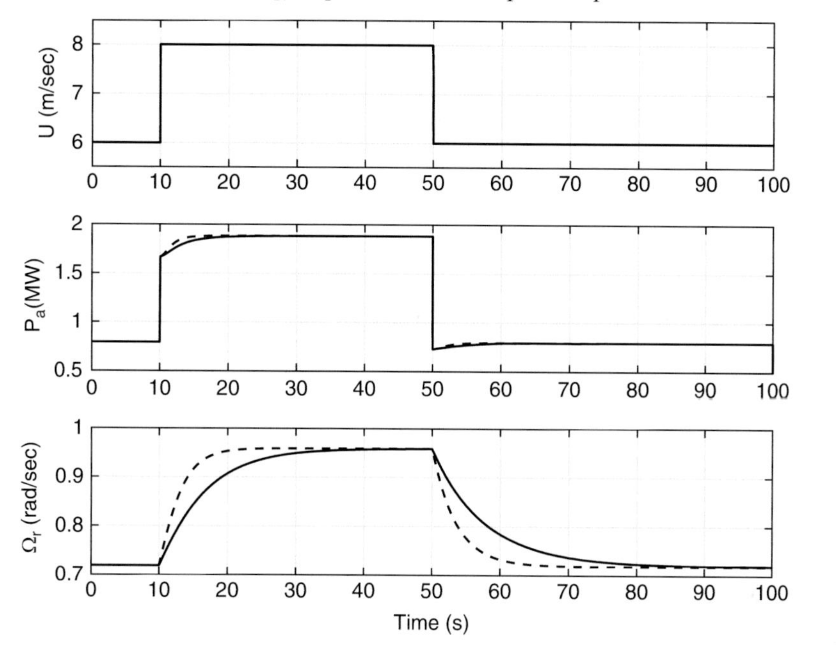

Figure 3.50 Aerodynamic power and rotor speed response due to steps in the wind speed. Solid lines: results from [3.101]. Dashed lines include the modified controller, [3.103], with $G^2B = 0.5I_d$.

3.9.3 Control above Rated Wind Speed

Above rated wind speed, both the rotational speed and the power output should be kept constant. In the NREL 5 MW case this corresponds to a rotor speed of 12.1 rev/min and a generator speed of 1173.7 rev/min. In this region a collective blade pitch controller may be used to control the generator speed.

As in [3.100], the difference between rotor torque and generator torque causes a change of the rotational speed. In Region 2, the rotational speed automatically tends toward the optimum value if the torque was controlled properly. Further, the large inertia of the rotor system secures a smooth variation in rotational speed. In Region 3, however, the aim is not to extract maximum power from the wind but to keep the power extraction at rated power and the rotational speed constant.

To maintain constant power, the generator torque in Region 3 must be inversely proportional to the generator rotational speed:

$$q_g = P_0/\omega_g. \tag{3.104}$$

where P_0 is the rated power and ω_g is the actual rotational speed of the generator. To control the rotational speed, a PID controller may be introduced to control the blade pitch angle. The deviation between the actual and rated rotational speed of the generator ω_0 is used as input to the blade pitch controller:

$$\Delta\beta = K_d \frac{d\Delta\omega_g}{dt} + K_p \Delta\omega_g + K_i \int_0^t \Delta\omega_g dt. \tag{3.105}$$

Here, $\Delta\omega_g = \omega_g - \omega_0$, the difference between actual and wanted rotational speed. To find proper values for the controller constants, Jonkman et al. (2009) use the following approach. Assume that the generator torque is controlled to be inversely proportional to the rotational speed. For a small deviation in rotational speed of the generator, $\Delta\omega_g$, the corresponding generator torque is thus obtained as:

$$q_g \simeq \frac{P_0}{\omega_0} + \frac{dq_g}{d\omega_g}\Delta\omega_g = \frac{P_0}{\omega_0} - \frac{P_0}{\omega_0^2}\Delta\omega_g. \tag{3.106}$$

Similarly, the aerodynamic torque after a small change in blade pitch angle, $\Delta\beta$, may be written as:

$$Q_r \simeq \frac{P_0}{\Omega_0} + \frac{1}{\Omega_0}\left(\frac{\partial P}{\partial \beta}\right)\Delta\beta. \tag{3.107}$$

Here, $\Omega_0 = \omega_0/G$ is the rated rotor speed. The dynamic equilibrium for the torque is given by [3.100]. The above expressions for blade pitch angle, generator torque and aerodynamic torque are now combined. It is convenient to replace the rotational velocity by the rotational angle by writing $\dot{\Phi} = \Omega_r$. The following expression is obtained as a result:

$$\ddot{\Phi}[I_d + \chi K_d] + \dot{\Phi}\left[\chi K_p - \frac{P_0}{\Omega_0^2}\right] + \Phi \chi K_i = 0, \qquad [3.108]$$

where $\chi = -\frac{G}{\Omega_0}\frac{\partial P}{\partial \beta}$. Equation [3.108] for the dynamic variation of the rotational angle with a PID controller behaves as a mass-spring-damper system. As observed from Figure 3.51, the pitch sensitivity $\partial P/\partial \beta$ is a negative quantity in the range considered. The generator torque relation as given from [3.106] introduces a negative contribution to the damping, $-\frac{P_0}{\Omega_0^2}$. Thus, the pitch proportional controller gain K_p must be sufficiently large to secure that the total damping becomes positive. The undamped natural frequency and damping ratio for the system in [3.108] are obtained as:

$$\omega_{\Psi 0} = \sqrt{\frac{\chi K_i}{I_d + \chi K_d}}$$
$$\zeta_\Psi = \frac{\chi K_p - \frac{P_0}{\Omega_0^2}}{2(I_d + \chi K_d)\omega_{\Psi 0}}. \qquad [3.109]$$

The derivative term K_d may be considered to be set to zero. Jonkman et al. (2009) found that the performance of the system was not improved by including the derivative term. The natural frequency should be chosen so that interaction with the structural frequencies is avoided; similarly, the damping should be sufficiently large that resonant oscillations are quickly reduced. $\omega_{\Psi 0} \simeq 0.6\ rad/\mathrm{s}$ and $\zeta_\Psi \simeq 0.6 - 0.7$ is recommended for the NREL 5 MW turbine (see example below).

The pitch sensitivity $\partial P/\partial \beta$ varies with the incident wind speed. Figure 3.51 shows the blade pitch angle and pitch sensitivity $\partial P/\partial \beta$ as a function of wind speed for the NREL 5 MW turbine. To account for the wind speed dependency of the pitch sensitivity, Jonkman et al. (2009) propose adjustment of the proportional and integral gain with the wind velocity. I.e., as the magnitude of the pitch sensitivity is increased with the wind speed, the gain is reduced. In the baseline controller, they add limitations to the torque rate and the pitch rate. They also set limits for the blade pitch angle.

The simple combined torque and blade pitch controller discussed above can be illustrated as shown in Figure 3.52. In Region 2.5, close to rated wind speed, one may implement a smooth transition between the control settings in

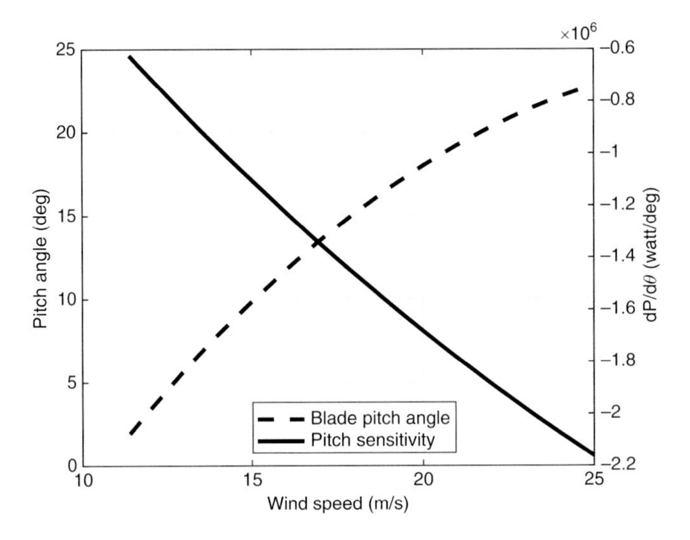

Figure 3.51 Blade pitch angle and pitch sensitivity $\partial P/\partial \beta$ for the NREL 5 MW turbine. Data adapted from Jonkman et al. (2009).

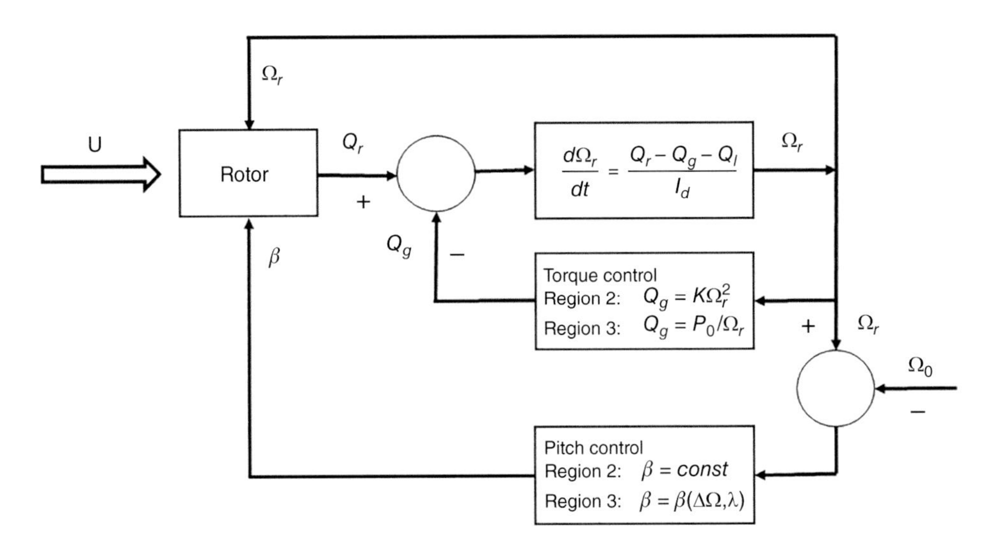

Figure 3.52 Simple torque and blade pitch controller for wind speeds in Regions 2 and 3. Q_r is the aerodynamic torque on the rotor, Q_g is the generator torque as experienced on the low speed side, Ω_r is the actual rotational speed and Ω_0 is the wanted rotational speed.

Regions 2 and 3. This will secure that the rotational speed is limited to the rated speed and that the peak in the turbine thrust as illustrated in Figure 3.46 is reduced.

Characteristics of the NREL 5 MW Baseline Controller

According to Jonkman et al. (2009), *the natural frequency of the first tower bending mode in fore-aft direction is 0.32 Hz. This is about the lowest natural frequency of the system. (The side-to-side natural frequency is marginally lower.) From the recommendations given by* Hansen et al. (2005), *they propose to set the controller constants to* $K_p(\beta = 0) = 0.01883$ s, $K_i(\beta = 0) = 0.008069$ *and* $K_d = 0$. *A gain scheduling is introduced to reduce the gains as the blade pitch angle increases. The gain scheduling function used is* $G_c = \frac{1}{1+\beta/\theta_k}$ *with* $\theta_k = 0.11$ *rad.*

The controller constants are thus varied with the blade pitch angle as $K(\beta) = K(0)G_c(\beta)$. *Using the above controller constants and* $I_d = 51.81 \cdot 10^6 kgm^2$, *the following controller natural frequency and damping are obtained by invoking [3.109] and assuming* $\beta = 0$: $\omega_{\Psi 0} = 0.58$ rad/s *and* $\zeta_\Psi = 0.62$.

$\partial P/\partial\beta_{\beta=0} = -28.24 \cdot 10^6$Watt/rad *has been assumed.*

Considering the case $\beta = 0.384$ rad (22 deg), *the gain control function becomes 0.2227 and the corresponding natural frequency and damping become* $\omega_{\Psi 0} = 0.27$ rad/s *and* $\zeta_\Psi = 0.20$.

It is observed that the ratio between the lowest natural frequency of the tower and the controller frequency is in the range 3.8–8.1. The controller is thus too slow to interact with the structural eigenmodes.

Control above Rated Wind Speed

Again, the NREL 5 MW reference turbine is used. A simple controller as illustrated in Figure 3.52 is used. The aerodynamic properties of the turbine are simulated using the power characteristics shown in Figure 3.48. The data used are computed for pitch angles less than 9 deg only. Above that value a simple spline extrapolation is used in the present example. The generator torque control is according to [3.104] and the pitch control according to [3.105]. The constants in the pitch controller as well as the factor for gain scheduling are as given in the example "Characteristics of the NREL 5 MW Baseline Controller." Further limits on the torque and pitch rates are introduced: $abs\left(\frac{dq_g}{dt}\right)_{\max} = 15\ kNm/\text{s}$, $abs\left(\frac{d\beta}{dt}\right)_{\max} = 8\ deg/\text{s}$. *Further, the maximum generator torque is set to* $47.4\ kNm$ *and the pitch angles are set to be within the range* $[0,\ 15]$ *deg. The turbine is exposed to a constant wind speed of 12 m/s, stepped up to 14 m/s and back to 12 m/s. The initial blade pitch angle is set to 0 deg. It should be noted that using the quasi-static aerodynamic characteristics of the turbine, no transient aerodynamic effects are accounted for. In Figures 3.53 and 3.54 the variation in aerodynamic power, rotational speed, power coefficient and pitch angle for this case are shown. The step change in power observed below is due to the fact that transient aerodynamic effects are ignored in this example.*

(cont.)

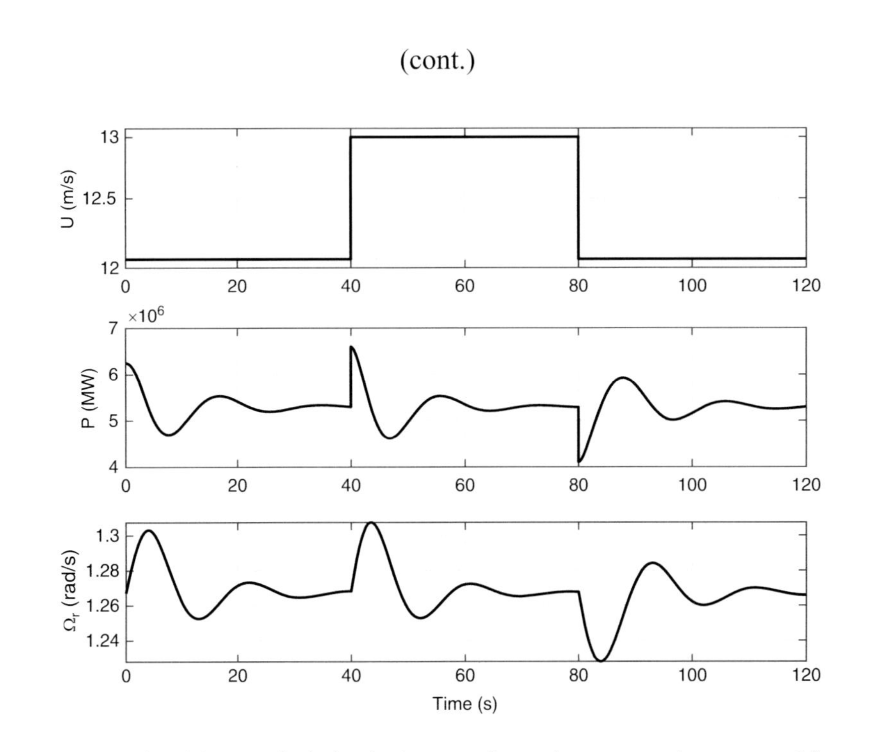

Figure 3.53 Time history of wind velocity, aerodynamic power and rotor speed for the above-rated case.

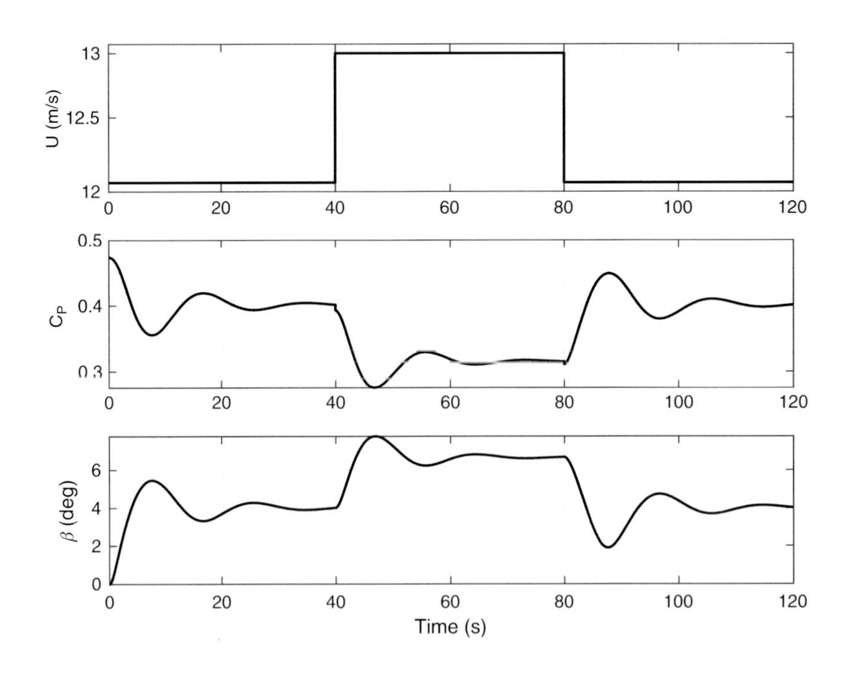

Figure 3.54 Variation of the power coefficient and blade pitch angle for the above-rated case.

3.9.4 Other Control Issues

In the previous chapters, simple controllers for regular operation below and above rated wind speed are discussed. However, others control functions exist. Various supervisory controls are needed. Supervisory control refers to a system that moves the turbine from one operational state to another. According to Burton et al. (2011) the operational states include:

- standby, when the turbine is available to run if external conditions permit;
- start-up;
- power production;
- shutdown;
- stopped, with fault.

The sequence of operations to bring a pitch-regulated turbine from rest to operation as the wind velocity increases from below 3 m/s might consist of (Burton et al. (2011) the following steps:

- Power up the pitch actuators and other subsystems.
- Release the shaft brake.
- Ramp the pitch position demand at a fixed rate to some starting pitch.
- Wait until the rotor speed exceeds a certain small value.
- Engage the closed-loop pitch control of speed.
- Ramp the speed demand up to the generator minimum speed.
- Wait until the speed has been close to the target speed for a specified time.
- Close the generator contactors.
- Engage the power or torque controller.
- Ramp the power/torque/speed set points up to the rated level.

Various control systems are to be implemented to ensure safety. These systems should operate independently of the main control system. In case of over-speeding, e.g., due to a fault on the electrical load side, an emergency shut-down procedure should be engaged. Emergency shutdown may also be engaged manually. Further, acceleration sensors may be used to warn if abnormal vibrations occur. This may be an indication of structural or mechanical failure and should initiate shutdown to avoid propagation of failure.

As observed in Section 3.6, the load on an aerofoil is delayed relative to a change in wind speed or pitch. All control systems that use the torque, rotational speed etc. as reference have thus a delayed action to compensate for the cause of the action. If the wind speed in front of the rotor could be measured continuously, the set-point for the torque could be adjusted to the actual wind speed in a more optimum manner. However, the inhomogeneous flow field experienced by large rotors causes a challenge to this approach.

By using individual blade pitching, it is possible to compensate for the variation in wind speed encountered by each blade as it moves from top position to lower position as well as the effect of the slow-down of the wind speed just in front of the tower. Such individual and active use of the blade pitch may call for a revised design of the pitch actuators as it implies much more wear of the system.

The first bending mode of the tower vibration may have very low damping, and active use of blade pitching may add damping to this mode of oscillation.

More details on the control systems of wind turbines are found in Burton et al. (2011) and Anaya-Lara et al. (2018).

Exercises Chapter 3

1. Derive the relation between the angular and axial induction factors as given in [3.24].
2. Show the derivation of the relation between axial and angular induction to obtain maximum power coefficient for at HAWT ([3.27]).
3. Show that for a 2D cylinder in a homogeneous ideal flow and with circulation, the force in the direction of the external flow, the drag force, is zero.
4. a. Plot the vortex distribution over the chord length for a parabolic aerofoil ([3.48]) at some angles of attack.
   b. Find a relation between angle of attack and camber that gives zero lift.
5. National Advisory Committee for Aeronautics (NACA) profiles are frequently used in aerofoils. In the four-digit NACA series, the aerofoil sections are defined by NACAxyzz. Here, the first digit, x, gives the maximum camber as a percentage of the chord length. The second digit, y, gives the distance of the maximum camber from the leading edge, in tenths of the chord length. The last two digits, zz, give the maximum thickness of the aerofoil as a percentage of the chord length. For example, NACA2412 means an aerofoil with maximum camber of 2%, located 0.4 of the chord length from the leading edge, with a maximum thickness of the aerofoil of 12% of the chord length. Expressions for the camber line and thickness distribution are found in literature; see, e.g., https://en.wikipedia.org/wiki/NACA_airfoil

(accessed March 2, 2023); https://m-selig.ae.illinois.edu/ads/coord_database.html (accessed March 2, 2023).

a. Select an aerofoil, e.g., NACA2412, and use [3.41]–[3.43] to compute the lift for various angles of attack.
b. Check the literature to compare your results with experimental or numerical results.

6. In [3.54] a tip-correction factor is given.
   a. Model a wing by a horseshoe vortex with straight-tip vortices and compute a tip-correction factor for the lift from the induced velocity due to the tip vortices.
   b. Compare your result with the given tip-correction factor.
7. Using 2D considerations, the lift on a wing formed as a flat plate with span S and chord c may be written as:

$$L = rU_0GS = rU_0^2 cp\alpha S.$$

   Here, α is the angle of attack. In Section 3.7.1, an example is given on how the downwash due to the trailing vortices may be computed. Assume $S/c = 10$, $c = 1\ m$, $\alpha = 5.7 deg.$ and $U_0 = 10\,m/s$.
   a. What is the lift on the wing using a pure 2D approximation?
   b. What is the effective angle of attack along the span when the effect of the downwash is included? Assume straight trailing vortices aligned with the incident flow starting at the ends of the span.
   c. Accounting for the downwash, what is the effect of the lift-distribution along the span and what is the total lift?
   d. Considering the answer in b), what is the consequence to the bound and trailing vortices along the span?
   e. How would you formulate a solution procedure to compute the lift along the length of the plate by using several horseshoe vortices to model the wing? Make a list of bullet points or a flow diagram.
   f. Make a computer code solving the problem in using one bound vortex for each y-position only (vortex line method).
8. Consider Figure 3.46. Discuss how the various parameters displayed vary with wind speed, which are controlled and how the parameters interrelate.
9. The dynamic equations for control of rotor speed below and above rated wind speed are given in [3.101] and [3.108]. Show how these are obtained.
10. Use the wind speeds given in the file Wind_vel.txt and the power curve for the NREL 5 MW reference turbine.

a. Interpolate the wind data with respect to height to find the wind velocity at nacelle height. Which interpolation method do you prefer?
b. Compute the yearly capacity factor of the turbine for the years available in the data. How large is the yearly variation in capacity factor and how does this compare with the yearly variation in mean wind speed?
c. Make similar comparisons as above, but on a seasonal basis.

4

Support Structures for Offshore Wind Turbines

4.1 Introduction

The main difference between land-based and offshore wind turbines is in the support structure that carries the rotor, electrical generator and related equipment. This chapter discusses some of the most common as well as proposed support structures for offshore wind turbines. The key requirements of the support structure are summarized, and the main principles, advantages and disadvantages of the various solutions are discussed.

As the offshore wind industry is not very old, a lot of support structures have been proposed and to some extent also designed. However, only a few designs have reached a mature level and mass production. Most bottom-fixed wind turbines use monopiles. Few floating wind farms exist, for which the only concepts in use are spar buoys and semisubmersibles. However, many floating concepts have been proposed, some of which have also been tested at various scales. This chapter discusses only support structures for Horizontal-Axis Wind Turbines (HAWTs). For vertical-axis wind turbines some additional challenges must be considered, e.g., how to withstand the torsional moment from the turbine.

The rated power of wind turbines is steadily increasing. The first offshore windfarm, installed at Vindeby in Denmark in 1991, was situated in water depth of approximately 4 m. The windfarm consisted of 11 turbines, each with a rated power of 0.45 MW. Thirty years later, we see projects underway with turbines with rated power exceeding 15 MW to be installed at water depths ten times as deep as the Vindeby turbines, or even as floating units.

The challenges faced can be illustrated by the following example.

Size of Support Structure Increases with Rated Power

Assume a turbine with power coefficient 0.48 at rated wind speed of 11 m/s (see Section 3.8). The extracted power at rated power is then 391 W/m^2. The diameters of a 5 MW and a 15 MW turbine with this power density become 128 m and 221 m respectively. With a clearance to the sea surface of 25 m, the rotor axis will be located at 89 m and 131.5 m above sea level in the two cases.

(cont.)

Considering that the weight of the rotor-nacelle assembly (RNA) in most cases increases more than proportional to the power, it is realized that the complexity in assembly and replacement of major components increases significantly with the size of the turbine. Also, the overturning moment at sea level due to the rotor thrust is 4.4 times larger for the 15 MW than the 5 MW turbine in this example.

The rated power of the turbine and the water depth are thus key parameters for choosing the support structure. Based upon the discussion of the atmospheric boundary layer in Chapter 2, the uncertainty related to the wind speeds at such heights should be kept in mind.

The functional requirement for the support structure can briefly be summarized as follows. The support structure should: carry the RNA; have sufficient strength to withstand the operational and extreme wind loads on the turbine and from waves; have sufficient fatigue resistance to withstand the dynamic loads during the lifetime of the wind turbine; and keep the motions of the wind turbine within acceptable limits. The last requirement is particularly relevant for floating wind turbines. The design lifetime for wind turbines is normally 20–25 years. Most support structures are made from steel, but composite materials and concrete are also used.

4.2 Components of an Offshore Wind Turbine

The most frequently used wind turbine is the "Danish design," a three-bladed, upwind HAWT. Horizontal-axis means that the rotor is mounted on an almost horizontal axis. This design has over time been developed to have very high efficiency as well as being robust. An illustration of an offshore HAWT of the Danish design is shown in Figure 4.1. Most designs are so-called upwind turbines. This implies that the rotor always works on the upwind side of the tower. Thereby, the shadow effect or wake effect from the tower is avoided by the rotor. If the blades should pass through such a wake at every rotation, they would induce large dynamic loads.

At the top of the tower we have the rotor-nacelle assembly (RNA). The blades are mounted to the rotor hub. The connection between the hub and the rotor blades is such that the blades may be pitched. The nacelle is the machinery house in which all the components transforming the rotor motion into electrical power are located. Some key components include a low-speed axle transmitting the rotor motion to the gear, which speeds up the rotational speed and transmits the high-speed rotation

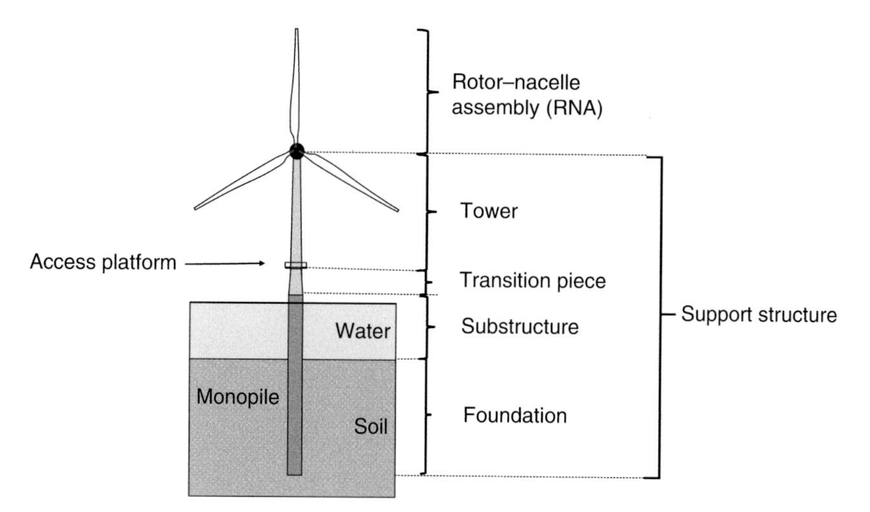

Figure 4.1 The main components of a bottom-fixed offshore wind turbine. Name convention according to DNV (2021a).

to the electrical generator. In addition, various control systems, brake systems, air-conditioning systems etc. are located in the nacelle.

Most towers are similar to those used for land-based turbines but designed to withstand the additional loads experienced in the offshore environment, such as wave loads and corrosion. Usually the tower is conical, with the largest diameter at the bottom.

The tower is mounted on top of a transition piece between the tower and the substructure. The transition piece is a consequence of the most common choice of substructure, the monopile, and its installation technique. The monopile is driven into the sea bottom and the transition piece is mounted on top. This makes it possible to install the tower vertically even if the verticality of the monopile is not exact. In newer concepts, where the complete support structure is an integrated design, the transition piece may be omitted.

Close to the bottom of the tower, at a height above the reach of the waves, a work platform or access platform is mounted. As the name indicates, this platform is used for accessing the wind turbine from boats. From the access platform a door gives access to the interior of the tower and the nacelle.

The structure below the transition piece and the sea floor is denoted the substructure. The most common design, the monopile, is a vertical steel cylinder. The part below the sea floor is denoted the foundation. How far down into the sediments the monopile is driven depends upon the geotechnical properties of the sediments. Other designs of substructures exist, which are discussed in more detail in the following sections.

All components below the RNA are known collectively as the support structure. For floating wind turbines, the floater constitutes the substructure.

4.3 Fixed Substructures

The support structure's primary function is to transfer weight and environmental loads to the sea floor. As explained in Chapter 3, the extraction of wind energy implies horizontal aerodynamic loads on the rotor. These loads cause bending loads in the support structure. The magnitude of the overturning moment increases with the distance downward from the nacelle. Therefore, a conical shape of the tower is a natural consequence.

The rotor blades bend downwind due to the aerodynamic loads. To avoid collision between the blades and the tower, the rotor blades are usually designed with some cone. The cone implies that the blades are not in the rotor plane when unloaded; instead, they have some forward cone angle. As seen from Table 2.2, the cone angle for the reference turbines is in the range −2.5 – −4.0 deg. (Negative cone angle implies that the blade tip is in front of the blade root). The rotor axis is not exactly horizontal but has some tilt in the order of 5 – 6 deg. Further, the blades are mounted in the hub some distance from the tower axis. The cone, tilt and overhang, together with a limited tower diameter, guarantee that the blades, even at maximum deflection, avoid collision with the tower.

The substructure should transfer the overturning moment and forces further down to the sea floor. At the same time, extra environmental loads on the substructure due to waves and current should be minimized. The design of the substructure must also take into consideration the conditions of the bottom sediments, i.e., how the loads are to be transferred from the substructure via the foundation to the bottom sediments or rock.

Bottom-mounted support structures may be classified according to how they are fixed to the sea floor (DNV, 2021a). The structures are:

- piled structures
- gravity-based structures
- skirt-and-bucket structures

Piled structures are presently the most frequently used solution. They are used for monopiles as well as for jacket structures (see Sections 4.3.1 and 4.3.2). Gravity-based structures are used where piling is not possible, for example, on a rocky sea floor. Skirt-and-bucket structures may be considered where the bottom conditions are very soft.

Further, the substructures may have various designs:

- monopile structures
- jacket (lattice/space-frame) structures
- tripod structures
- gravity structures

The monopile, jacket and tripod solutions are illustrated in Figure 4.2. In addition to these main categories, hybrid solutions exist.

To fulfil the fatigue life requirements, the support structure should not be excited at its natural frequencies. Due to low damping, excitation of natural frequencies may cause large dynamic responses and stresses, reducing the fatigue life considerably. For this reason, computation of the system's natural frequencies is an important part of the design process for the support structure. This issue is discussed in further detail in Chapters 5 and 6.

In the actual design of the support structure, several practical issues are to be considered. For example, practical limits exist with respect to manufacturing very thick steel plates. If large thicknesses are required (e.g., above 100 mm), few manufacturing sites may be available, and the production of tubular elements may be difficult. Lifting, transport and installation procedures may also put restrictions on physical dimensions as well as the weight of components.

4.3.1 Monopiles

The monopile is the most frequently used substructure for bottom-fixed wind turbines. The monopile is a tubular structure of constant diameter, as illustrated in Figure 4.2. It is normally installed in sea floors consisting of sand or clay-like sediments. The

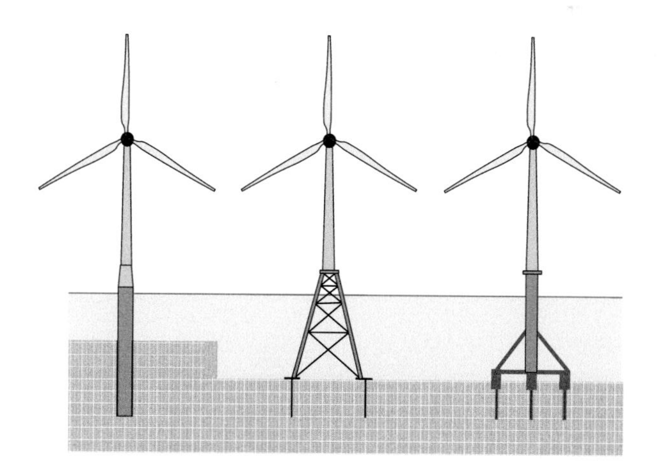

Figure 4.2 Monopile (left), jacket substructure (middle) and tripod substructure (right).

monopile is lifted by a crane vessel into vertical position and subsequently lowered to the sea floor. It is then driven to the required depth by hammering. The tolerances on verticality are strict (typically less than 1 deg). A transition piece is used to correct the verticality prior to installing the tower.

The overturning moment in the monopile due to wind and wave loads on the turbine increase as the water depth increases. Also, the natural frequency of the first bending mode decreases (see Chapter 5). To fulfil the design criteria, increased water depth thus calls for larger diameter and/or larger wall thickness, both driving weight and costs. In addition, an increased diameter increases the wave loads. Monopiles are therefore best suited for shallow water applications, even if the limits both with respect to turbine size and water depth are continuously challenged. Manufacturing and installation issues limit the size of the monopile, e.g., wall thickness and weldability, equipment for lifting and piling/hammering. The IEA 15 MW reference turbine is proposed with a monopile diameter of 10 m and a monopile mass of $1.318 \cdot 10^6$ kg (Gaertner et al., 2020). The length of the monopile depends upon the site-specific sediment conditions and may thus vary throughout a wind farm. In 2019, monopiles were used for 70% of the newly installed substructures in Europe. Jackets were the second-most used substructure (WindEurope, 2020).

An obvious advantage of the monopile is the simple geometry, which is well suited to automated manufacture. Examples of use and sizing of monopiles are given in, e.g., Negro et al. (2017) and Arany et al. (2017).

4.3.2 Jackets

To avoid the increased diameter and steel thickness of the monopile, the jacket substructure represents an alternative. As the jacket widens toward the sea floor, the forces in the legs do not need to increase, even if the overturning moment increases downward. Further, the jacket has only small structural volumes exposed to waves. The wave forces are thus lower on a jacket than on a monopile. The jacket is normally fixed to the sea floor by one or more piles at each corner. Most jackets have four legs (corners), but three-legged versions exist.

A disadvantage of the jacket substructure is the more complicated geometry, involving higher manufacturing costs per unit mass of steel than monopiles.

4.3.3 Tripods

Tripods represent an intermediate solution between the monopile and the jacket structure (see Figure 4.2). They consist of a large-diameter central tubular element, which is stiffened toward the bottom with inclined legs in three directions. The tripod is secured to the sea floor by piles. However, both the jacket and the tripod can use gravity or suction foundations as an alternative. Tripods have been used for some offshore development, but not in more recent years. An example of structural analysis of a tripod substructure is found in Chen et al. (2013). A variant, the "quadropod," using stiffeners in four directions, is also a possible solution. Both the jacket and the tripod can carry the increased moment close to the sea floor without increasing the structural dimensions close to the sea surface, causing increased wave loads.

4.3.4 Gravity-Based Substructures

Gravity-based substructures are of particular interest if the bottom sediment conditions are not suited for piling. The overturning moment is then transferred to the sea floor by a sufficiently heavy and wide substructure. The gravity-based substructure may be formed as a caisson with open compartments. After the substructure is placed on location, the compartments are filled with rock, iron ore or similar heavy, low-cost materials. Depending upon the sediment conditions, gravity-based substructures may be fitted with skirts penetrating the upper sediment layer.

Some gravity-based substructures are designed to be floated out to the installation site, then submerged and filled with ballast. Examples of gravity-based substructures are given in Peire, Nonnemann and Bosschen (2009) and Attari and Doherty (2015).

4.4 Floating Substructures

Wind turbines with fixed substructures will experience dynamic motions due to elastic deformation of the support structure, in particular bending deformation of the support structure. The first couple of bending modes (see Chapter 5) may have low natural frequencies. A design requirement is that these frequencies should not be in the range of wave load excitation frequencies, nor should they interfere with the blade passing frequencies. Floating substructure will, in addition to the elastic deformations, move as a rigid body in up to six degrees of freedom. These degrees of freedom are frequently denoted "free" modes, as opposed to the "restricted" modes that have a stiff connection to the sea floor and thus very limited dynamic deformations. Further to the additional modes of motion, the floaters must also satisfy requirements related to buoyancy, stability and maximum static and dynamic motions, in particular platform pitch. These issues are discussed in more detail in Chapter 7. The commonly used names for the six rigid-body modes of motion of a floating structure are given in Figure 7.2.

Floating substructures are applied in waters too deep for fixed substructures, either due to costs, loads or natural frequency issues. The floating structure is allowed to move freely in the "free modes," where the natural frequency is lower than the frequencies of the excitation loads. The dynamic motion response is thus limited mainly by inertia effects, while the mean position is kept by "soft springs," typically mooring lines. As for most dynamic systems, the designer has the option to either allow for large dynamic motions or to accept large restoring forces (see Chapter 5 for details).

In the design of mooring systems, one must fulfil requirements related to strength, flexibility and fatigue. If very small motions are required, large mooring line loads causing increased requirements for ultimate strength and fatigue capacity must be expected. If larger motions can be accepted, lower extreme and fatigue loads can be obtained. However, the dynamic deformations of the power transfer cable, frequently called the "dynamic cable," will increase. The directional control (yaw) of the substructure may also be worse to handle. Design of floaters for shallow water (less than approximately 50 m) and severe wind and wave conditions may be a challenging task. The mooring system must be sufficiently strong and at the same time flexible.

In the following sections, four main groups of floating substructures are discussed: semisubmersibles, spars, tension leg platforms (TLP) and barge-like structures. Most of the proposed solutions are in one of these categories. Most concepts are designed to have one turbine per support structure. Solutions using multiple turbines on one support structure are also proposed. In most cases these use variants of the above substructures, except spars.

4.4.1 Semisubmersibles

The semisubmersible substructure is in most cases designed with three or four vertical columns and either horizontal pontoons connecting the columns or horizontal plates at the lower end of the columns (see the principles illustrated in Figure 4.3).

To obtain favorable dynamic motion response of a semisubmersible , the designer plays with parameters like column diameter, draft and column spacing, height and width of pontoons, and size of bottom plates. The area of the water plane together with the mass of the complete structure plus the vertical hydrodynamic mass determines the natural period in heave. Similarly, the roll and pitch restoring forces together with the inertia in rotation determine the natural periods in roll and pitch. Definition of the modes of motion and examples of how to compute hydrostatic and hydrodynamic quantities are given in Chapter 7. All six rigid-body modes of motion for a semisubmersible are free modes. The natural periods in the free

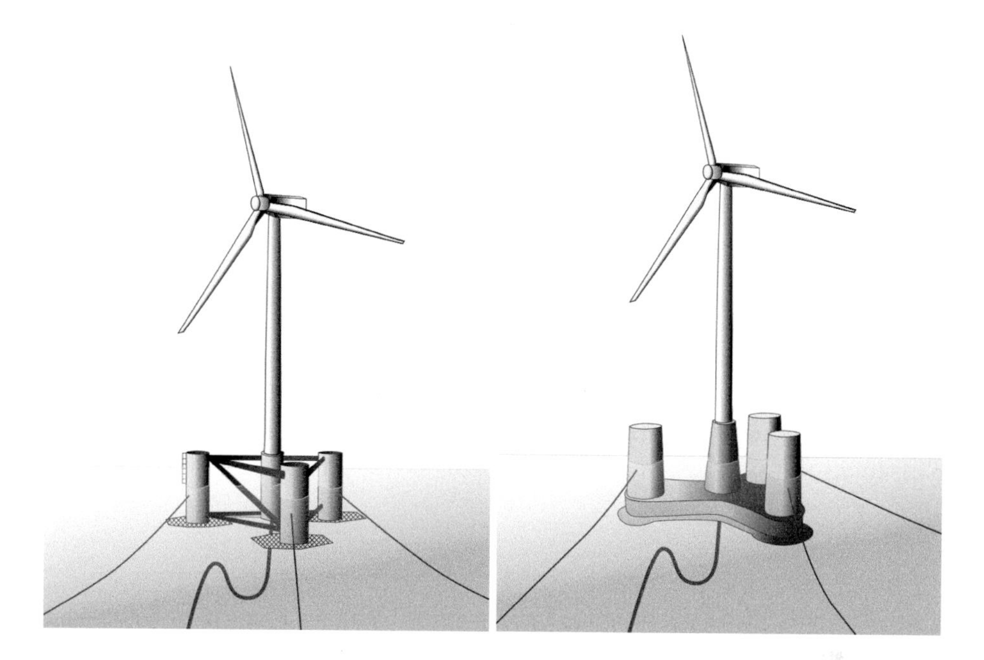

Figure 4.3 Examples of semisubmersible substructures: Left: vertical columns with perforated bottom plates and connected by a space-frame structure. Right: vertical columns connected with large horizontal pontoons.

modes are to be located outside the range of wave excitation periods. By playing with the dimensions for the columns and the pontoons, one may almost cancel out the wave excitation forces at certain wave periods. This is illustrated in Figure 4.5 for a spar-like substructure. The characteristics of the heave and roll/pitch motions of a semisubmersible are similar. Examples are given in Section 7.5. The turbine tower may either be located on top of one of the corner columns or on top of a separate center column.

As state-of-the-art wind turbines are designed for fixed foundations, little information is available about the maximum acceptable for roll and pitch motions. The roll and pitch angle and corresponding non-vertical loading in the machinery is one design issue; another is the accelerations due to pitch and roll combined with the effect of gravity.

The restoring forces in surge, sway and yaw are obtained by the mooring lines. The mooring lines will represent soft springs, securing that the floater is kept within a certain limited excursion radius to avoid overstressing the power cable, and to avoid yaw motions that are too large. Large yaw motions of the floater will challenge the yaw mechanism on top of the tower.

Concepts also exist where the semisubmersible is moored at one point only ("single-point mooring"). In such cases the whole support structure will rotate around the point of anchoring.

The challenge with the semisubmersible is, among others, the static tilt due to the thrust on the wind turbine. To avoid a too-large static tilt, the roll and pitch stiffnesses must be sufficiently large. This may challenge the requirement of keeping the natural period in roll and pitch outside the range of wave excitation. To avoid this problem, the size of the floater may be increased, or one may use active ballasting, i.e., pumping water between compartments in the substructure to compensate for the static wind overturning moment.

A favorable characteristic of the semisubmersible is the shallow draft. This allows for assembly of the complete wind turbine at a quayside and tow-out in shallow water areas. The complex geometry may on the other hand be a cost driver in manufacturing.

4.4.2 Spars

A spar-like substructure is illustrated in Figure 4.4. The requirement to the dynamic behavior of the spar substructure is similar as for the semisubmersible. However, due to the simpler geometry, there are fewer parameters to play upon to fulfil the various requirements. As shown in Figure 4.4, the diameter of the spar at the waterline is smaller than that of the main hull. By tuning the ratios between these two diameters as well as the draft of the substructure, the desired heave natural period is obtained. By proper ballasting, the moment of inertia in roll and pitch as well as the center of gravity (CG) are controlled and the desired natural periods in roll and pitch can be obtained.

A typical heave motion response characteristic of a wind turbine upon a spar substructure is illustrated in Figure 4.5. The cancellation at about 22 s is due to the change in diameter along the spar. To further increase the vertical inertia of the structure, a horizontal circular plate with diameter exceeding that of the column may be placed at the bottom of the column. By such an arrangement the natural period in heave will increase and extra damping will be introduced. The plate will have minor impact on the wave loads.

The large draft of the spar structure (typically in the range 50–100 m) causes a large moment of inertia in roll and pitch and makes it thus easy to obtain large natural periods in these modes of motion. Further, a large distance between CG and center of buoyancy (CB), together with a large displacement, implies that the restoring forces in roll and pitch are sufficiently large to ensure small trim and heel.[1]

[1] In the terminology of marine structures, roll and pitch are used for the rotational dynamic motion around the two horizontal axes (x and y), while the corresponding static displacements are denoted trim and heel or list.

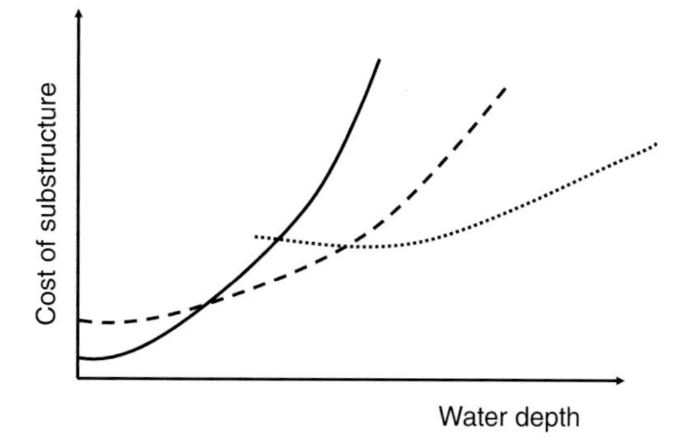

Figure 4.9 Qualitative illustration of cost of support structures versus water depth. Solid line: monopiles; dashed line: jacket/tripod solution; dotted line: floaters.

has been favorable to use fewer large units than many smaller to reduce the cost of electricity.

Schematically, the cost of a wind turbine substructure as a function of water depth is illustrated in Figure 4.9. In shallow waters, the monopile is the preferred solution if the bottom sediment conditions allow for driving the monopile to a sufficiently large depth. As the water depth increases, the cost of the monopile increases significantly and jackets and tripods become more favorable solutions. Presently typical transitional depth from monopiles to jackets or tripods is in the range of 30–50 m. Damiani, Dykes and Scott (2016) find that for conditions in the US Gulf of Mexico, the cost of monopiles exceeds that of jacket substructures for water depths above approximately 40 m. A 5 MW wind turbine was used in the study. Such absolute limits are, however, continuously challenged. As the water depth increases further, the cost of the jacket structures also becomes prohibitive. As there are many floater concepts and the technology is not as mature as for jacket structures, it is hard to state at which water depth the cost of a jacket exceeds that of a floater. Also, the cost of the mooring system for shallow-water floaters is frequently underestimated and may be higher than in deeper waters. There may thus be a range of water depths in which neither the jacket/tripod solution nor the floater solution is attractive. The cost of floaters increases more slowly with water depth than for fixed bottoms as the substructure does not need to be modified as the water depth increases, and thus the extra costs are related to the increased length of mooring lines and electrical cables only.

5
Linear Dynamics

This chapter gives a short review of the theory of linear dynamics of single-body, multiple-body and continuous systems. In the analysis of offshore wind turbines, a thorough understanding of dynamic properties is important for the avoidance of excitation of natural frequencies and to understand how the various modes of motion are coupled and excited.

It is convenient to consider a linear dynamic system in the frequency domain. However, even if the structural system may be assumed to be a linear system, the forces acting on the system may be nonlinear. Thus, most offshore wind turbines need to be analyzed in the time domain. For this purpose, the state-space formulation of dynamic equations is very useful. In this chapter, the state-space formulation will be demonstrated for a system with two degrees of freedom. The formulation can easily be extended to a system with multiple degrees of freedom.

5.1 SDOF System: Free Oscillations

As an introduction to the dynamic response of complex systems, the key features of a linear, single-degree-of-freedom (SDOF) system are summarized. The system is illustrated in Figure 5.1. The linear dynamic equilibrium equation for the system is written as:

$$m\ddot{x} + b\dot{x} + kx = F(t). \qquad [5.1]$$

Here, x is the displacement of the body, m is the body mass, b is the linear damping, k is the stiffness and $F(t)$ is a time-dependent external force. To find the natural frequency of the system, the external force is set to zero and it is assumed that the solution may be written in the form $x = x_A e^{\lambda t}$. x is now the displacement from the mean position. [5.1] may then be written as:

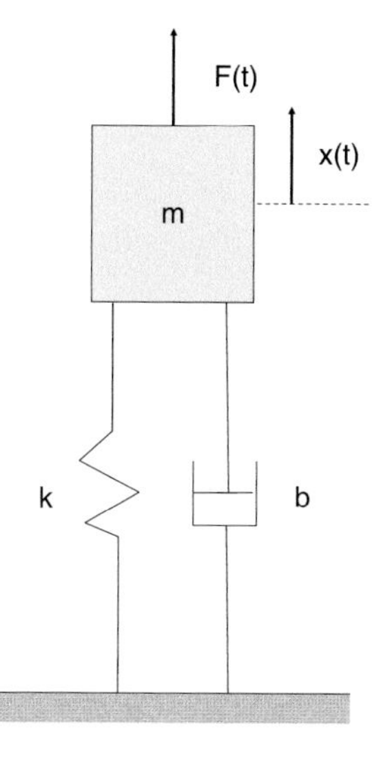

Figure 5.1 Illustration of the SDOF system considered.

$$[\lambda^2 m + \lambda b + k]x = 0. \quad [5.2]$$

The solution of [5.2] is given by:

$$\lambda = -\beta \pm \sqrt{\beta^2 - \omega_0^2} = -\beta \pm i\sqrt{\omega_0^2 - \beta^2}. \quad [5.3]$$

Here, $\omega_0^2 = k/m$, the undamped natural frequency, and $\beta = b/(2m)$, have been introduced. The change of sign and introduction of $i = \sqrt{-1}$ in the last expression of [5.3] is done to be able to handle cases where $\omega_0^2 > \beta^2$. In this case the damping is called subcritical, and the system is oscillating around a mean position. If $\omega_0^2 < \beta^2$, the system has supercritical damping, and the motion is non-oscillatory.

The nontrivial solution for x in the subcritical damped case can be written as:

$$\begin{aligned} x(t) &= e^{-\beta t}\left[C_1^* e^{+i\left(\omega_0\sqrt{1-\zeta^2}\right)t} + C_2^* e^{-i\left(\omega_0\sqrt{1-\zeta^2}\right)t}\right] \\ &= e^{-\beta t}\left[C_1\cos(\omega_d t) + C_2\sin(\omega_d t)\right]. \end{aligned} \quad [5.4]$$

$\zeta = \beta/\omega_0 = b/(2m\omega_0)$ is denoted the relative damping and $\omega_d = \omega_0\sqrt{1-\zeta^2}$ is the modal frequency or the damped natural frequency. ζ is a positive quantity as both mass, damping and stiffness are assumed to be positive. The coefficients C_1, C_1^* and C_2, C_2^* are given from the initial conditions. The initial conditions are the displacement and velocity at $t = 0$, denoted x_0 and $\dot{x}_0$. Working out the derivatives and inserting for the initial conditions, the following values are obtained for the coefficients:

$$C_1 = x_0$$
$$C_2 = \frac{\dot{x}_0 + \omega_0 \zeta x_0}{\omega_0 \sqrt{1 - \zeta^2}} \,. \qquad [5.5]$$

Frequently, it is convenient to use complex quantities in solving the dynamic equations. Using the complex expressions in [5.4] and solving for C_1^* and C_2^*, the following is obtained:

$$C_1^* = \frac{1}{2}\left[x_0 + \frac{\dot{x}_0 + \beta x_0}{i\omega_d}\right]$$
$$C_2^* = \frac{1}{2}\left[x_0 - \frac{\dot{x}_0 + \beta x_0}{i\omega_d}\right]. \qquad [5.6]$$

The solution for x thus becomes:

$$x = \frac{1}{2} x_0 [e^{i\omega_d t} + e^{-i\omega_d t}] e^{-\beta t} + \frac{1}{2} \frac{\dot{x}_0 + \beta x_0}{i\omega_d} [e^{i\omega_d t} - e^{-i\omega_d t}] e^{-\beta t}. \qquad [5.7]$$

Utilizing the definitions $\cos\alpha = \frac{1}{2}(e^{i\alpha} + e^{-i\alpha})$ and $\sin\alpha = \frac{1}{2i}(e^{i\alpha} - e^{-i\alpha})$, [5.7] can thus be written as:

$$x = \left[x_0 \cos\omega_d t + \frac{\dot{x}_0 + \beta x_0}{\omega_d} \sin\omega_d t\right] e^{-\beta t}, \qquad [5.8]$$

or

$$x = x_A \cos(\omega_d t + \phi) = x_A \mathrm{Re}\left\{e^{i(\omega_d t + \phi)}\right\}, \qquad [5.9]$$

with

$$x_A = \sqrt{x_0^2 + \left(\frac{\dot{x}_0 + \beta x_0}{\omega_d}\right)^2} \quad \text{and}$$
$$\phi = \arctan\left[\frac{\dot{x}_0 + \beta x_0}{\omega_d x_0}\right]. \qquad [5.10]$$

It is understood that it is the real part of the complex expressions that has physical meaning. In the subcritical case already discussed, $\zeta < 1$, and the modal frequency ω_d is positive and real. The terms in brackets at the first line in [5.4] thus represent an oscillatory term.

Cases with $\zeta > 1$ are denoted as supercritical damping. In these cases the solution for λ in [5.3] becomes real and the solution for x can be written as:

$$x = [\widetilde{C}_1 e^{\omega_0\left(-\zeta+\sqrt{\zeta^2-1}\right)t} + \widetilde{C}_2 e^{\omega_0\left(-\zeta-\sqrt{\zeta^2-1}\right)t}], \qquad [5.11]$$

where the coefficients again are given from the initial conditions. For $\zeta > 1$, both $-\zeta + \sqrt{\zeta^2 - 1}$ and $-\zeta - \sqrt{\zeta^2 - 1}$ are less than zero. That implies that x decays toward zero without crossing zero. The case with $\zeta = 1$ is denoted as critical damping and represents the limiting case between oscillatory and non-oscillatory behavior.

Free decay of a dynamic system with an initial displacement and zero initial velocity is illustrated in Figure 5.2. Here, critical damping as well as subcritical damping and zero damping cases are illustrated. Note the period elongation for the highly damped cases.

The damping level for the free decay is frequently expressed through the logarithmic decrement. Consider the ratio between two following amplitudes in the subcritical case, $x(t_p)$ and $x(t_p + T_d)$. Here, t_p indicate a peak value in the oscillations and $T_d = 2\pi/\left(\omega_0\sqrt{1-\zeta^2}\right)$ is the damped natural period of oscillation (see Figure 5.3). The relation between the logarithmic of the ratio between two following amplitudes δ and the relative damping is given by:

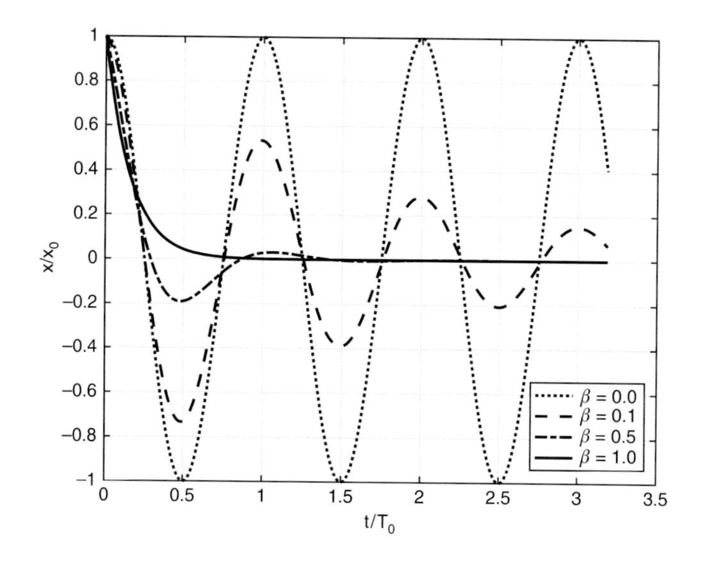

Figure 5.2 Free decay response, with initial conditions $x(t = 0) = x_A$ and $\dot{x}(t = 0) = 0$.

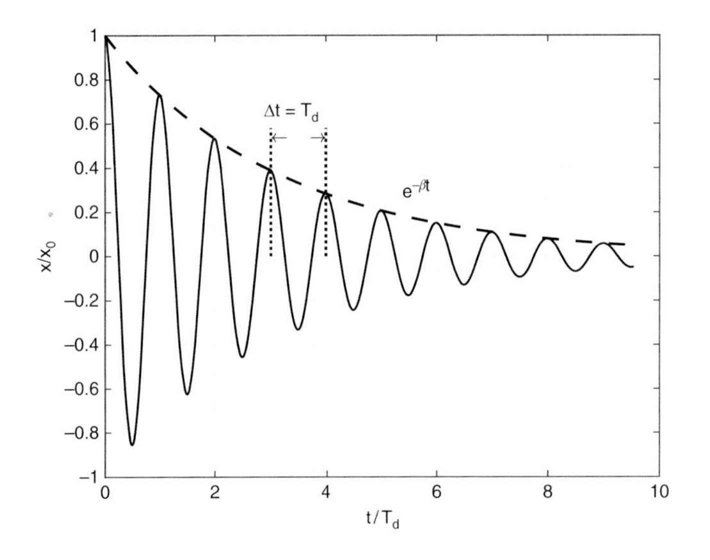

Figure 5.3 Damped, free-decay oscillations.

$$\delta = \ln\left[\frac{x(t_p)}{x(t_p + T)}\right] = \ln[e^{\beta T_d}] = \beta T_d = \frac{2\pi\zeta}{\sqrt{1-\zeta^2}}. \qquad [5.12]$$

This relation is useful in estimating damping from free-decay experiments.

5.2 SDOF System: Forced Oscillations

Assume that the force term in [5.1] can written as $F(t) = F_A \cos(\omega t)$. If such a force is suddenly introduced to the system at $t = 0$, a transient response is experienced. With a positive damping in the system, the transient will die out over time and finally a pure oscillatory response with the same frequency as the load will remain. The solution for this oscillatory response is obtained as:

$$x(t) = x_A \cos(\omega t + \phi),$$
$$x_A = \frac{F_A}{k\sqrt{\left(1-(\omega/\omega_0)^2\right)^2 + (2\zeta\omega/\omega_0)^2}}, \quad \phi = \arctan\left(\frac{-2\zeta\omega/\omega_0}{1-(\omega/\omega_0)^2}\right). \qquad [5.13]$$

In complex form, the oscillatory response can be written as:

$$x(t) = \mathrm{Re}\left\{\frac{F_A e^{i\omega t}}{k\left(1-(\omega/\omega_0)^2 + 2i\zeta\omega/\omega_0\right)}\right\}. \qquad [5.14]$$

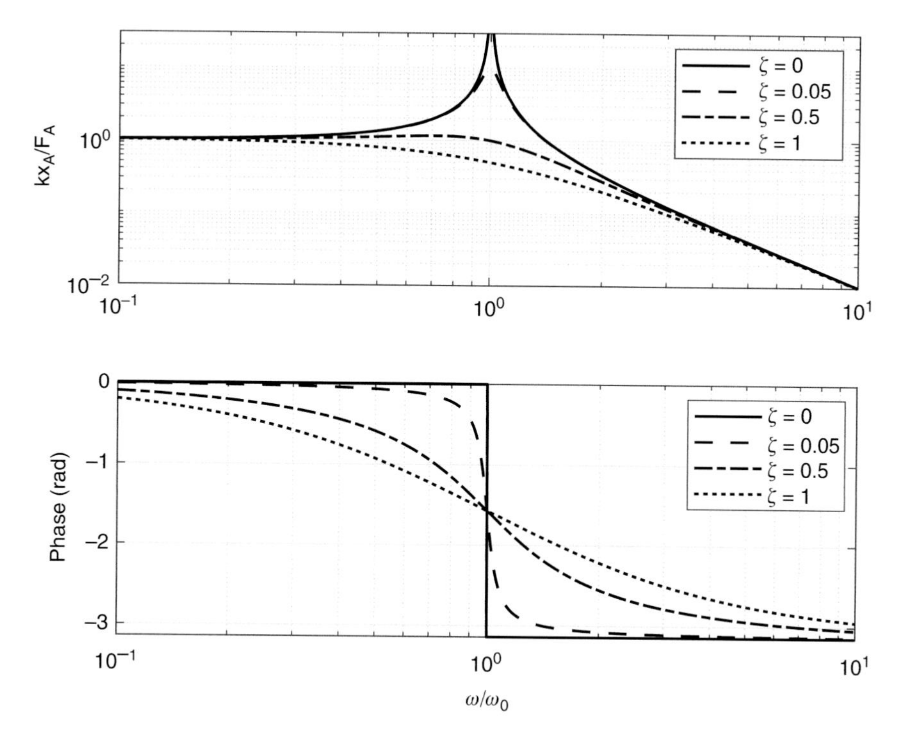

Figure 5.4 Example of stationary response of a SDOF system as a function of excitation frequency and damping. Mass $m = 1$, and stiffness $k = 1$.

Examples of the computed response for various levels of damping are shown in Figure 5.4. The peak response frequency is shifted toward lower frequencies as the damping increases. The phase between the response and the forcing is shifted 180 deg as the excitation frequency is increased from below to above resonance. Note that the phase shift takes place more gradually as the damping is increased. For frequencies much lower than the natural frequency, the response becomes almost quasi-static and almost frequency-independent. A quasi-static response implies that the motion amplitude is close to the ratio between the force amplitude and the stiffness. On the contrary, for high frequencies the response is controlled by the inertia and the response amplitude becomes inversely proportional to $\omega^2 m$.

5.3 System with Multiple Degrees of Freedom

In most practical cases the system considered has more than one degree of freedom. A general linear system with N degrees of freedom should thus be considered. For the sake of simplicity let us initially assume N = 2. A possible two-degrees-of-freedom

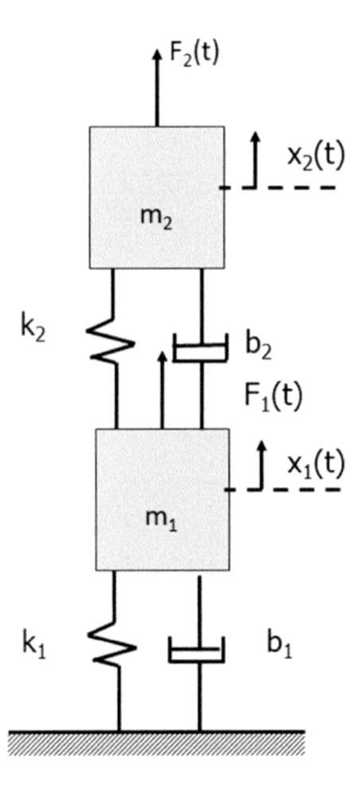

Figure 5.5 Example of a two-degrees-of-freedom dynamic system.

(2DOF) system is illustrated in Figure 5.5. Considering the forces acting on each of the two bodies m_1 and m_2, the following two equations are obtained:

$$\begin{aligned} m_2\ddot{x}_2 &= b_2(\dot{x}_1 - \dot{x}_2) + k_2(x_1 - x_2) + F_2 \\ m_1\ddot{x}_1 &= -b_1\dot{x}_1 - k_1x_1 - b_2(\dot{x}_1 - \dot{x}_2) - k_2(x_1 - x_2) + F_1. \end{aligned} \quad [5.15]$$

Rearranging these equations, we may write the coupled dynamic equilibrium as:

$$\begin{aligned} m_1\ddot{x}_1 + (b_1 + b_2)\dot{x}_1 - b_2\dot{x}_2 + (k_1 + k_2)x_1 - k_2x_2 &= F_1\cdot \\ m_2\ddot{x}_2 - b_2\dot{x}_1 + b_2\dot{x}_2 - k_2x_1 + k_2x_2 &= F_2. \end{aligned} \quad [5.16]$$

On matrix form the above equations can be written as:

$$\mathbf{M}\ddot{\mathbf{x}} + \mathbf{b}\dot{\mathbf{x}} + \mathbf{k}\mathbf{x} = \mathbf{F}, \quad [5.17]$$

where:

$$\mathbf{M} = \begin{pmatrix} m_1 & 0 \\ 0 & m_2 \end{pmatrix}, \quad \mathbf{x} = \begin{pmatrix} x_1 \\ x_2 \end{pmatrix}, \mathbf{F} = \begin{pmatrix} F_1 \\ F_2 \end{pmatrix},$$

$$\mathbf{b} = \begin{pmatrix} b_1 + b_2 & -b_2 \\ -b_2 & b_2 \end{pmatrix}, \quad \mathbf{k} = \begin{pmatrix} k_1 + k_2 & -k_2 \\ -k_2 & k_2 \end{pmatrix}.$$

This matrix formulation can be extended to an arbitrary number of freedoms. The linear system is straightforward to solve in the frequency domain. The nice feature about the linearity is that the system may be solved for simplified load cases and then a summation of the responses for several load cases provides the solution for a more complicated load case.

In the case of nonlinear equations, when some of the coefficients may be time-, velocity- or position-dependent, a time domain solution of the system response is required. To perform efficient time domain integration of the coupled system of equations, it is useful to replace the N second-order differential equations with 2 N first-order differential equations, a so-called state-space representation of the equations. As an example, let us consider the 2DOF system studied above. A new variable $y = (x_1\ x_2\ \dot{x}_1\ \dot{x}_2)^T$ is introduced. Here, the superscript T denotes the transposed matrix. Rearranging [5.16] and introducing y, the following expressions are obtained:

$$\begin{aligned} \ddot{x}_1 &= \frac{1}{m_1}[-(k_1+k_2)x_1 + k_2x_2 - (b_1+b_2)\dot{x}_1 + b_2\dot{x}_2 + F_1]. \\ \ddot{x}_2 &= \frac{1}{m_2}[k_2x_1 - k_2x_2 + b_2\dot{x}_1 - b_2\dot{x}_2 + F_2]. \end{aligned} \qquad [5.18]$$

In matrix form this may be written as:

$$\begin{pmatrix} \dot{x}_1 \\ \dot{x}_2 \\ \ddot{x}_1 \\ \ddot{x}_2 \end{pmatrix} = \mathbf{A}\begin{pmatrix} x_1 \\ x_2 \\ \dot{x}_1 \\ \dot{x}_2 \end{pmatrix} + \mathbf{B}\begin{pmatrix} F_1 \\ F_2 \end{pmatrix} \quad \text{or} \quad \dot{\mathbf{y}} = \mathbf{A}\mathbf{y} + \mathbf{B}\mathbf{F}. \qquad [5.19]$$

Here,

$$\mathbf{y} = \begin{pmatrix} \mathbf{x} \\ \dot{\mathbf{x}} \end{pmatrix}, \quad \mathbf{A} = \begin{pmatrix} \mathbf{0} & \mathbf{I} \\ -\mathbf{M}^{-1}\mathbf{K} & -\mathbf{M}^{-1}\mathbf{b} \end{pmatrix}, \quad \mathbf{B} = \begin{pmatrix} \mathbf{0} \\ \mathbf{M}^{-1} \end{pmatrix}.$$

In the general N-degrees-of-freedom system, **0** is the N x N zero matrix; **I** is the N x N unit diagonal matrix; **M** is the N x N mass matrix; **K** is the N x N stiffness matrix; and **b** is the N x N damping matrix. **F** is the N x 1 force vector.

The simplest way to solve this system of equations numerically is by the forward Euler integration method. Using a discrete time step of length Δt, the solution for time $t + \Delta t$ is found from the state at time t as:

$$\mathbf{y}(t+\Delta t) = \mathbf{y}(t) + \Delta t\Big(\mathbf{A}\mathbf{y}(t) + \mathbf{B}\mathbf{F}(t)\Big). \qquad [5.20]$$

The integration starts by stating the initial positions and velocities in $\mathbf{y}(0)$ and the initial force vector $\mathbf{F}(0)$. The forward Euler method works well for simple dynamic systems but should be used with care. If the system exhibits high natural frequencies, the integration scheme may diverge as the damping level is low. For complicated systems more advanced integration schemes, such as for example the fourth-order Runge–Kutta scheme, are recommended (see, e.g., Press et al., 1989; Dahlquist and Björck, 1974).

Figure 5.6 shows an example of the frequency-dependent response of a 2DOF system. The force vector is in this example $\left(0 \ F_A\cos(\omega t)\right)^T$. For very low frequencies a quasi-static response is observed, $x_1 \simeq F_A/k_1$ and $x_2 \simeq F_A(1/k_1 + 1/k_2)$.

To interpret the response at higher frequencies it is very useful to find the undamped eigenfrequencies and eigenmodes. A stationary solution in the form $\mathbf{x} = \psi e^{i\omega t}$ is assumed. The undamped eigenvalue problem then reads:

$$\left(-\omega^2\mathbf{M} + \mathbf{K}\right)\psi = 0 \qquad \text{or} \qquad \lambda\psi = \mathbf{M}^{-1}\mathbf{K}\psi. \qquad [5.21]$$

To each eigenvalue $\lambda_i = \omega_{0i}^2$ there is a corresponding eigenvector ψ_i. ψ_i defines the contribution from each of the different modes of motions, in our case given by x_1 and x_2, to the specific undamped resonant mode of motion, given by ψ_i. In the case of no damping, ψ_i is real. For floating vessels, the six rigid modes of motion are denoted as surge, sway, heave, roll, pitch and yaw, and defined according to a chosen coordinate system. However, due to coupling effects, off-diagonal terms in the mass and restoring matrices, the eigenmodes seldom correspond to these modes. For example, a "heave resonant motion" will in most cases also contain some contribution from pitch. Details about multibody dynamics and eigenvalues can be found in standard textbook on dynamics (e.g., Irgens, 1999). More examples on multibody dynamics are found in Chapter 8.

2DOF System, Eigenmodes and Frequency Response

Consider the example shown in Figure 5.5 and use the following numerical values:

$$\begin{aligned} m_1 &= 2\ \text{kg} & m_2 &= 1\ \text{kg} \\ b_1 &= 0.1\ \text{Ns/m} & b_2 &= 0.1\ \text{Ns/m} \\ k_1 &= 1\ \text{N/m} & k_2 &= 2\ \text{N/m}. \end{aligned}$$

Solving for the eigenvalues and the eigenvectors, the following values are obtained:

$$\omega_i = (0.560,\ 1.785)\ \text{rad/s} \quad \psi_1 = (0.645\ \ 0.765)^T \quad \psi_2 = (-0.510\ \ 0.860)^T. \qquad [5.22]$$

(cont.)

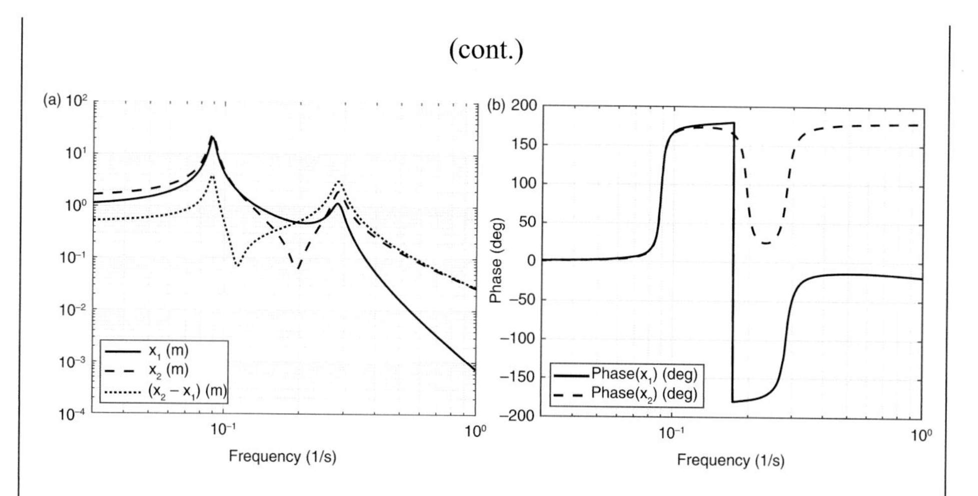

Figure 5.6 Amplitude and phase of a 2DOF system described in Figure 5.5 with characteristics as given in the above example and excited harmonically by a force $f(t) = F_A \cos(\omega t)$, $F_A = 1\ N$ acting at the top of mass m_2.

The elements of ψ_i *tell how much each of the displacements* x_1 *and* x_2 *contribute to the eigenvector. Let the system be excited by a harmonic force acting at* m_2:
$F = \begin{bmatrix} 0 & f_A \cos(\omega t) \end{bmatrix}^T$ *with* $f_A = 1\text{N}$. *The response characteristics are shown in Figure 5.6, where it is observed that the two eigenfrequencies correspond to maxima of the amplitudes of motion. For frequencies below the first eigenfrequency,* $\omega_{01} = 0.56\ \text{rad/s} = 0.0892\ \text{Hz}$, *the motion is dominated by mode 2. The two masses move in phase,* m_2 *with larger amplitude than* m_1. *In the low frequency limit the response is quasi-static. When the excitation frequency is close to* ω_{01}, *both* m_1 *and* m_2 *exhibit large oscillations. The same is the case close to the second natural frequency* $\omega_{02} = 1.79\ \text{rad/s} = 0.284\ \text{Hz}$. *In the range between the two eigenfrequencies, the responses of the two masses are partly in phase and partly out of phase. For frequencies far above* ω_{02}, *the response is again dominated by the motion of* m_2 *and the two masses are oscillating almost out of phase.*

2DOF System, Transient Response

Figure 5.7 shows the transient displacements of the system displayed in Figure 5.5. The initial displacements are $x_1(0) = 0$, $x_2(0) = 1$ *and zero initial velocities. No external forces are acting. It is observed how the transient is damped out over time, converging toward a zero displacement for both masses. It is also observed that initially both modes of motion oscillate with comparable amplitudes, while after a while the first mode, corresponding to a period of 11.22 s, dominates. From the right-hand part of the figure, it is observed that initially there is a large force in spring 2, while later the masses move almost in phase, causing a small force in spring 2 and a dominating force in spring 1.*

In this example, a simple forward Euler integration scheme was used to step the equations in [5.19] forward in time.

(cont.)

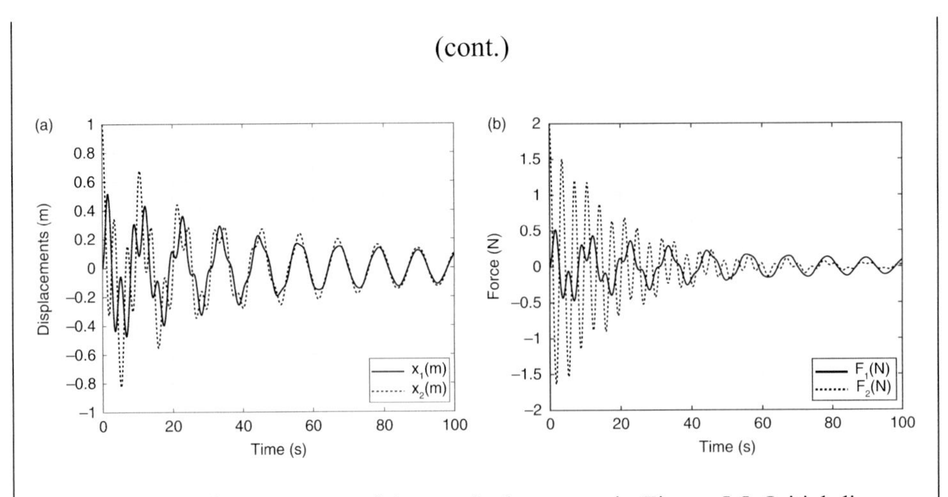

Figure 5.7 Transient response of the two-body system in Figure 5.5. Initial displacements $x_1(0) = 0$, $x_2(0) = 1$ m. Zero velocities at $t = 0$. Left: displacements of the two masses; right: forces in the two springs.

5.4 Continuous System

SDOF and 2DOF systems have so far been considered. Continuous systems such as, for example, a beam with distributed mass have an infinite number of degrees of freedom. For some simple systems, analytical solutions of the eigenmodes and eigenfrequencies exist. For more complicated systems one has to use numerical methods as finite element techniques (see, e.g., Næss and Moan, 2013).

Consider the vertical beam illustrated in Figure 5.8. The beam may resemble a simplification of a wind turbine tower with constant diameter and mass per unit length. The beam has a rigid fixture at x = 0, implying that it cannot rotate at this level. The mass per unit length of the beam is m. The damping is set to zero. Consider the equilibrium of a short vertical section of length dx. The horizontal force acting on the lower end of the section is denoted V. Similarly, the force at the top of the segment may be written as $V + \frac{\partial V}{\partial x}dx$. Without any external force acting on the segment the difference between these two forces will cause an acceleration of the segment:

$$mdx\frac{\partial^2 u}{\partial t^2} = \frac{\partial V}{\partial x}dx. \qquad [5.23]$$

Here, $u(x, t)$ is the horizontal deflection of the beam in $y-$ direction. The moment acting at the lower end of the segment is denoted M. Considering moment equilibrium about the upper end of the segment and ignoring rotation of the segment, the following equilibrium equation is obtained:

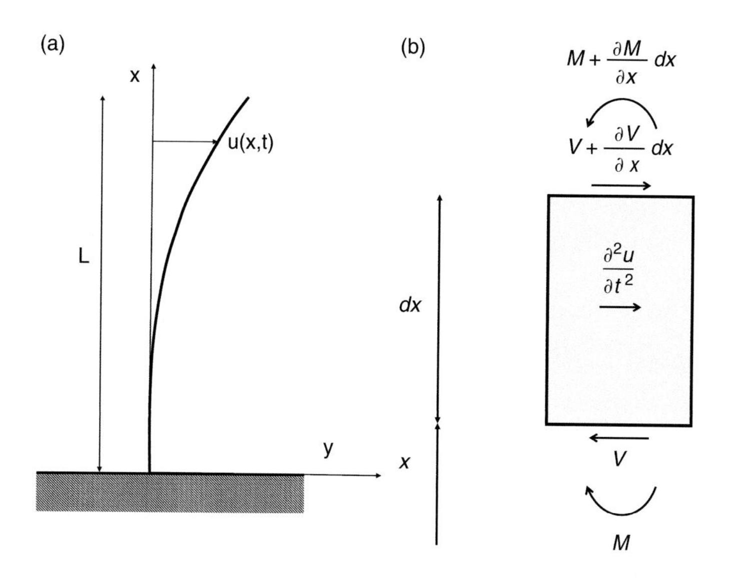

Figure 5.8 Illustration of a flexible beam with homogeneous mass distribution. The beam is restrained from rotation in x = 0. Right: element of the beam with forces V and moments M acting on it.

$$Vdx - \frac{\partial M}{\partial x} dx = 0. \qquad [5.24]$$

Further, the following relation exists between the bending moment and the curvature of the beam:

$$\frac{\partial^2 u(x,t)}{\partial x^2} = -\frac{M(x,t)}{EI}. \qquad [5.25]$$

Here, E is the module of elasticity of the material and $I = \int y^2 dA$ is the second moment of the cross-sectional area of the beam. EI is denoted the bending stiffness. Combining these relations and ignoring damping, the differential equation for the dynamic equilibrium for the beam without external forcing can be written as:

$$\frac{\partial^2 u}{\partial t^2} + \frac{EI}{\rho A} \frac{\partial^4 u}{\partial x^4} = 0. \qquad [5.26]$$

Here, the density of the beam material ρ has been introduced so $m = \rho A$. To solve this equation, the technique of separation of variables is used, writing the solution as a product of a time-dependent and a space-dependent function:

$$u(x,t) = X(x)T(t). \qquad [5.27]$$

It can be shown that the general, stationary solution of [5.27] is given (see, e.g., Irgens, 1999) by:

$$X(x) = K_1 \sin(cx) + K_2 \cos(cx) + K_3 \sinh(cx) + K_4 \cosh(cx).$$
$$T(t) = T_0 \cos(\omega t - \phi), \qquad \omega = c^2 \sqrt{\frac{EI}{m}}. \qquad [5.28]$$

The coefficients $K_1 - K_4$ and c are given from the boundary and initial conditions of the beam. The boundary conditions require:

no deflection at $x = 0: u(0,t) = 0, \;\; X(0) = 0$
no rotation at $x = 0$: $\frac{\partial u(0,t)}{\partial x} = 0, \;\; \frac{dX(0)}{dx} = 0$
zero moment at $x = L$, i.e., no curvature at $x = L$: $M(L,t) = 0, \frac{\partial^2 u(L,t)}{\partial x^2} = 0, \frac{d^2 X(L)}{dx^2} = 0$
no load at $x = L$, i.e., no gradient in moment at $x = L$: $\frac{\partial M(L,t)}{\partial x} = 0, \frac{\partial^3 u(L,t)}{\partial x^3} = 0, \frac{d^3 X(L)}{dx^3} = 0$

Inserting these boundary conditions into [5.28], we obtain:

$$\begin{aligned} K_1 + K_3 &= 0 \\ K_2 + K_4 &= 0. \end{aligned} \qquad [5.29]$$

Then we are left with the following two equations for K_1 and K_2:

$$\begin{bmatrix} (\cosh cL + \cos cL) & (\sinh cL + \sin cL) \\ (\sinh cL - \sin cL) & (\cosh cL + \cos cL) \end{bmatrix} \begin{pmatrix} K_2 \\ K_1 \end{pmatrix} = \begin{pmatrix} 0 \\ 0 \end{pmatrix}. \qquad [5.30]$$

To avoid a nontrivial solution, the determinant of the 2 x 2 matrix must be zero, giving the requirement that:

$$1 + \cosh cL \cos cL = 0. \qquad [5.31]$$

This equation has an infinite number of solutions; the smallest, corresponding to the lowest natural frequencies, are: $(cL)_i = 1.875,\ 4.694,\ 7.855,\ 10.996,\ 14.137,\ 17.279, \ldots$. The relation between $(cL)_i$ and ω_i is given in [5.28].

The corresponding mode shapes are obtained as:

$$\psi_i(x) = A_i \left[\cosh\left(\frac{(cL)_i}{L} x\right) - \cos\left(\frac{(cL)_i}{L} x\right) \right.$$
$$\left. - \frac{\sinh(cL)_i + \sin(cL)_i}{\cosh(cL)_i + \cos(cL)_i} \left(\sinh\left(\frac{(cL)_i}{L} x\right) - \sin\left(\frac{(cL)_i}{L} x\right) \right) \right]. \qquad [5.32]$$

A_i is a scale factor, in the following determined by the choice $|\psi_i(L)| = 1$. The four mode shapes corresponding to the lowest natural periods are displayed in Figure 5.9.

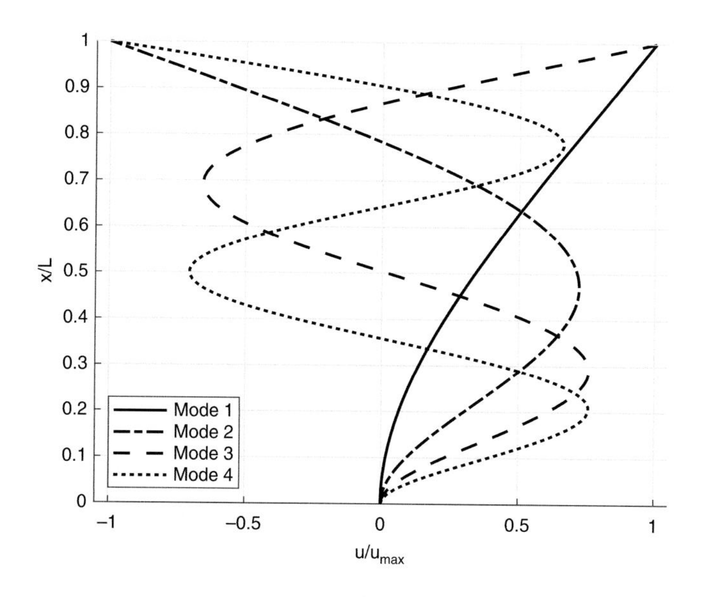

Figure 5.9 First four mode shapes for a cantilever beam with uniform mass and stiffness.

If a continuous system, such as the above cantilever beam, is exposed to a transient load or some initial displacements, several natural modes may be excited. This is similar to what was observed in the two-body example shown above. The natural modes are orthogonal, formally expressed as:

$$\int_0^L \psi_i(x)\psi_j(x)dx = 0 \quad \text{for } i \neq j. \qquad [5.33]$$

This implies that oscillations in one mode shape do not couple into other mode shapes. Further, a general oscillation may be solved by considering one mode shape at the time and then making a summation of the modal responses to find the total response. The dynamic response for each of the mode shapes may be considered as an SDOF system, with unique modal mass, damping and stiffness. The initial displacements and external forces are weighted by the modal shape functions to establish the dynamic response of each mode. For example, if the cantilever beam above has an initial displacement $X_0(x)$ and is excited by an external force $F(x, t)$, then the initial displacement for each mode and the modal excitation forces are given by:

$$\psi_{0i}(x) = \int_L \boldsymbol{\psi}_i X_0(x)dx$$

$$F_{\psi_i}(x,t) = \int_L \boldsymbol{\psi}_i F(x,t)dx. \qquad [5.34]$$

Similarly, the modal mass and stiffness for a beam with non-constant mass and stiffness properties becomes:

$$m_{ij} = \int_L \psi_i^T m(x) \psi_j dx$$

$$k_{ij} = \int_L \frac{d^2\psi_i^T}{dx^2} EI(x) \frac{d^2\psi_i}{dx^2} dx. \qquad [5.35]$$

From these expressions it is observed that the modal masses and stiffnesses using orthogonal mode shapes are diagonal. For details about dynamic analysis using modal superposition, see, e.g., Næss and Moan, 2013).

In analyzing wind turbines, a modal representation of the structure is frequently used. Using a modal representation of the tower, only a few modes will be required to obtain a reasonable representation of the dynamics, while if a finite element representation is applied, many more degrees of freedom are needed. Analyzing the dynamics of rotor blades involves special challenges, as the coordinate system of the blade is rotating, and the centrifugal forces influence the stiffness and geometry. The consequence is natural frequencies depending upon the rotational speed (Skjoldan and Hansen, 2009).

Exercises Chapter 5

1. In Section 5.3, a dynamic system with two degrees of freedom (2DOF) is discussed.
   a. Make a routine to solve the dynamics in the time domain. Use the data from the example in Section 5.3. Animate the results and observe how the motions of the two masses behave as the frequency of excitation varies.
   b. Introduce a third mass into the system and repeat the analysis and discussion above.
2. Write down the low-frequency and high-frequency asymptotic limits for the 2DOF system in the example in Section 5.3 and check the responses towards the response characteristics shown in Figure 5.6.
3. Consider a vertical steel tower of a length of 100 m, an outer radius of 2 m and a wall thickness of 3 cm.
   a. Assume modal shapes as given by [5.32] and compute the modal mass for the first and second mode.

b. Add a point of mass M = 200 Mg at the top. What is now the modal mass? Assume the same mode shapes as above. Are the two modes considered above eigenmodes in this case?
c. Simplify the first mode shape by $\psi_1(x) = 1 - \cos\left(\frac{\pi}{2}\frac{x}{L}\right)$ and $\psi_1(x) = \frac{x}{L}$. Compute the modal mass in these cases. Disregard the top mass. Compare the results with the result in Exercise 3a.
d. Assume the mass of the beam (tower) is much smaller than the top mass and that the bending stiffness along the tower EI is constant. What is the lowest natural frequency of the system? How many natural frequencies exist in this case?

6

Wave Loads on Fixed Substructures

This chapter discusses various ways to estimate the wave loads on offshore wind turbines with bottom-fixed substructures. The most common wind turbine substructure currently uses monopiles, i.e., a vertical cylinders with constant diameter. However, jacket (space-frame) structures consisting of several inclined circular cylinders are also in use. Therefore, the general concept of computing wave loads on such structures is discussed as well. The foundation of bottom-fixed structures is addressed in short, in particular the various ways to include the forces from the sea floor in the structural model.

Both fixed and floating substructures may be constructed from other-than-cylindrical shapes. Therefore, methods for computing the wave forces on more general geometries are addressed. These methods are examined further in Chapter 7 when discussing the dynamics of floating substructures.

6.1 General Principles of Computing Wave Loads on Small Bodies

Consider a body fully submerged in water, as illustrated in Figure 6.1. The force in x-direction is obtained by integrating the pressure at the surface of the body times the $x-$ component of the surface normal:

$$F_x = \int_{S_b} p n_x dS. \qquad [6.1]$$

The integral must be taken over the complete wetted surface. Note that under the assumption of an ideal fluid, the pressure only acts perpendicular to the wetted surface. No shear forces are present. The pressure over the body may be obtained by the Bernoulli equation. Ignoring hydrostatic effects and linearizing, the pressure is obtained as:

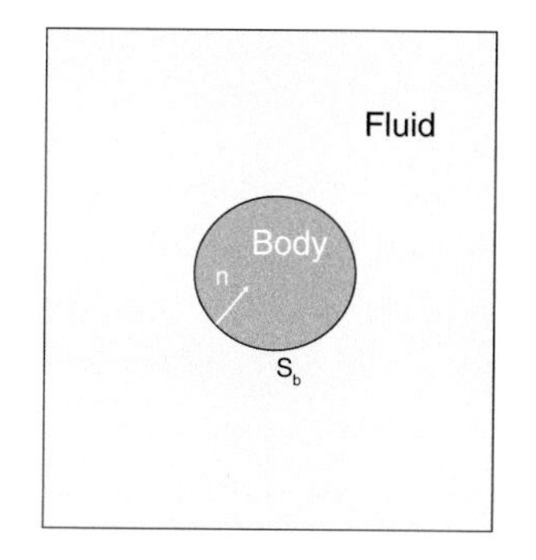

Figure 6.1 A fully submerged body with wetted surface S_b and unit surface normal n pointing out of the fluid volume.

$$p = -\rho\left[\frac{\partial\phi}{\partial t} + \frac{1}{2}\nabla\phi\cdot\nabla\phi\right] \simeq -\rho\frac{\partial\phi}{\partial t}. \qquad [6.2]$$

Here, ϕ is the velocity potential of the fluid flowing around the body. It is observed that the velocity-squared terms disappear due to the linearization, thus there is no (dynamic) pressure on the body unless there is a time variation in the velocity potential.

Consider the two-dimensional (2D) flow around a circular cylinder. The free flow far from the cylinder is in $x-$ direction and has magnitude U. If the cylinder has radius R, the velocity potential can be written as:

$$\phi = U\left(r + R^2/r\right)\,\cos\theta. \qquad [6.3]$$

Here, $r = \sqrt{x^2+y^2}$ and $\theta = \arctan(x/y)$. Using [6.1] and [6.2], the following is obtained for the force in $x-$ direction:

$$F_x = -\rho\int_{S_b}\frac{\partial\phi}{\partial t}n_x dS = -2\rho R\dot{U}\int_{S_b}\cos\theta(-\cos\theta)Rd\theta = 2\rho\pi R^2\dot{U}. \qquad [6.4]$$

It is observed that the force is twice the force needed to accelerate the mass of fluid displaced by the fluid. Considering the force in $y-$ direction, zero total force is obtained.[1] For geometries other than a cylinder, the factor will not be 2 and in the general case the force may be written as:

[1] This contrasts with the lifting surfaces discussed in Chapter 3, where a circulation was introduced to satisfy the Kutta–Joukowski condition and a lift force perpendicular to the incident flow was obtained.

$$F_x = (1 + C_m)\rho\pi R^2 \dot{U}. \qquad [6.5]$$

The factor 1 corresponds to the force obtained by integrating the pressure in the fluid undisturbed by the body. In the case of a cylinder in a uniform flow in $x-$ direction, this corresponds to the contribution from the potential $\phi_0 = Ur\ \cos\theta = Ux$. When considering wave-induced forces, the contribution to the force from the pressure in the undisturbed wave field is frequently denoted as the Froude–Krylov force. C_m is denoted as the added mass coefficient. This can be explained by considering the acceleration of the body in $x-$ direction in a fluid at rest. The force needed to accelerate the body is obtained as:

$$F_{xa} = \left(m + C_m\rho\pi R^2\right)\ddot{x}. \qquad [6.6]$$

Here, m is the mass of the body and $\ddot{x}$ is the acceleration of the body in $x-$ direction. Thus, in addition to the force needed to accelerate the mass of the body itself, the dry mass, an additional force is needed. This is related to the pressure distribution at the body surface caused by the acceleration of the fluid surrounding the body.

The result in [6.5] was obtained by considering a flow field with a homogeneous velocity distribution. In considering bodies in a wave field, the above expression may be used under the approximation that the acceleration in the undisturbed wave field is approximately constant over the volume of the body. This means that the characteristic cross-sectional dimension of the body should be small compared to the wavelength. For a circular cylinder, it is engineering practice to assume that this criterion is fulfilled if the cylinder diameter is less than 1/5 of the wavelength.

6.2 Wave Forces on Slender Structures

6.2.1 Wave Forces on Vertical Cylinders: The Morison Equation

To arrive at the force in [6.5], an ideal, nonviscous fluid was assumed. Morison et al. (1950) wanted to estimate the force on a vertical pile in waves. They proposed a simple approach to account for the viscous effects[2] by adding to [6.5] a term proportional to the velocity squared. This formula was later named the "Morison equation":

$$F_{x2D} = (1 + C_m)\rho\pi\left(\frac{D}{2}\right)^2 \dot{U} + \frac{1}{2}\rho C_d D U|U|. \qquad [6.7]$$

[2] The term "viscous effects" is sometimes used for the effects related to the shear forces along the body surface only. The reduced pressure on the backside of a cylinder due to separated flow is frequently denoted as separation effects. However, the origin of both is the viscosity. The viscous effects in the Morison equation are dominated by separation effects.

Here, D is the cylinder (pile) diameter and C_d is the drag coefficient. F_{x2D} is the 2D force on the cylinder, or the force per unit length in the 3D case. The absolute value of the velocity is introduced to secure that the force is acting in the direction of the fluid velocity. In the case of a cylinder oscillating in a fluid at rest, the force needed to drive the oscillation is similarly given by:

$$F_{xa2D} = \left(m_{2D} + C_m \rho \pi \left(\frac{D}{2}\right)^2\right)\ddot{x} + \frac{1}{2}\rho C_d D \dot{x}|\dot{x}|. \quad [6.8]$$

Here, m_{2D} is the mass per unit length. One may say that Morison equation adds two asymptotic expressions, the force on the body in a fluid of zero velocity and finite acceleration and the force in steady flow with no acceleration. As shown above, $C_m = 1$ and $C_d = 0$ for a circular cylinder deeply submerged in an ideal fluid. However, in a real fluid, the value of the two coefficients depends on the flow conditions. In a fluid at rest and harmonic oscillations of the body, the key parameters defining the flow condition are the Reynolds number $\mathrm{Re} = \dot{x}_A D/v$; the Keulegan–Carpenter number $KC = \dot{x}_A T/D = 2\pi x_A/D$; and the surface roughness expressed by k/D. Here, v is the kinematic viscosity of the fluid; the index A denotes amplitude; T is the period of oscillation; x_A is the amplitude of oscillation; and k is a characteristic roughness of the cylinder surface. The Reynolds number is a measure of the turbulence level in the boundary layer at the cylinder. In a steady flow, the flow separation and thus the drag coefficient are very sensitive to the Reynolds number. At the critical Reynolds number in the range 10^5–10^6, depending upon surface roughness, large changes in the drag coefficient are experienced. The Keulegan–Carpenter number is an expression of the amplitude of oscillation versus the diameter of the cylinder. It can either be an oscillation of the cylinder in a fluid at rest or an oscillation of fluid around a body at rest. How well the separated flow is developed depends upon the Keulegan–Carpenter number. For large Keulegan–Carpenter numbers the separated flow is similar to steady flow conditions. For further details and advice on drag and inertia coefficients for slender structures, textbooks and standards should be consulted (e.g., Sarpkaya and Isaacson, 1981; Faltinsen, 1990; DNV, 2021c).

Consider a vertical cylinder mounted at the sea floor and extending through the sea surface, as illustrated in Figure 6.2. The diameter of the cylinder may be assumed to be small relative to the wavelength. Thus, the Morison equation should apply. However, the amplitude of the wave particle velocities decays with depth. The approach used to handle this case is called "strip theory." In strip theory, we consider a small vertical section of the cylinder of length dz. In the Morison equation only the flow perpendicular to the cylinder is considered. Thus, the

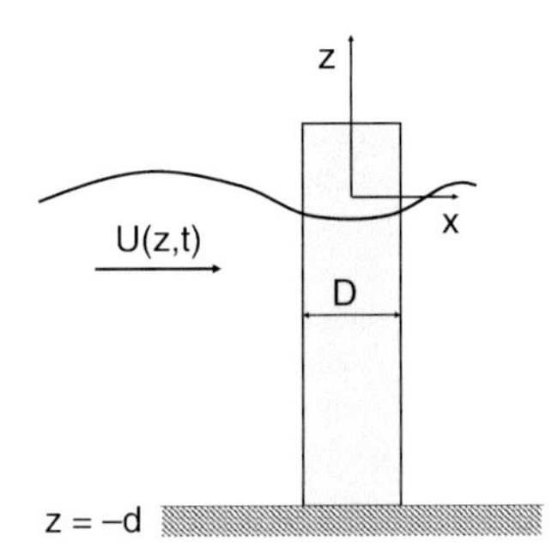

Figure 6.2 A vertical cylinder extending from the sea floor through the sea surface.

force on this small section is given by $F_{x2D}dz$. The total force on the cylinder in $x-$ direction is obtained by integrating from the sea floor to the mean free surface:

$$F_x = \int_{-d}^{0} F_{x2D}dz. \qquad [6.9]$$

Similarly, the overturning moment about the sea floor is obtained as:

$$M_d = \int_{-d}^{0} F_{x2D}(z+d)dz. \qquad [6.10]$$

In [6.9] and [6.10] the integration is performed up to $z = 0$. This is consistent with linear wave theory. For the acceleration term, this is fine as the horizontal fluid acceleration is maximum as the wave elevation is zero. However, the horizontal velocity in the wave is maximum when the wave elevation is maximum. Thus, there may be a significant difference in the horizontal force, and in particular the overturning moment, if the drag force is integrated to $z = 0$ or to the instantaneous free surface elevation. Integrating beyond $z = 0$ implies that a choice of kinematic model must be done. This is in particular an issue in irregular waves (see Chapter 2).

6.2.2 Wave Forces on Inclined Cylinders

The Morison equation is used also for inclined cylinders, e.g., to compute the wave forces on jacket ("space-frame") structures, as illustrated in Figure 6.3. The Morison equation gives the forces perpendicular to the cylinder only. The forces in the axial direction due to the viscous forces of the flow in the axial direction are normally ignored. The approach used is to consider the components of the fluid

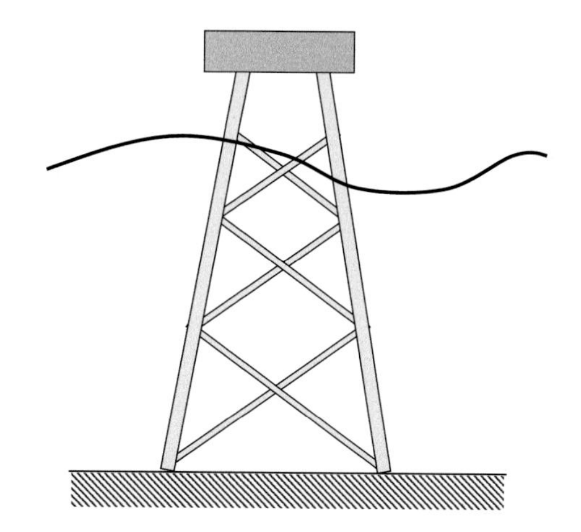

Figure 6.3 Illustration of jacket structure.

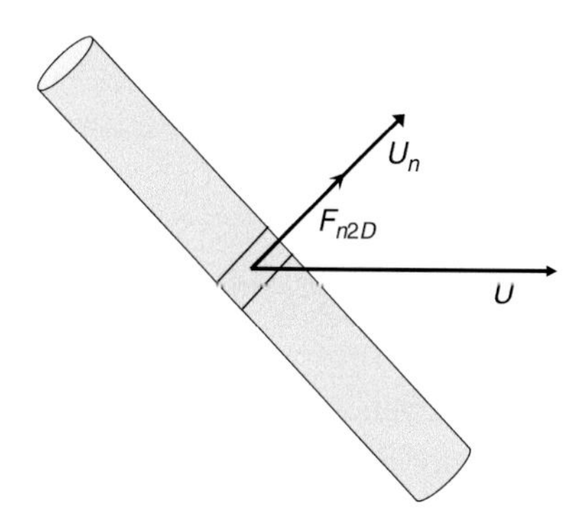

Figure 6.4 The component of the fluid flow perpendicular to the axis of an inclined cylinder. U is the fluid velocity, U_n is the normal component. F_{n2D} is the 2D normal force per unit length.

acceleration and velocities perpendicular to small sections ("strips") of the cylinder (see Figure 6.4), compute the normal force and integrate over the length of the cylinder. This approach is considered valid if the flow direction does not deviate too much from perpendicular to the axis. If the deviation is large, however, the normal forces are small anyhow.

In integrating the force along the inclined cylinder, one must account for variation in acceleration and velocities due to the position dependence of the wave kinematics in $x-$ direction as well as the depth dependence.

The axial force on a cylinder may be estimated by using a Froude–Krylov approach. I.e., the axial force is estimated by assuming the pressure in the undisturbed wave is acting in the axial direction on each end surface, given that the end surface is wetted. If the end surface is not wetted, the axial force is zero.

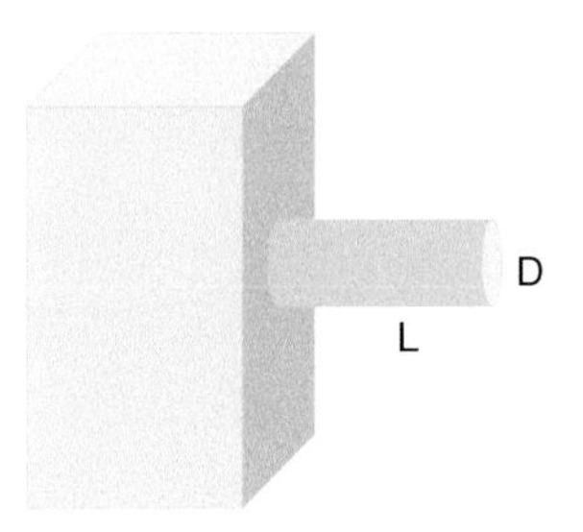

Figure 6.5 Cylinder attached to a larger body.

6.2.3 Effect of Finite Length of a Cylinder

In the above implementation of the Morison equation, it was assumed that the flow was 2D over every cross-section of the cylinder. However, if the cylinder has finite length and one or both ends are in the fluid, the flow close to the ends will have a three-dimensional (3D) component. This effect is frequently expressed through a 3D correction to the added mass coefficient; see, e.g., DNV (2021c). Considering an acceleration of a cylinder of length L perpendicular to the axis, the corresponding inertia force may be written as:

$$F_x = \left(m_{2D} + \alpha\left(\frac{D}{L}\right) C_m^{(2D)} \rho\pi \left(\frac{D}{2}\right)^2 \right) L\ddot{x}. \qquad [6.11]$$

The value of α as function of the diameter over length ratio is illustrated in Figure 6.6. Note that by this correction, only the total force integrated over the length of the cylinder is obtained. To obtain the sectional forces, other 3D methods are needed; see Section 6.4. If the fluid is free to flow around both ends of the cylinder, the length to be used in Figure 6.6 is equal to the physical length of the cylinder. If, however, one end is connected to a structure such that the flow around the end of the cylinder is restricted, a "flow mirror effect" is obtained. This mirror effect causes the hydrodynamic length to be used in Figure 6.6 to be twice the physical length. E.g., when considering the added mass for the cylinder in Figure 6.5, the correct α-value is obtained by considering $D/2L$ in Figure 6.6. If both ends of the cylinder are attached to large bodies, the 2D value of the added mass applies corresponding to $D/L \simeq 0$.

6.3 Wave Force on Non-Slender Vertical Cylinders: MacCamy and Fuchs Theory

It has so far been assumed that the cylinder is slender in the sense that the undisturbed fluid acceleration and velocity can be assumed to be constant over the diameter of the section considered. If the diameter of the cylinder increases, this assumption fails.

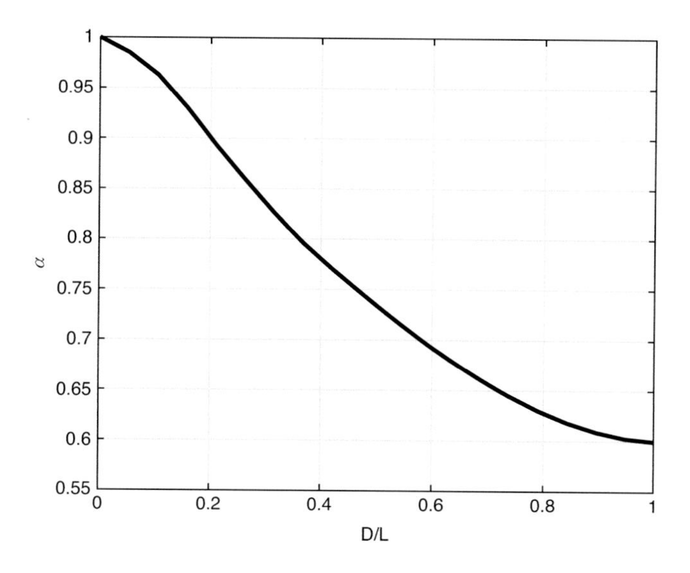

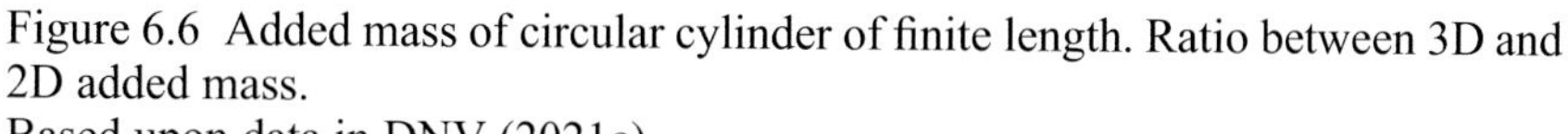

Figure 6.6 Added mass of circular cylinder of finite length. Ratio between 3D and 2D added mass.
Based upon data in DNV (2021c).

However, under the assumption of ideal flow, the horizontal force and overturning moment on a vertical, circular cylinder can be derived analytically in the linear wave case. This was done by MacCamy and Fuchs (1954).

Consider a linear harmonic wave propagating in the positive $x-$ direction and a vertical cylinder with diameter $D = 2R$ located at $x = y = 0$, as illustrated in Figure 6.2. The velocity potential as function of horizontal radius from the cylinder axis, depth and time can be written (MacCamy and Fuchs, 1954; Sarpkaya and Isaacson, 1981) as:

$$\phi(r,z,t) = \frac{ig\zeta_A}{\omega}\frac{\cosh\left(k(z+d)\right)}{\cosh(kd)}$$

$$\left[\sum_{m=0}^{\infty}\beta_m\left(J_m(kr) - \frac{J_m{}'(kR)}{H_m^{(1)\prime}(kR)}H_m^{(1)}(kr)\right)\cos(m\theta)\right]e^{i\omega t}$$

$$\beta_m = \begin{cases} 1 \text{ for } m = 0 \\ 2\mathrm{i}^{\mathrm{m}} \text{ for } m \geq 1 \end{cases}. \qquad [6.12]$$

Here, ζ_A is the wave amplitude and ω is the wave circular frequency. J_m is the Bessel function of first kind and order m; similarly, $H_m^{(1)} = J_m + iY_m$ is the Hankel function of first kind and order m. θ is the direction in the $x - y$ plane such that $x = r\cos\theta$ and $y = r\sin\theta$. The first term in the summation represents the undisturbed incident wave. In cartesian coordinates the incident wave potential contains the term e^{ikx}. In cylindrical coordinates this may be rewritten as:

$$e^{-ikx} = e^{ikr\cos\theta} = \cos(kr\cos\theta) + i\sin(kr\cos\theta). \qquad [6.13]$$

The right-hand side of [6.13] may be replaced by an infinite sum of Bessel functions (see, e.g., Abramowitz and Stegun, 1970), resulting in the first term of the summation in [6.12].

The second term in the summation in [6.12] represents the diffraction potential, i.e., the waves diffracted from the cylinder. Note that the expression contains the derivative of the Bessel and Hankel functions. The derivative of the Bessel function of first kind and order m is given by $J_m'(x) = \frac{m}{x}J_m(x) - J_{m+1}(x)$. For the Bessel function of second kind and the Hankel function the same relation is valid (Abramowitz and Stegun, 1970).

The dynamic force acting on the cylinder is obtained by integrating the pressure over the cylinder surface:

$$F_x = \int_{-d}^{0}\int_{0}^{2\pi} pn_x dz d\theta = \int_{-d}^{0}\int_{0}^{2\pi} -\rho\frac{\partial\phi}{\partial t}(-\cos\theta)dzd\theta. \qquad [6.14]$$

As the potential contains terms of the form $\cos(m\theta)$, all terms with $m\neq 1$ give zero contribution when integrating over the circumference. The horizontal force is thus obtained as:

$$F_x = \pi\rho g R^2\zeta_A C_M \tanh(kd)e^{i(\omega t+\delta)}$$

with

$$C_M = \frac{4}{\pi(kR)^2}\left\{J_1'^2(kR) + Y_1'^2(kR)\right\}^{-1/2}$$

$$\delta = \arctan\left\{\frac{Y_1'(kR)}{J_1'(kR)}\right\}J_1'. \qquad [6.15]$$

Frequently the overturning moment about a vertical location at the sea floor is wanted, which is obtained in a similar way to the horizontal force in [6.14], while accounting for the contribution to the overturning moment:

$$\begin{aligned} M_{(z=-d)} &= \int_{-d}^{0}\int_{0}^{2\pi} pn_x(z+d)dzd\theta. \\ &= \pi\rho g R^2\zeta_A C_M\frac{1}{k}\left\{\frac{1 + kd\sinh(kd) - \cosh(kd)}{\cosh(kd)}\right\}e^{i(\omega t+\delta)}. \end{aligned} \qquad [6.16]$$

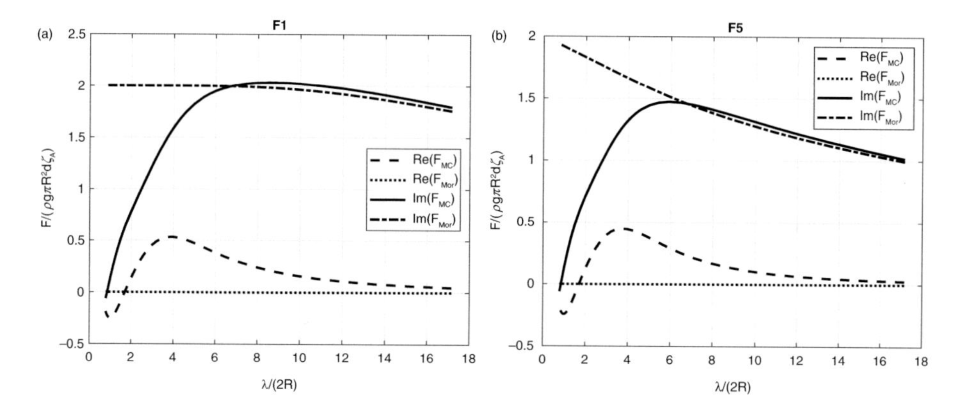

Figure 6.7 Comparison of horizontal force (left) and overturning moment about the sea floor (right) as computed by long wavelength theory (the Morison equation with zero drag and added mass coefficient 1.0 ("Mor")) and MacCamy and Fuchs (1954) theory ("MC"). Water depth to diameter ratio is 3.75. The forces and moments are made dimensionless by $\rho g R^2 \zeta_A$ and $\rho g R^2 d \zeta_A$. $R = 4\ m$ is the cylinder radius, ζ_A is the wave amplitude and $d = 30\ m$ is the water depth. Real and imaginary part of the forces.

In Figure 6.7, the horizontal wave force and overturning moment is computed on a vertical circular cylinder with diameter 8 m. A water depth of 30 m is used. It is observed that for wavelength to diameter ratio $\lambda/D > 5$, the long wavelength theory compares well with the MacCamy and Fuchs theory. For shorter waves, the deviation is, however, significant. If drag forces are added, the force using the Morison equation depends upon the drag coefficient used as well as how the wave kinematic is extrapolated above the mean free surface.

In comparing the long wavelength limit for the MacCamy and Fuchs theory with the long wavelength approach, it is observed that for long wavelengths $C_M \simeq 2 = \left(C_m^{(2D)} + 1\right)$ and $\delta \simeq \pi/2$.

6.4 Bodies of General Shape

Fixed-substructure wind turbines in most cases are mounted upon monopiles or various kinds of slender space-frame structures as jackets. The wave forces are thus estimated using the Morison equation or the MacCamy and Fuchs method, as has already been discussed. However, a great variety of substructures have been proposed, as illustrated in Chapter 4. In most cases floating substructures are assembled from slender, horizontal pontoons and vertical columns. Both the pontoons and the columns may have a cross-sectional area that varies along its length. In addition, flat solid or perforated plates may be introduced to obtain the wanted dynamic characteristics of the floater (see Chapter 7). Some floating substructures have a barge-like shape. To estimate the wave forces on a long, slender structure of general cross-section, "strip theory" methods as discussed in Chapter 7 can be used. These methods have some similarities with the beam element momentum (BEM) methods used in aerodynamics, as discussed in Chapter 3. Both assume a local 2D flow over the structure. By the "strip theory" approach, forces on the total structure as well as sectional forces may be obtained. The sectional forces are needed for computing local bending moments and shear forces. The strip theory assumes no hydrodynamic interaction between neighboring strips or between the various structural components.

To compute the linearized wave forces on a body of general shape, where neither the small body approximation nor the slender body approximation will work, linear potential theory methods based upon the application of Green's identity, so-called panel methods as described by Newman and Sclavounos (1988), are usually applied. The basic principles for these methods are based upon the panel methods developed by Hess and Smith (1962) for the aero industry. However, the presence of free surface waves adds some complications. Details on the 3D panel method are described by Newman and Sclavounos (1988) and Faltinsen (1990).

The basic concept is based upon linear wave-body interaction assuming potential flow and stationary conditions. The problem is solved in the frequency domain. The boundary problem considered is illustrated in Figure 6.8. The velocity potential in the fluid surrounding the body considered is denoted $\Phi = \mathrm{Re}\{\varphi e^{i\omega t}\}$. As all terms contain the factor complex $e^{i\omega t}$, only the complex amplitude of the potential φ is considered in this section. The fluid is assumed to be ideal, irrotational and incompressible, and the potential thus fulfils the Laplace equation in the fluid domain:

$$\nabla^2 \varphi = 0. \qquad [6.17]$$

Further, as the problem is assumed to be linear, the velocity potential describing the flow around a 3D rigid body oscillating in an incident wave field may be divided into three main components: one potential describing the undisturbed incident waves; one potential describing the disturbance, the scattering, due to the presences of a rigid body at rest; and one potential describing the flow effect due to motion of the body, the radiation:

$$\varphi = \varphi_I + \varphi_S + \varphi_R. \qquad [6.18]$$

The scattering potential φ_S and the incident potential φ_I are frequently summed and denoted as the diffraction potential, $\varphi_D = \varphi_I + \varphi_S$. The incident potential is discussed in Chapter 2. The radiation potential may be split into six contributions, one

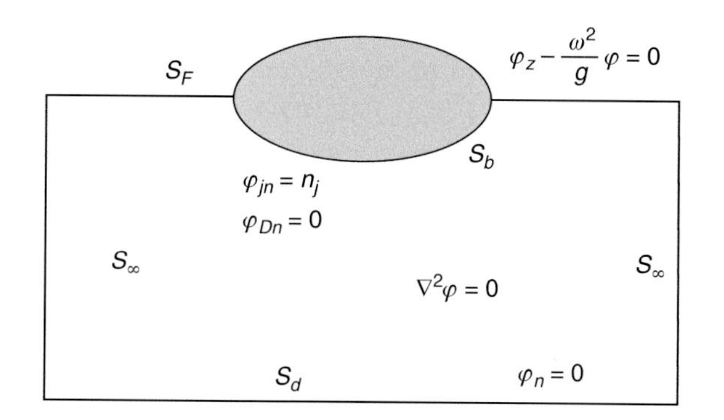

Figure 6.8 The boundary problem considered. The partial derivative, $\partial\varphi/\partial n$, is written as φ_n.

from each of the six rigid-body modes of motion. As the potential for each mode of motion is proportional to the velocity of the body in that mode, it is convenient to write:

$$\varphi_R = i\omega\sum_{i=1}^{6}\eta_i\varphi_i. \quad [6.19]$$

Here, η_i is the complex amplitude of the motion in mode i. Referring to Figure 6.8, the boundary conditions for the velocity potentials are obtained as follows. At the flat bottom the normal velocity must be zero, i.e.:

$$\frac{\partial\varphi}{\partial n} = \frac{\partial\varphi}{\partial z} = 0 \quad \text{at } z = -d. \quad [6.20]$$

As the water depth tends toward infinity, this boundary condition is replaced by the requirement that the gradient should tend to zero as z tends toward minus infinity. On the body surface S_b a similar zero normal velocity condition must be required for the diffraction potential:

$$\frac{\partial\varphi_D}{\partial n} = 0, \quad [6.21]$$

while for the radiation potential it must be required that the normal velocity of the water at S_b must equal the normal velocity of the body. This is required for each of the six modes of motion:

$$\frac{\partial\varphi_i}{\partial n} = n_i \quad i = 1 - 6. \quad [6.22]$$

$\mathbf{n}$ is the unit surface normal pointing out of the fluid domain with the three components $\mathbf{n} = \{n_1, n_2, n_3\}$. The rotational components are defined by $(n_4, n_5, n_6) = \mathbf{x} \times \mathbf{n}$, where $\mathbf{x} = (x, y, z)$ is the coordinates of the body surface point considered. At the free surface S_F the same linearized boundary condition is valid as for the incident waves (see Chapter 2):

$$\frac{\partial \varphi}{\partial z} - \frac{\omega^2}{g}\varphi = 0 \quad \text{at} \quad z = 0. \qquad [6.23]$$

Far away from the body, at S_∞ a radiation requirement is imposed on the scattering and radiation potentials. This condition implies that the potentials should represent outgoing waves far away from the body.

In state-of-the-art computer programs, e.g., WAMIT (2016), the potentials are solved by invoking Green's second identity, relating integrals over the fluid volume and the surface enclosing the volume:

$$\iiint_\Omega \left(\psi\, \nabla^2\varphi - \varphi\, \nabla^2\psi\right) d\Omega = \oiint_S \left(\varphi\frac{\partial \psi}{\partial n} - \psi\frac{\partial \varphi}{\partial n}\right) dS. \qquad [6.24]$$

Ω is the fluid volume considered and S is the surface enclosing the volume. Ψ is a "help function," normally denoted as Green's function, which also satisfies the Laplace equation except in one singular point P_1 . In the case of an infinite fluid volume without boundaries, except at the body surface, $\psi = 1/r$ with r equal to the distance from P_1 to the point entering into the integration. $1/r$ is the potential for a source with strength 4π in an infinite fluid, the "Rankine source." If the fluid volume is a half-space, e.g., an infinite fluid volume above a flat bottom located at $z = 0$, a Green's function $\psi = 1/r + 1/r'$ may be used. Here, $r' = [(x-\xi)^2 + (y-\eta)^2 + (z+\zeta)^2]^{1/2}$; see Figure 6.9. By introducing this Green's function, it is observed that the boundary condition $\partial\psi/\partial z = 0$ at $z = 0$ is satisfied. By constructing a Green's function satisfying all the boundary conditions except at the body considered, it can be shown that the contributions to the surface integral in [6.24] from all boundaries except the body surface S_b vanish. An artificial small surface (sphere) around the singularity P_1 is introduced to ensure the validity of the volume integral. The radius of the sphere is infinitesimal. The result from [6.24] is obtained as:

$$\varphi(P_1) = \frac{1}{4\pi}\iint_{S_b} \left(\varphi\frac{\partial \psi}{\partial n} - \psi\frac{\partial \varphi}{\partial n}\right) dS. \qquad [6.25]$$

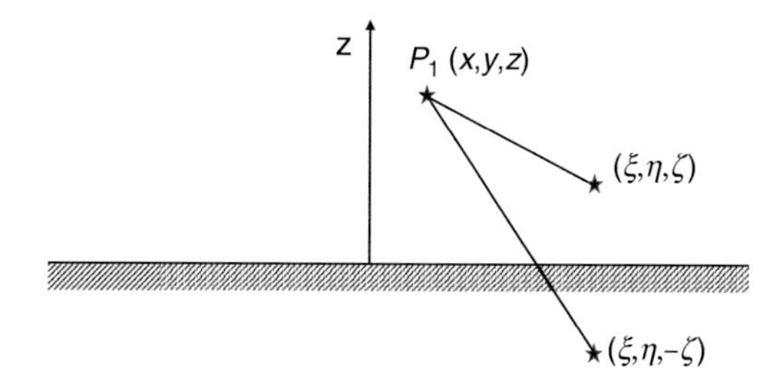

Figure 6.9 Illustration of the concept of mirroring.

When a free surface is present, the Green's function ψ becomes much more involved than the potential corresponding to the pure Rankine source. In addition to the Rankine part, a singular integral over all wavenumbers is to be included. For the infinite water depth case the Green's function may be written (Newman and Sclavounos,1988) as:

$$G(\boldsymbol{\xi};\mathbf{x}) = \frac{1}{r} + \frac{1}{r'} + \frac{2\omega^2}{\pi g}\int_0^\infty \frac{e^{k(z+\zeta)}}{k-\omega^2/g} J_0(kR)dk. \qquad [6.26]$$

Here, r and r' are given above; R is the horizontal distance between the source and P_1. J_0 is the Bessel function of first kind and order zero. Great care must thus be taken in implementing this Green's function to ensure accuracy as well as reasonable computational times. For details on the free surface Green's function and its numerical implementation, see Wehausen and Laitone (1960), Newman (1985) and Faltinsen (1990).

With P_1 located in the fluid domain, the potential in the point $P(\mathbf{x})$ is obtained by introducing [6.26] into [6.25]. To do so, the potential distribution over the body surface is needed. If $P(\mathbf{x})$ is at the body surface, the following result is obtained for the potentials at this location:

$$\varphi_i(\mathbf{x}) + \frac{1}{2\pi}\iint_{S_B} \varphi_i(\boldsymbol{\xi})\frac{\partial G(\boldsymbol{\xi};\mathbf{x})}{\partial n_\xi}d\boldsymbol{\xi} = \frac{1}{2\pi}\iint_{S_B} n_j G(\boldsymbol{\xi};\mathbf{x})d\boldsymbol{\xi}$$

$$\varphi_D(\mathbf{x}) + \frac{1}{2\pi}\iint_{S_B} \varphi_D(\boldsymbol{\xi})\frac{\partial G(\boldsymbol{\xi};\mathbf{x})}{\partial n_\xi}d\boldsymbol{\xi} = 2\varphi_I(\mathbf{x}). \qquad [6.27]$$

Here, $\varphi(\mathbf{x})$ is the potential at point $\mathbf{x}$ at the body surface. $G(\boldsymbol{\xi};\mathbf{x})$ is the Green's function computed between the integration point $\boldsymbol{\xi}$ at the body surface and $\mathbf{x}$. $\partial/\partial n_\xi$ is the derivative in the surface normal direction at $\boldsymbol{\xi}$. Note that when P_1 is located at the body surface, the factor $1/4\pi$ in [6.25] is replaced by $1/2\pi$. [6.27] implies that potential at all points along the body surface interacts. The right-hand side of the equations in [6.27] are known, as is the derivative of the Green's function on the left-hand side. To solve for the total potential, the integrals may be replaced by a summation over the body surface.

The body may be divided into N flat, quadrilateral panels as illustrated in Figure 6.10. If the potential is assumed to be constant over each panel, and the distance is calculated to a central point at each panel, [6.27] may be reformulated to a set of equations with N unknowns:

$$\varphi_j + \frac{1}{2\pi}\sum_{k=1}^{N} D_{jk}\varphi_k = \frac{1}{2\pi}\sum_{k=1}^{N} S_{jk}\left(\frac{\partial \varphi}{\partial n}\right)_k. \qquad [6.28]$$

$$\varphi_{Dj} + \frac{1}{2\pi}\sum_{k=1}^{N} D_{jk}\varphi_{Dk} = 2\varphi_{Ij}$$

Here, the index i is omitted for the radiation potentials. D_{jk} and S_{jk} are matrices given from the integrals over each panel, including the Green's function or its normal derivative to all panels, i.e.:

$$D_{jk} = \iint_{S_k} \frac{\partial G(\boldsymbol{\xi}, \mathbf{x}_j)}{\partial n_{\xi}} d\xi$$

$$S_{jk} = \iint_{S_k} G(\boldsymbol{\xi}, \mathbf{x}_j) d\xi. \qquad [6.29]$$

The integration takes place over each panel surface S_k located at $\boldsymbol{\xi}$ and the functions are evaluated at the collocation points, representing geometrical mid-points, of the panel j, located at $\mathbf{x}_j$. In most implementations, the Rankine part of the Green's function is integrated analytically over the flat, quadrilateral panels, while the wave part of the Green's function is evaluated between the collocation points.

To achieve high accuracy in this approach, a large number of panels may be needed. Close to sharp corners and close to the free surface special care should be used to ensure sufficiently small panels to represent the rapidly changing strength of the potential. The panel size has also to be small relative to the wavelength considered. Figure 6.10 shows an example of the discretization of a spar-floating substructure. A large number of panels increases the computational time significantly. To improve the accuracy without increasing the number of panels too much,

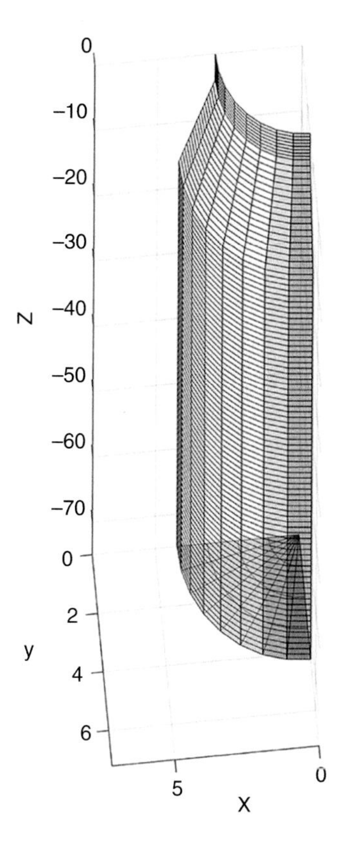

Figure 6.10 Illustration of discretization of a spar platform into quadrilateral panels. Only one quarter of the model is modeled. The remaining three-quarters are modeled by mirroring about the planes $x = 0$ and $y = 0$.

higher-order panel methods[3] may be applied. Such methods may allow for non-flat panels and a variation of the potential strength over the panels. Details of one such method may be found in the WAMIT Manual (2016).

It has so far been shown how the velocity potential in the fluid may be found. This is a convenient approach when the pressure distribution is sought for as the dynamic pressure is proportional to the velocity potential:

$$p = -\rho \frac{\partial \Phi}{\partial t} = -i\omega\rho\Phi = -i\omega\rho\varphi e^{i\omega t}. \qquad [6.30]$$

[3] "Higher-order panel methods" must not be confused with "higher-order solutions." The former refers to the representation of the panel surface as well as the distribution of the potential over the panel. Thus, the panels may not be flat and the potential may not be constant over each panel. The problem considered is still linear from a hydrodynamic point of view. However, higher-order hydrodynamic problems and solutions (in particular the second-order problem) may also be solved by boundary integral methods. The higher-order solutions use a Taylor expansion to account for finite body motions and wave heights. In this case integration over the free surface is also required, increasing the complexity of the numerical problem considerably.

However, if accurate estimates of the fluid velocities are of importance, it may be preferred to solve for the source strength at each panel. The approach is similar to solving for the velocity potential (Newman and Sclavounos, 1988).

One of the nice things about using the boundary element method is that only the wetted body boundary needs to be discretized. However, solving for the problem in the exterior fluid domain, an internal, artificial problem is solved at the same time. For bodies piercing the water surface, this may cause some challenges. The internal problem is a homogeneous "Dirichlet" problem corresponding to $\varphi = 0$ at the internal surface. This problem exhibits eigenfrequencies that may disturb the results for the exterior problem. These frequencies are denoted as "irregular frequencies." The location of the irregular frequencies for a general body cannot be found analytically. However, for a vertical circular cylinder and a rectangular barge, simple expressions exist (see Hong, 1987; Korsmeyer, Newman and Sclavounos, 1988). For the vertical cylinder, the irregular frequencies are located at $J_n(kR) = 0$, where R is the cylinder radius and k is the wave number corresponding to gravity waves at the depth of the cylinder. J_n is the Bessel function of first kind and order n. For $n = 0,\ 1,\ 2, 3$ the zeros of the Bessel function of first kind are found at $kR = 2.4048.., 3.8317.., 5.1356.., 5.5200...$

For the rectangular barge with constant draft h the irregular frequencies are found approximately at wave numbers:

$$k = \pi\sqrt{\left(\frac{n}{L}\right)^2 + \left(\frac{m}{B}\right)^2}, \quad n = 1, 2, 3, \ldots \text{ and } m = 1, 2, 3, \ldots. \qquad [6.31]$$

Here, L is the length of the barge, B is the width of the barge and k is the wave number corresponding to gravity waves at depth h, so the corresponding wave frequencies are given from $\omega = \sqrt{gk\coth(kh)}$. Methods exist to remove the irregular frequencies. One such method is by introducing a lid at the internal volume.

6.5 Effects of Steep Waves

6.5.1 Drag Forces

The above derivations of wave forces are based upon linear wave theory. That implies that the structure considered is assumed to be wetted to the mean water only level even in a wave crest. Thus, when integrating the pressure to obtain the forces, the upper limit is taken at the mean free surface level. For vertical columns with inertia-dominated loads, where the acceleration term dominates over the drag or wave radiation terms, the approach of integrating to the mean free surface is good as the maximum load occurs as the wave elevation is close to zero. However, for drag-dominated structures, the maximum force occurs as the wave amplitude is at the maximum, and the contribution from the drag force in the region from the mean free surface level to the actual wave elevation becomes significant. An approximate way of estimating this contribution to the drag force can be made as follows. Consider the drag term in [6.7]. Integrated from the mean free surface level to the actual wave elevation, the contribution to the horizontal force becomes:

$$F_{FS} = \int_0^{\zeta(t)} \frac{1}{2}\rho C_D D U(z,t)|U(z,t)|dz. \qquad [6.32]$$

The maximum horizontal wave particle velocity coincides with the maximum wave elevation. Using the linear expressions for the velocity and wave elevation, we obtain the following approximate value of the maximum force:

$$F_{FS(\max)} \simeq \frac{1}{2}\rho C_D D \omega^2 \zeta_A^3. \qquad [6.33]$$

Here, ζ_A is the wave amplitude. It is observed that this force contribution is proportional to the third power of the wave amplitude. It is thus of increasing importance as the wave height increases. In [6.33] the linear value of the maximum horizontal wave particle velocity is used. As has been discussed in Section 2.2, the real maximum velocity in steep waves exceeds the linear value. [6.33] thus underestimates the real maximum drag force. Further, the choice of a proper drag coefficient in the splash zone is a nontrivial issue.

6.5.2 Slamming

When the wave becomes very steep, the particle velocity in the wave crest may exceed the phase speed of the wave and the wave breaks. Such a development is simulated by, e.g., Vinje and Brevig (1980) and is illustrated in Figure 6.11. From a force perspective, the situation with an almost vertical wall of water hitting the column represents the most severe situation. In this case, water with high velocity hits a large length of the column almost simultaneously. As breaking starts (see the lower two frames in Figure 6.11), the local particle velocities may become even larger, but the water hits the column on a more limited area. Further, after breaking, air will normally mix into the water, contributing to a reduced impact pressure.

The situation with an almost vertical "wall of water" hitting the column is called slamming and resembles the more common slamming situation when an (almost) flat body hits a horizontal water surface at high speed. A classical problem considered is a horizontal circular cylinder moving toward a calm free surface at constant speed, as illustrated in Figure 6.12. In this situation,

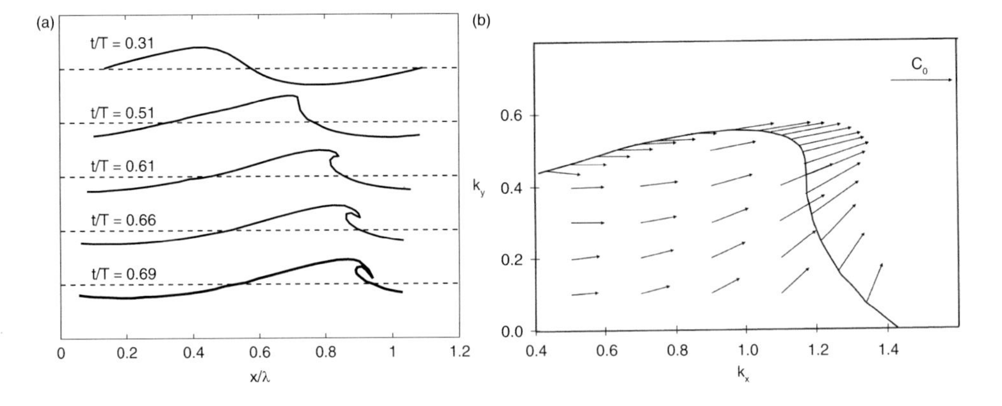

Figure 6.11 Development of a steep breaking wave in deep water as computed by Vinje and Brevig (1980). Left: development of a plunging breaker, T is the wave period. Right: velocity field in the wave when the wave front is approximately vertical. C_0 is the linear phase speed of the wave. $k = 2\pi/\lambda$ is the wave number. Reproduced with permission by SINTEF OCEAN.

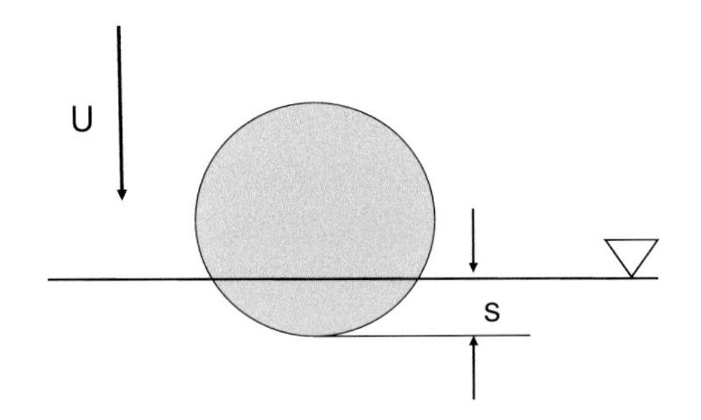

Figure 6.12 Horizontal cylinder impacting a calm free water surface.

the submerged volume and thus the added mass of the system varies with time. The vertical hydrodynamic force on the cylinder, ignoring viscous effects, can then be expressed by:

$$F_z(z,t) = \frac{d}{dt}[A^{(2D)}(t)U] = \frac{d}{ds}[A^{(2D)}(s)]\frac{ds}{dt}U + A^{(2D)}(s)\dot{U}$$
$$\simeq \frac{dA^{(2D)}(s)}{ds}U^2 \qquad [6.34]$$

U is the vertical velocity, positive downward, and $A^{(2D)}(s)$ is the 2D added mass for the cylinder as function of the submergence s. In the initial phase of the impact, it is assumed that U is almost constant and thus the acceleration dependent force is much smaller than the slamming force, related to the time derivative of the added mass. The key challenge is to estimate the added mass as function of submergence. For practical purposes, the derivative of the added mass is frequently expressed by a slamming coefficient:

$$C_s = \frac{dA^{(2D)}/ds}{\frac{1}{2}\rho D}, \qquad [6.35]$$

where D is the cylinder diameter. Classical theoretical derivations (see, e.g., Faltinsen, 1990), give the initial value of C_s, i.e., the value just after touching water. The so-called von Kármán (1929) solution gives $C_s(s=0)=\pi$. In the von Kármán approach the deformation of the free surface is ignored. In the Wagner (1932) approach, the deformation of the free surface, and thus the increased wetted length, is accounted for and $C_s(s=0)=2\pi$ is obtained. Most experiments result in values between these two theoretical values. A frequently used relation for the slamming coefficient is proposed by Campbell and Weynberg (1980):

$$C_s = 5.15\left[\frac{D}{D+19s} + \frac{0.107s}{D}\right]. \qquad [6.36]$$

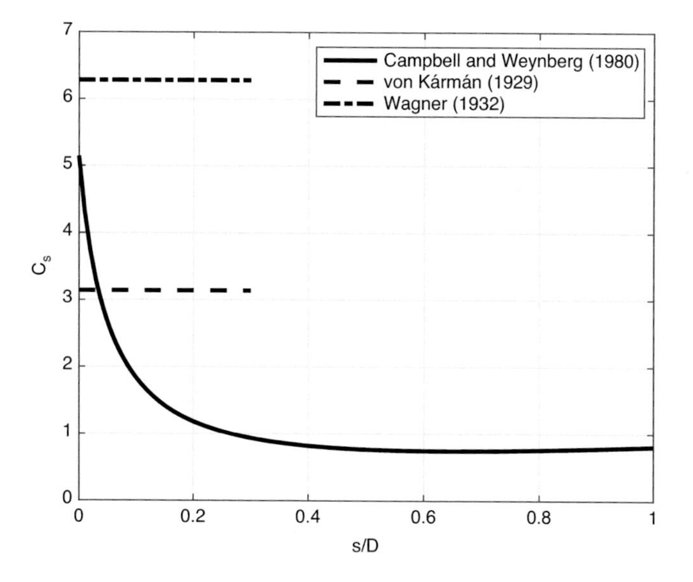

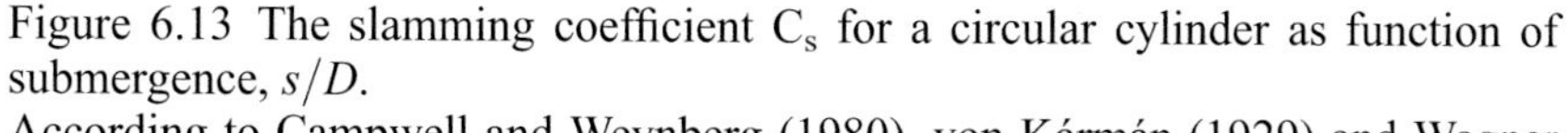

Figure 6.13 The slamming coefficient C_s for a circular cylinder as function of submergence, s/D.
According to Campwell and Weynberg (1980), von Kármán (1929) and Wagner (1932).

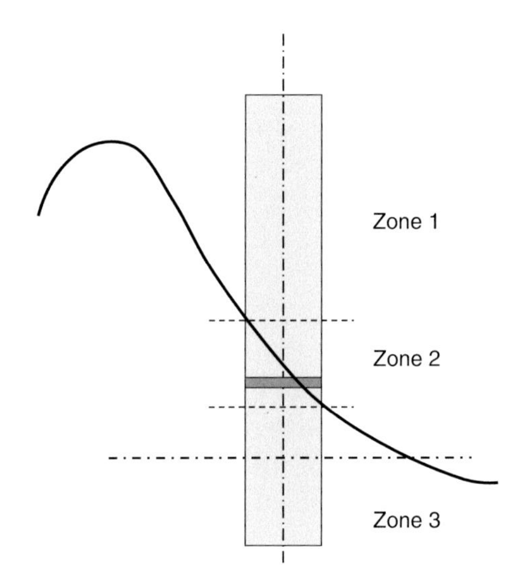

Figure 6.14 The three zones used in computing wave loads on a vertical, slender cylinder in steep waves. A narrow strip is indicated. The load on each strip is computed for each time instant and summed to establish the total load at a certain time instant. Based upon Kalleklev and Nestegård (2005).

The three expressions for the slamming coefficient are illustrated in Figure 6.13. Other expressions for the slamming coefficient have also been proposed. For an overview, see, for example, DNV (2021c).

To compute the horizontal slamming load of a steep wave impacting upon a vertical column, Kalleklev and Nestegård (2005) recommend using the slamming results for calm water, as illustrated in [6.36]. They use a strip theory approach and divide the vertical column into three zones, as illustrated in Figure 6.14. Strips in Zone 1 are not exposed to water and have thus zero load. Strips in Zone 2 are partially submerged, and the slamming formulation in [6.34] is used together with a proper slamming coefficient. For strips in Zone 3 the Morison equation is applied. As discussed in Chapter 2, great care should be used in estimating the wave particle kinematics. This is in particular the case for waves close to breaking (see Figure 6.11). The approach proposed by Kalleklev and Nestegård (2005) assumes that the horizontal component of the wave particle velocity is constant during the wetting phase. The effect of vertical run-up, a 3D effect, is ignored. This effect increases as the diameter of the cylinder increases.

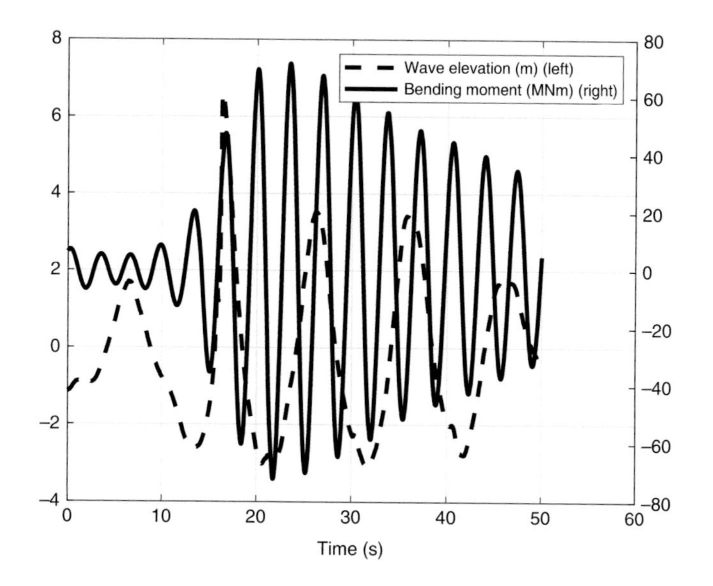

Figure 6.15 Example of ringing response of the first bending mode for a monopile. Results from model tests in scale 1:30.6. The plots are based upon data obtained from tests analyzed by Suja-Thauvin et al. (2017).
Courtesy: Suja-Thauvin, NTNU.

6.5.3 Ringing and Springing

Another transient wave load phenomenon that may excite resonant response of a vertical column is ringing. Ringing is caused by nonlinear effects in waves that typically are much longer than the column diameter. Perturbations up to third order in the wave steepness are used for estimating the wave kinematics to be input to computation of the ringing loads (see, e.g., Faltinsen et al., 1995; Krokstad et al., 1998; Grue and Huseby, 2002). Slamming causes a classical transient dynamic response in which the first transient response amplitude is the largest, and the following response amplitudes have gradually decaying amplitudes. The ringing-induced dynamic response, on the contrary, typically builds up over a few resonant cycles to a maximum response, and thereafter decays gradually. An example of measured ringing response is shown in Figure 6.15, which illustrates the response of the first bending mode of a wind turbine mounted on a monopile in 27 m water depth and exposed to steep waves. The natural frequency of the first bending mode is 0.29 Hz, while the wave spectrum has a peak frequency of 0.1 Hz. Figure 6.15 shows the characteristic ringing response triggered by a steep wave, which builds up over a few cycles before it dies out slowly. The natural frequency of the response is close to three times the peak frequency of the waves. For a further discussion of the experiments, see Suja-Thauvin et al. (2017).

Both slamming and ringing are caused by steep waves and may cause large resonant response amplitudes. However, the response dies out after a few cycles. Springing is a stationary response phenomenon experienced in long structures such as large ships. The bending modes of the structure may be excited by both linear and nonlinear wave loads. For the excitation to take place, both the wavelength and the frequency of the excitation forces must correspond to the mode shape and natural frequency of the structural mode. For offshore wind turbines, springing may thus be relevant for long horizontal structures such as pontoons or stiffeners.

6.6 Modal Loads

The dynamics of continuous systems are discussed briefly in Chapter 5. As most bottom-fixed wind turbines are mounted on monopiles, the contribution to inertia, damping and excitation forces due to waves may be obtained from the MacCamy and Fuchs solution or the Morison equation. The horizontal force and the overturning moment from waves are given in [6.14] and [6.15]. If the excitation of elastic mode-shapes is to be considered, the following approach may be used.

Denote the mode shape considered $\Psi_j(z)$. This may be elastic bending mode j of the monopile-tower structure. The modal inertia for mode j due to the surrounding water may then be written as:

$$A_{jj} = \rho\pi R^2 \int_{L_W} \Psi_j^2(z) C_m^{(2D)} dz. \qquad [6.37]$$

Here, L_w is the length of the wetted part of the monopile. A constant radius R is assumed. Using a long-wavelength approximation, the 2D added mass coefficient $C_m^{(2D)} = 1$ may be used. Similarly, the modal wave excitation force is obtained as:

$$F_j = \int_{L_w} \Psi_j(z) F^{(2D)}(z) dz. \qquad [6.38]$$

Using the Morison equation and ignoring the drag force, the long-wavelength approximation for the modal excitation force becomes:

$$F_j = \frac{ikg\zeta_A}{\cosh(kd)} \rho\pi R^2 \left(1 + C_m^{(2D)}\right) \int_{-d}^{0} \Psi_j(z) \cosh\left(k(z+d)\right) dz \, e^{i\omega t}. \qquad [6.39]$$

Using the MacCamy and Fuchs solution, the modal force may similarly be written as:

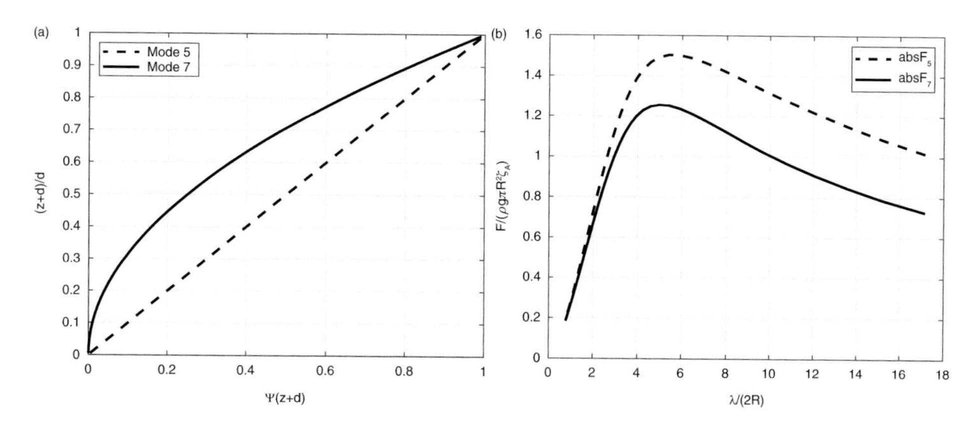

Figure 6.16 Left: mode shape for mode 5 (pitch) and a simplified bending mode (7). Right: absolute values of the modal loads as function of wavelength. A water depth of 30 m and radius of 4 m are used in the example.

$$F_j = \frac{kg\zeta_A}{\cosh(kd)}\rho\pi R^2 C_M \int_{-d}^{0} \Psi_j(z)\cosh\Big(k(z+d)\Big)dz\ e^{i(\omega t+\delta)}. \qquad [6.40]$$

Here, d is the draft of the monopile; R is the radius; ζ_A is the wave amplitude; and C_M is given by [6.15]. For modes 1 and 5 (overturning moment about the sea floor), the mode shapes may be written as $\Psi_1(z) = 1$ and $\Psi_5(z) = \alpha_5(z+d)$ respectively. The mode shape of an elastic beam fixed at the sea floor may be approximated by a parabola:[4] $\Psi_7(z) = \alpha_7(z+d)^2$. The α-values must be chosen so that the mode shape has a proper value at its maximum. If the first elastic mode is chosen to have a unit displacement at the nacelle level, $\Psi_7(z_{nacelle}) = 1$, the scaling factor becomes $\alpha_7 = 1/(z_{nacelle} - z_{bottom})^2$. Figure 6.16 displays the mode shapes for rigid-body pitch and simplified elastic bending modes. Both are given a unit displacement at the waterline, $z = 0$. In the same figure, the computed modal loads are also shown as function of wavelength over diameter. The wave loads are computed by the MacCamy and Fuchs approach. It is observed that except for cases with very short waves, the modal load in pitch is larger than the modal load in bending. This is explained by the fact that the bending mode shape function is less than the pitch mode shape function for all vertical positions. For very short waves, the modal loads are almost equal. This is because short waves act close to the waterline only and here both modal functions have a value close to unity.

[4] Elastic modes are number 7 onward. This is to reserve modes 1 to 6 for the rigid-body modes.

Effect of Water on the Modal Mass

Assume the monopile as well as the tower may be modeled as a vertical cylinder with radius 2.5 m and wall thickness 0.055 m. The water depth is $d = 22m$ and the vertical coordinate of the nacelle is at $z_n = 75m$. The elastic bending mode shape is approximated by a parabola and unit displacement at nacelle level, i.e.:

$$\Psi_7(z) = \left(\frac{z+d}{z_n+d}\right)^2. \qquad [6.41]$$

The mode is denoted mode 7. The structural mass of the tower plus monopile down to the sea floor becomes $m = 2\pi Rt\rho_{st}(z_n + d) = 657.8 \cdot 10^3 kg$. The density of steel is set to 7850 kg/m^3. The modal mass of the tower plus monopile for the above elastic bending mode becomes:

$$m_{7t} = \rho_{st} 2\pi Rt \int_{-d}^{z_n} \Psi_7^2(z)dz = \rho_{st} 2\pi Rt \frac{d+z_n}{5} = 131.6 \cdot 10^3 kg. \qquad [6.42]$$

With a mass of the nacelle and rotor assembly equal to $250 \cdot 10^3 kg$, the total modal mass of the structure becomes $381.6 \cdot 10^3 kg$. The monopile may be assumed to be filled with water up to the still water level; further, using a 2D added mass coefficient of 1.0, the contribution from external and internal water to the modal mass becomes:

$$m_{7w} = \rho\pi R^2 \left(1 + C_m^{(2D)}\right) \int_{-d}^{0} \Psi_7^2(z)dz = 2\rho\pi R^2 \frac{d^5}{5(d+z_n)^4} = 469\ kg. \qquad [6.43]$$

I.e., the contribution from the internal and external water is only 0.12% of the contribution from the steel and the top head mass. This is explained by the fact that the modal displacement at the waterline is only 5.1% of the top displacement.

Even if the hydrodynamic contribution is negligible to the modal mass, the wave excitation force may be significant, in particular if the wave frequency is close to the modal natural frequency.

6.7 Modeling of Bottom Sediments

The purpose of a monopile is to secure the wind turbine to the sea bottom. Monopiles are driven, most frequently by hammering, down into the bottom sediments. The required diameter and the length the monopile to be driven into the sediments depends upon the stiffness and strength characteristics. The sediments may consist of layers of silt, clay and sand, all with different

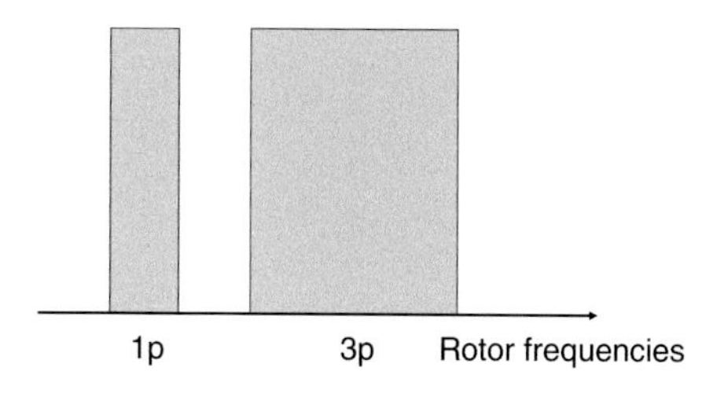

Figure 6.17 Illustration of ranges of the rotor frequency (1p) and three times the rotor frequency (3p) for a variable-speed HAWT.

properties. The stiffness properties with respect to horizontal displacement are essential in determining the eigenfrequencies for the elastic bending modes of the wind turbine. For a modern HAWT with a variable rotor speed control, one should avoid eigenfrequencies coinciding with the rotational frequency (1p) and three times this frequency (3p) for a three-bladed rotor. As the variable speed range increases, the gap between the highest values of 1p and the lowest values of 3p decreases, as illustrated in Figure 6.17. The rotational speed for modern 6–8 MW turbines may be in the range of 5–12 rev/min, corresponding to 1p in the range of 0.08–0.2 Hz.

Frequently the design of the monopile-tower structure is such that the eigenfrequency of the first bending mode is located in the gap between the 1p and 3p ranges. Using the above approximate numbers, this means that the first elastic eigenfrequency should be within the range 0.2–0.24 Hz (see, e.g., Kallehave et al., 2015). Thus, very accurate estimates of the eigenfrequencies are required, with a corresponding requirement for an accurate estimate of the sediment stiffness. The second eigenfrequency is normally designed to be above the 3p range.[5]

In industry it is common to denote the tower as "stiff" or "stiff-stiff" if the first eigenfrequency is above the 3p range. If the first eigenfrequency is below the 1p range, the tower is denoted as "soft" or "soft-soft." With the first eigenfrequency in the range between 1p and 3p the tower is denoted "soft-stiff." As the size of the turbines increases, the 1p frequency is reduced. Thus, using a "soft-stiff" design of the tower, the first eigenfrequency may be in the wave frequency range.

Three ways of modeling the pile–sediment interaction are used in design, as illustrated in Figure 6.18. The three options are as follows.

[5] This discussion considers only deformation in one plane. In a full 3D case there will be bending modes both in the direction of the rotor axis and perpendicular to the rotor axis. The mode shapes and eigenfrequencies in the two directions may differ a little due to differences in the inertial contribution of the rotor in the two directions.

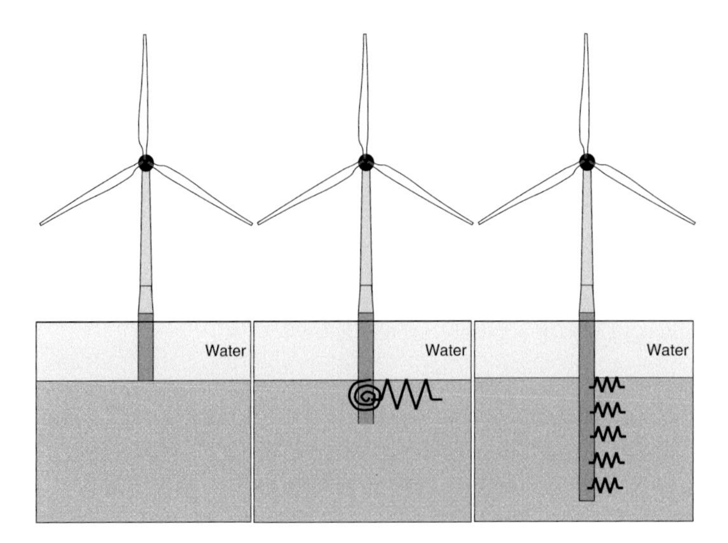

Figure 6.18 Three options for modeling the soil. Left: the pile is assumed to have a rigid fixation at the sea floor; middle: the soil stiffness is modeled by a horizontal and rotational spring at soil level; right: the soil stiffness is modeled by a vertical distribution of springs of variable strength, representing the various soil layers.

Rigid fixation at sea-bottom level disregards the stiffness and damping effects of the sediments. The modal shape will have zero displacement and inclination at the sea floor (see Section 5.4). This approach will overestimate the stiffness corresponding to the first eigenfrequency. The overestimation of the stiffness may be corrected for by moving the point of fixation to a level below the sea floor.

Another option is to represent the sediments by a linear and a rotational spring at the sea floor level. This approach accounts for the effect of the sediments in an integrated manner. To determine the magnitudes of the springs, the soil properties along the pile must be known. Damping may be introduced in a similar way.

A more complete representation of the sediments is by a distribution of springs (and dampers) along the full length of the pile. The stiffness at each level should represent the stiffness of the sediments at that location. For that purpose, data based upon a 2D analysis of a cylinder in the soil may be used, i.e., a strip approach. The continuity between the layers is accounted for by the elastic deflection of the pile. This approach requires detailed knowledge of the sediments at all levels, and accurate mode shapes and eigenfrequencies may be obtained. For further details, see, e.g., Aasen et al. (2017).

The material properties of the sediments have a very nonlinear behavior. The relation between load and deflection in the material may be represented by so-called "p-y curves." The "p-y curve" represents in this case the relation between the force

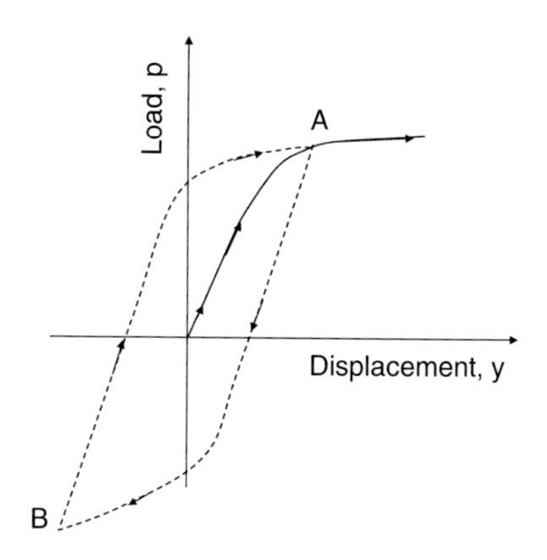

Figure 6.19 Schematic load displacement curve for bottom sediments. Solid line: increased, static forced displacement; dashed line: cyclic displacement.

on a 2D section of the pile and a sideways displacement. A typical shape of the "p-y curve" is shown in Figure 6.19. The solid line shows the load deflection for a monotonic increasing displacement. It starts out from zero with almost a linear relationship, then gradually the stiffness of the material is reduced, and the slope of the curve decreases. At some point, "A" in the figure, the displacement is assumed to be reversed, and an almost linear curve is followed until the nonlinearities again become significant. At "B," the displacement again is reversed.

A cyclic deformation pattern will thus follow a curve like the dashed curve in Figure 6.19. From this curve a hysteretic behavior is observed, causing energy dissipation during cyclic loading. The area of the loop increases as the amplitude of the motion increases. This implies increased damping as the motion amplitude is increased. If the displacements are within the almost linear range of the "p-y curve," the damping contribution is almost negligible. For further discussion of the importance of soil modeling and damping contribution from the soil, reference is made to Aasen et al. (2017) and Carswell et al. (2015).

Exercises Chapter 6

1. Consider a monopile with diameter D = 6 m. The pile is installed in d = 30 m water depth. We want to compute the wave loads on the monopile. Assume the maximum regular wave height at the site is limited by $H/\lambda < 1/10$ and $H < 12\ m$. Assume a drag coefficient of 0.7 and an added mass coefficient of 1.0.
   a. The waves on the location may have periods in the range of 3–15 s. In which range of wave periods may you consider the waves to be "deep-water waves"?

b. Use the Morison equation to compute the horizonal force and overturning moment about the level of the sea floor. Use regular waves in the range of 3–15 s and the maximum wave height as given above. Show the contribution from inertia and drag effects as well as the total load. For simplicity, integrate the loads to the still water level only.
c. Discuss the error in Exercise 1b due to not integrating to the real physical wave elevation.
d. Make an estimate of the error discussed in Exercise 1c by estimating the "missing" force contribution at maximum and minimum wave elevation.

2. Assume you are going to compute the wave load on a gravity foundation with a cylindrical substructure on top. You want to use a computer program based upon the boundary element method. The cylindrical structure has a diameter of D = 8 m at the water line. Check if the issue of irregular frequencies may cause any problems in this case.
3. Consider a wind turbine mounted upon a monopile in d = 30 m water depth. The diameter of the pile is D = 6 m. The distance from the sea floor to the center of the rotor is L = 130 m. Assume the nacelle mass $M_c = 500$ Mg and neglect the mass of the tower. The lowest natural period for tower bending is 3 s. Assume a modal shape of this mode given by $\Psi_1(x) = 1 - \cos\left(\frac{\pi}{2}\frac{x}{L}\right)$ where x is the vertical distance from the sea floor.
   a. What is the modal stiffness for the given modal shape?
   b. Compute the modal wave excitation force on the pile at a range of wave frequencies assuming a unit wave amplitude and ignoring the drag forces.
   c. Compute the corresponding displacement of the top of the tower due to wave loading.
4. Consider a steep, close-to-breaking wave with a vertical wave front as illustrated in Figure 6.11. Assume the horizontal velocity of the wave front is equal to the linear phase speed of the wave. Deep-water waves may be assumed. Consider a wave with a period of 7 s and a vertical pile with a diameter of D = 8 m.
   a. What is the horizontal slamming force per unit length of the pile? Plot the force as a function of time.
   b. How large is the slamming force compared to the linear inertia force and the drag force at the water line for a regular wave with steepness 1/7?

7
Floating Substructures

As energy production from offshore wind expands, new and deeper ocean areas are being considered for development. As discussed in Chapter 4, floating support structures should be considered for water depths beyond 50 m. Floating support structures introduce several new aspects with respect to dynamic behavior compared to bottom-fixed support structures. These aspects will be discussed in more detail in this chapter.

The starting point is equations of motion for a rigid body in six degrees of freedom (6DOF). The forcing mechanisms from waves are addressed as well as the inertia effects due to the surrounding fluid, the added mass, hydrostatics and the effect of mooring. The effect of wind forces is discussed in Chapter 3. This chapter further discusses the combined effect of wind forces and the motion control system.

Floating support structures can take several geometric shapes. Various methods for computing the wave loads on rectangular pontoons, barges etc. will therefore be outlined in more detail. In Chapter 6, the boundary element method for computing wave loads on a 3D body of general shape was discussed. This method is well suited also for floating bodies. However, simpler and computationally faster methods are useful in the design process, in particular for optimization purposes. Therefore, strip theory methods are outlined in some detail. Most of the derivations in this chapter are based upon linear methods. This implies that forces are computed at the initial or mean position of the structure, and that inertia, damping and restoring effects are also linearized and referred to the initial or mean position. The linearization also implies that all dynamic rotations of the support structure are assumed to be small. The linearization makes the computations efficient and allows for solving the dynamics in the frequency domain. However, in real design processes the importance of various nonlinear effects must be assessed.

For floating support structures, a great variety of shapes have been proposed, as illustrated in Chapter 4. In most cases the floater is assembled of slender horizontal pontoons and vertical columns. Both the pontoons and the columns may have a cross-sectional area that varies along the length. In addition, flat solid or perforated plates may be introduced to obtain the wanted dynamic characteristics of the floater. Some floating foundations have a barge-like shape; thus, the applicability and accuracy of the various methods must be evaluated for each case. For example, in a preliminary design phase involving an optimization process, strip theory methods may be applied. Having concluded on a geometry, the results obtained by strip theory should be compared to results obtained by 3D methods.

7.1 Wave-Induced Motions: Equations of Motion

Considering the six rigid-body degrees of motion, the dynamic equations may be written as:

$$(\mathbf{M}+\mathbf{A})\ddot{\boldsymbol{\eta}}+(\mathbf{B_v}+\mathbf{B_r})\dot{\boldsymbol{\eta}}+(\mathbf{C_m}+\mathbf{C_h})\boldsymbol{\eta}=\mathbf{F_{wa}}+\mathbf{F_{wi}}+\mathbf{F_{cu}}+\mathbf{F_{wt}}. \qquad [7.1]$$

Here, $\boldsymbol{\eta}$ is the vector of the six degrees of motion, as illustrated in Figure 7.1. The figure also shows the common naming convention for the motions. The linear motions in direction (x, y, z) are denoted (η_1, η_2, η_3) and the rotations about the (x, y, z) axes are denoted (η_4, η_5, η_6). It is here is assumed that the (x, y)-plane coincides with the mean water surface and that z is vertical, zero at the mean free surface and positive upward. $\mathbf{M}$ is the 6 x 6 dry mass matrix of the complete wind turbine and $\mathbf{A}$ is the hydrodynamic mass matrix. The damping matrix is split into two parts, the radiation part, $\mathbf{B_r}(x, y, z)$, related to wave generation, and the remaining damping, $\mathbf{B}_v$, mainly linearized viscous damping from water and air. The damping could also contain effects due to the control of the wind turbine, but these effects may also be included in the forcing term. The restoring matrix is split into a hydrostatic part, $\mathbf{C_h}$, and a mooring part, $\mathbf{C_m}$. The four excitation force vectors are the wave force vector; the wind force on the structural parts; the current force; and the force due to the action of the wind turbine.

If the equations are linearized, [7.1], and a stationary, dynamic response is considered, the force vector may be written as $\mathbf{F}=\mathbf{F_A}e^{i\omega t}$ and the response as $\boldsymbol{\eta}=\boldsymbol{\eta}_\mathbf{A}e^{i\omega t}$, where ω is the frequency of oscillation and $\boldsymbol{\eta}_\mathbf{A}$ is the complex response vector. The equation of motion in frequency domain may thus be written as:

$$\{-\omega^2[\mathbf{M}+\mathbf{A}(\omega)]+i\omega\mathbf{B}(\omega)+\mathbf{C}\}\boldsymbol{\eta}(\omega)=\mathbf{F}(\omega). \qquad [7.2]$$

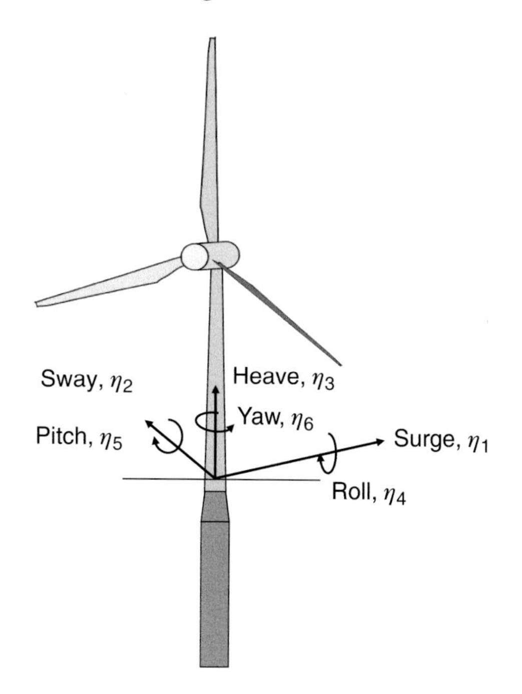

Figure 7.1 The six rigid-body motion degrees of freedom for a floating wind turbine. Surge is in direction of the wind, perpendicular to the rotor plane. The (x, y) plane is located at the mean water surface. z is vertical and positive upward.

Here, it is indicated that in the general case, the added mass as well as the damping are frequency-dependent. The frequency domain format of the equation of motions is useful when wave forces dominate the excitation. If significant nonlinear effects are present, which is the case for wind turbines during operation and active control functions, the equations must be written and solved in time domain. If the hydrodynamic forces are assumed to be linear but frequency-dependent, a convolution integral is needed in the time domain version of the equations to account for the frequency dependence. In time domain the frequency dependence represents a memory effect. In the 1D case the equation of motion in time domain may then be written as:

$$(M + A_\infty)\ddot{\eta} + \int_0^t h(t-\tau)\dot{\eta}(\tau)d\tau + C\eta = F(t). \qquad [7.3]$$

The convolution term now accounts for the frequency dependency of added mass and damping (these are related) and $\boldsymbol{A}_\infty$ is the high-frequency limit of the added mass. Further discussion of time domain formulation of the equation of motion with frequency-dependent coefficients is found in, e.g., Falnes (2002). Further details are given in Section 7.4.8.

7.2 The Mass Matrix

7.2.1 The Dry Mass Matrix

The mass matrix for the dry body can be written as:

$$\mathbf{M} = \begin{bmatrix} M & 0 & 0 & 0 & Mz_G & 0 \\ 0 & M & 0 & -Mz_G & 0 & 0 \\ 0 & 0 & M & 0 & 0 & 0 \\ 0 & -Mz_G & 0 & M_{44} & 0 & 0 \\ Mz_G & 0 & 0 & 0 & M_{55} & 0 \\ 0 & 0 & 0 & 0 & 0 & M_{66} \end{bmatrix}. \quad [7.4]$$

Here, it assumed that the center of gravity (CG) is located at $(0, 0, z_G)$ and that the (x, z) and the (y, z)-planes are planes of symmetry, which frequently is the case for floating bodies. M is the mass of the body, and the moments of inertia are given by:

$$\begin{aligned} M_{44} \equiv I_{11} &= \int_M (z^2 + y^2)dm = I_{11G} + z_G^2 M \\ M_{55} \equiv I_{22} &= \int_M (x^2 + z^2)dm = I_{22G} + z_G^2 M. \\ M_{66} \equiv I_{33} &= \int_M (y^2 + x^2)dm = I_{33G} \end{aligned} \quad [7.5]$$

Here, I_{ii} refers to the mass moment of inertia about axis i and I_{iiG} refers to the mass moment when the axis has origin in CG.

In the more general case without symmetry and where the CG is located in (x_G, y_G, z_G), the mass matrix may be obtained by:

$$\mathbf{M} = \begin{bmatrix} M\mathbf{I}_{3*3} & -M\mathbf{S} \\ M\mathbf{S} & \mathbf{I_{bb}} \end{bmatrix}, \quad [7.6]$$

where

$$\mathbf{I}_{3*3} = \begin{bmatrix} 1 & 0 & 0 \\ 0 & 1 & 0 \\ 0 & 0 & 1 \end{bmatrix}, \mathbf{S} = \begin{bmatrix} 0 & -z_G & y_G \\ z_G & 0 & -x_G \\ -y_G & x_G & 0 \end{bmatrix} \text{ and } \mathbf{I_{bb}} = \int_M \begin{bmatrix} y^2 + z^2 & -xy & -xz \\ -yx & z^2 + x^2 & -yz \\ -zx & -zy & x^2 + y^2 \end{bmatrix} dm.$$

For further details, see Perez and Fossen (2007).

7.2.2 The Added Mass Matrix

In many of the proposed designs for offshore wind support structures, the floater is composed of slender horizontal pontoons and vertical columns. Both the pontoons and the columns may have a cross-sectional area that varies along its length. In addition, flat solid or perforated plates may be introduced to obtain the wanted dynamic characteristics of the floater.

There are two main options to obtain the added matrix for such structures: strip theory or 3D ideal fluid theory based upon, e.g., boundary element techniques, as discussed in Chapter 6. Strip theory approach will be addressed here.

7.2.2.1 Vertical Columns

Consider a slender, circular and vertical column of constant radius R and extending from z_b to z_t, where $z_b < z_t \leq 0$. The cylinder axis is located at (x_c, y_c). The added mass for linear motion in the horizontal direction can then be approximated by:

$$A_h = A^{(2D)}(z_t - z_b) = \pi\rho R^2 C_{ah} L, \quad [7.7]$$

where the length of the column is L and C_{ah} is the 2D added mass coefficient for the cylinder. The added mass for oscillation in the vertical direction can similarly be written as:

$$A_v = (C_{avb} + C_{avt})\pi\rho R^3. \quad [7.8]$$

Here, the indices b and t refer to the bottom and top of the column respectively. If the column pierces the free surface, $C_{avt} = 0$, and if the column is sitting on top of a pontoon, $C_{avb} = 0$. If two columns are sitting on top of each other, an approximate value for the added mass contribution at the junction may be applied; see Section 7.4.2. A 3 x 3 added mass matrix for linear motions is obtained as:

$$\mathbf{A_c} = \begin{bmatrix} A^{(2D)}L & 0 & 0 \\ 0 & A^{(2D)}L & 0 \\ 0 & 0 & A_v \end{bmatrix}. \quad [7.9]$$

As compared to the dry mass matrix, it is observed that the mass values differ between the three directions. $\mathbf{A_c}$ will now constitute the new submatrix corresponding to the upper-left part of [7.6]. Similarly, the submatrix $m\mathbf{S}$ is replaced by $\mathbf{S}_{Ac} = \mathbf{A_c} * \mathbf{S_c}$, where:

$$\mathbf{S_c} = \begin{bmatrix} 0 & -\frac{1}{2}(z_b + z_t) & y_c \\ \frac{1}{2}(z_b + z_t) & 0 & -x_c \\ -y_c & x_c & 0 \end{bmatrix}. \quad [7.10]$$

The rotational coupling terms are obtained as:

$$I_{11} = \int_L \left(y^2 + z^2\right) dm = y_c^2 A_v + \frac{1}{3} A^{(2D)} \left(z_t^3 - z_b^3\right)$$

$$I_{12} = \int_L -xydm = -x_c y_c A_v$$

$$I_{13} = \int_L -xzdm = -x_c \frac{1}{2} A^{(2D)} \left(z_t^2 - z_b^2\right)$$

$$I_{22} = \int_L \left(x^2 + z^2\right) dm = x_c^2 A_v + \frac{1}{3} A^{(2D)} \left(z_t^3 - z_b^3\right)$$

$$I_{23} = \int_L -yxdm = -y_c \frac{1}{2} A^{(2D)} \left(z_t^2 - z_b^2\right)$$

$$I_{33} = \int_L \left(y^2 + x^2\right) dm = \left(y_c^2 + x_c^2\right) A^{(2D)} \left(z_t - z_b\right). \quad [7.11]$$

With $I_{ij} = I_{ji}$, the rotational submatrix becomes:

$$\mathbf{I_{bbc}} = \begin{bmatrix} I_{11} & I_{12} & I_{13} \\ I_{21} & I_{22} & I_{23} \\ I_{31} & I_{32} & I_{33} \end{bmatrix}. \quad [7.12]$$

The full added mass matrix for one vertical column thus becomes:

$$\mathbf{A_{col}} = \begin{bmatrix} \mathbf{A_c} & -\mathbf{S_{Ac}} \\ \mathbf{S_{Ac}} & \mathbf{I_{bbc}} \end{bmatrix}. \quad [7.13]$$

It should be kept in mind that in this derivation it has been assumed that the 2D added mass is equal at all sections, i.e., no end effects are accounted for when integrating the 2D added mass along the column. If end effects are to be accounted for, the various terms involved should be obtained from [7.7] and [7.11] by performing integration along the axis and accounting for variation in $A^{(2D)}$.

The added mass related to the end surfaces of a long slender cylinder is frequently taken to be the mass of a half-sphere with the same radius as the column, i.e., $C_{av} = 2/3$. If two columns are located on top of each other, a rough estimate of the vertical added mass can be obtained by setting the vertical added mass for the surface of the column with the smallest diameter to zero and for the column with the largest diameter to the difference between two half-spheres. I.e., $A_v \simeq \frac{2\pi}{3}\left(R_2^3 - R_1^3\right)$, where the indices 2 and 1 refer to the largest and smallest radius respectively. Experience has shown that this approach may overestimate the vertical added mass; however, it provides the correct results in the limits of $R_1 = R_2$ and $R_1 = 0$.

7.2.2.2 Horizontal Pontoons

Consider a horizontal pontoon of rectangular cross-section extending from (x_1, y_1, z_1) to (x_2, y_2, z_2), see Figure 7.8. As the pontoon is horizontal, $z_1 = z_2 = z_p$. To establish the added mass matrix in this case, we employ strip theory once more. The pontoon is split into short transverse sections over which the flow is assumed to be 2D. It is assumed that the 2D added mass in the horizontal and vertical direction differs, i.e., $A_h^{(2D)} \neq A_v^{(2D)}$. Further, the added mass in the axial direction due to the end surfaces of the pontoon may be included. Consider a section of length ΔL of the pontoon. The mid-point of the center axis through the section is located at (x, y, z). The pontoon axis forms an angle α with the x-axis. Considering an acceleration in x direction $\ddot{\eta}_1$, the forces acting on the fluid in direction 1, 2 and 3 due to this acceleration are:

$$\begin{aligned} \Delta F_{11} &= a_n A_h^{(2D)} \sin\alpha = \ddot{\eta}_1 \sin\alpha \, A_h^{(2D)} \sin\alpha = \ddot{\eta}_1 A_h^{(2D)} \sin^2\alpha. \\ \Delta F_{21} &= a_n A_h^{(2D)} \cos\alpha = \ddot{\eta}_1 A_h^{(2D)} \sin\alpha \cos\alpha \\ \Delta F_{31} &= 0. \end{aligned} \quad [7.14]$$

The same procedure applies for the two other directions. Integrating over the length of the pontoon thus gives the following added mass matrix for linear translations in (x, y, z):

$$\mathbf{A}_p = \begin{bmatrix} A_{hx} & A_{hxy} & 0 \\ A_{hxy} & A_{hy} & 0 \\ 0 & 0 & A_v \end{bmatrix}, \quad [7.15]$$

where:

$$\begin{aligned} A_{hx} &= A_h^{(2D)} L \sin^2\alpha + 2A_e \cos^2\alpha. \\ A_{hy} &= A_h^{(2D)} L \cos^2\alpha + 2A_e \sin^2\alpha \\ A_{hxy} &= \left(-A_h^{(2D)} L + 2A_e\right) \cos\alpha \sin\alpha \\ A_v &= A_v^{(2D)} L. \end{aligned} \quad [7.16]$$

Here, the end surfaces are also accounted for, the added mass due to an acceleration in axial direction is $2A_e$. The angle of the pontoon relative to the x-axis is given by $\alpha = \arctan\left(\frac{y_2 - y_1}{x_2 - x_1}\right)$ and L is the length of the pontoon.

Considering rotational acceleration around the x-axis, and the resulting moments about the other axes ΔL, the following contributions are obtained:

$$\begin{aligned}\Delta M_{11} &= -\Delta F_y z + \Delta F_z y = -\Delta F_h \cos\alpha\ z + \Delta F_z y \\ &= -a_h A_h^{(2D)} \Delta L \cos\alpha\ z + a_v A_v^{(2D)} \Delta L y \\ &= \ddot{\eta}_4 z \cos\alpha\ A_h^{(2D)} \Delta L \cos\alpha\ z + \ddot{\eta}_4 y A_v^{(2D)} \Delta L y \\ &= [z^2 A_h^{(2D)} \cos^2\alpha + y^2 A_v^{(2D)}] \ddot{\eta}_4 \Delta L.\end{aligned} \tag{7.17}$$

Here, ΔM_{11} is the moment around the x-axis from a small section of the pontoon of length ΔL located at (x, y, z) due to an acceleration around the x-axis, $\ddot{\eta}_4$. Similar considerations are made for the other moments. The contributions from each section are integrated over the length of the pontoon. The result is a symmetric rotational inertia matrix, $I_{ij(L)}$ due to the sectional added mass of the pontoon (details are given in Appendix D):

$$I_{11(L)} = A_h^{(2D)} L\ z_p^2 \cos^2\alpha + A_v^{(2D)} L \left(y_p^2 + \frac{1}{12} L^2 \sin^2\alpha \right) \tag{7.18}$$

$$I_{21(L)} = -A_v^{(2D)} L \left(x_p y_p + \frac{1}{12} L^2 \cos\alpha \sin\alpha \right) + A_h^{(2D)} L z_p^2 \cos\alpha \sin\alpha \tag{7.19}$$

$$I_{31(L)} = -A_h^{(2D)} L\ z_p \cos\alpha \left(y_p \sin\alpha + x_p \cos\alpha \right) \tag{7.20}$$

$$I_{22(L)} = A_v^{(2D)} L \left(x_p^2 + \frac{1}{12} L^2 \cos^2\alpha \right) + A_h^{(2D)} L z_p^2 \sin^2\alpha \tag{7.21}$$

$$I_{32(L)} = -A_h^{(2D)} L\ z_p \sin\alpha \left(y_p \sin\alpha + x_p \cos\alpha \right) \tag{7.22}$$

$$I_{33(L)} = A_h^{(2D)} L \left[\left(x_p \cos\alpha + y_p \sin\alpha \right)^2 + \frac{1}{12} L^2 \right]. \tag{7.23}$$

(x_p, y_p, z_p) is the volume center of the pontoon. The end surfaces will only experience pressure in axial direction and only if they are wetted. However, if the pontoon is attached to a column, as illustrated in Figure 6.5, one will normally ignore the effect of the pontoon when considering the added mass of the column. Thus, one should evaluate if the total added mass is better represented by considering the pontoon ends to be wet or dry. This may be done by using a 3D panel method. The contributions from a wetted end to the rotational inertia are given in Appendix D. The total rotational added mass matrix thus becomes:

$$\mathbf{I}_{p3*3} = \mathbf{I}_{(L)} + \mathbf{I}_{(e)} \ . \tag{7.24}$$

The total 6 x 6 added mass matrix for one horizontal pontoon becomes thus:

$$\mathbf{A}_{pon} = \begin{bmatrix} \mathbf{A}_p & -\mathbf{A}_p{}^* \mathbf{S}_p \\ \mathbf{A}_p{}^* \mathbf{S}_p & \mathbf{I}_{p3*3} \end{bmatrix}, \tag{7.25}$$

with:

$$\mathbf{S}_p = \begin{bmatrix} 0 & -z_p & y_p \\ z_p & 0 & -x_p \\ -y_p & x_p & 0 \end{bmatrix}. \qquad [7.26]$$

Added Mass of a Horizontal Pontoon

Consider a horizontal pontoon with length $L = 30\ m$, *width* $B = 5\ m$ *and height* $H = 3\ m$. *The axis of the pontoon is 8.5 m below the free surface. The 2D added mass in vertical and horizontal direction for the pontoon section is estimated to be* $A_v^{(2D)} = 1.67HB$ *and* $A_h^{(2D)} = 0.72HB$ *respectively. The added mass of the end sections,* A_e, *is ignored in this example. The pontoon is located with one end of the axis at* $(x = 10\ m,\ \ y = 0\ m,\ \ z = -8.5\ m)$. *The angle between the pontoon axis and the x-axis is 30 deg; see Figure 7.2.*

The 6 × 6 added mass matrix is computed using the strip theory approach as well as using a 3D boundary element method as described in Section 6.4. The distribution of the quadrilateral, constant potential boundary elements are shown by the black lines in Figure 7.2. Note that the panel sizes are reduced toward the edges of the pontoon. This is to improve the computational accuracy. The 3D method accounts for the free surface effect. The added mass thus becomes frequency-dependent. In Table 7.1, the added mass matrix as obtained by strip theory as well as 3D results at a low frequency (0.087 Hz) and a high frequency (0.5 Hz) are presented. The matrix is symmetric. In general, the strip theory method and 3D results do not differ much. One exception is A_{11}, *which is sensitive to the added mass related to the end surfaces. This effect was ignored in the strip theory method example.*

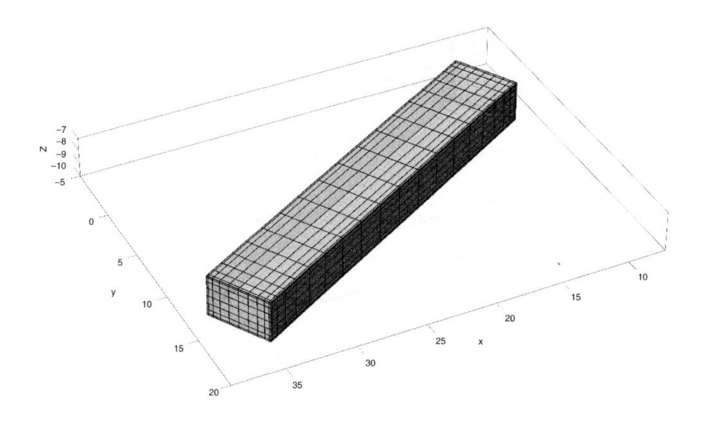

Figure 7.2 The horizontal pontoon used in the example. Quadrilateral panels as used in the 3D boundary element method.

(cont.)

Table 7.1 *Nondimensional added mass for the pontoon shown in Figure 7.2 as computed by strip theory and at a low frequency (0.087 Hz) and a high frequency (0.5 Hz) using a 3D panel method. The added mass values are made dimensionless in the following way:* $\widetilde{A}_{ij} = \frac{A_{ij}}{\rho B^{\gamma}}$, *with* $\gamma = 3$ *for* $(i,j) = (1:3, 1:3)$, $\gamma = 4$ *for* $(i,j) = (1:3, 4:6)$ *and* $(i,j) = (4:6, 1:3)$ *and* $\gamma = 5$ *for* $(i,j) = (4:6, 4:6)$. $B = 5$ *m.*

	i / j	1	2	3	4	5	6
Strip	1	0.650	-1.126	0.000	-1.913	-1.105	-6.150
3D high	1	0.871	-0.959	0.000	-1.648	-1.448	-5.717
3D low	1	0.937	-1.015	0.000	-1.696	-1.647	-6.071
Strip	2		1.949	0.000	3.314	1.913	10.652
3D high	2		1.979	0.000	3.351	1.648	10.537
3D low	2		2.109	0.000	3.605	1.696	11.218
Strip	3			6.001	9.002	-27.594	0.000
3D high	3			5.740	8.610	-26.394	0.000
3D low	3			6.381	9.571	-29.340	0.000
Strip	4				23.637	-45.934	18.108
3D high	4				22.254	-42.810	17.878
3D low	4				24.382	-47.567	19.120
Strip	5					142.260	10.455
3D high	5					134.417	9.749
3D low	5					149.038	10.268
Strip	6						66.000
3D high	6						63.498
3D low	6						67.291

7.2.2.3 Horizontal Disks

In some cases, the substructures are equipped with horizontal plates of almost circular shape and with small thickness (as discussed in Section 4.4.1). The reason for using such plates is to tune the dynamic behavior of the platform. The plates will add inertia to the system, thus moving the natural periods in heave, roll and pitch to higher values. At the same time, plates with sharp edges will contribute to viscous damping and thus reduce the motion response in the resonant domain. To improve the damping properties, perforation of the plates is an option. A perforation will, however, reduce the added mass effect of the plate (Molin and Nielsen, 2004).

The added mass of a circular disk with radius R oscillating in infinite fluid is given by Lamb (1975, 144):

$$A_n = \frac{8}{3}\rho R^3. \qquad [7.27]$$

In most cases, the plate will be located at the bottom of a vertical column. In such cases the added mass will be somewhat smaller, depending upon the ratio of the disk radius to the column radius (see discussion on vertical columns in Section 7.2.2.1).

Figure 7.3 shows examples of the importance of the perforation to the added mass and linearized damping. The figures are from Molin and Nielsen (2004). The nondimensional added mass and damping is presented as a function of the "porous Keulegan–Carpenter number":

$$KC_{por} = \frac{1-\tau}{2\mu\tau^2}\frac{A}{R}. \qquad [7.28]$$

Here, τ is the perforation ratio (open area divided by total area of disk) and μ is the "discharge ratio", relating the pressure drop over the disk and the relative fluid velocity through the disk. It is thus related to the flow resistance through the disk, which again is dependent upon the local geometry of the perforation. μ usually has a

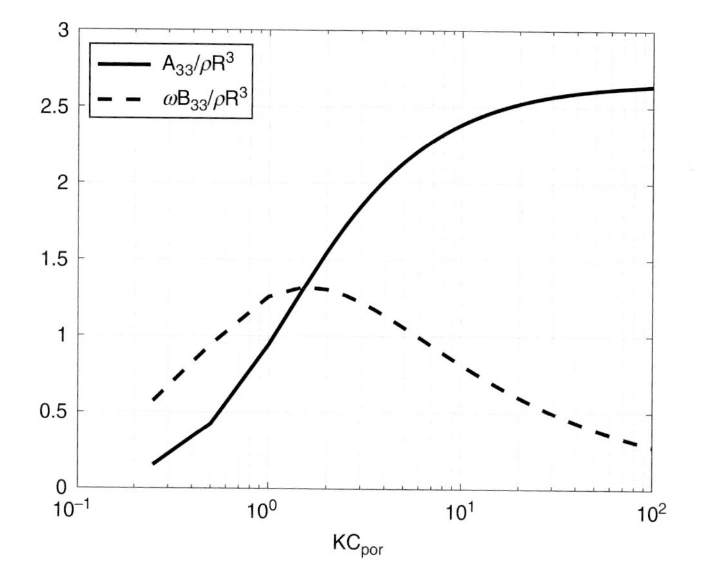

Figure 7.3 Added mass and linearized damping for a perforated disk as a function of the "porous Keulegan–Carpenter number," KC_{por}. Period of oscillation 20 s, water depth 100 m, radius of disk 10 m and submergence of disk 20 m. According to theory as described by Molin and Nielsen (2004).

value between 0.5 and 1.0. Molin (2011) discusses various approaches to estimate the discharge ratio. It is observed from Figure 7.3 that for small KC_{por}, the added mass as well as the damping tends to zero. This case corresponds to a situation with a very large perforation area, $\tau \rightarrow 1$. On the one hand, as $\tau \rightarrow 0$ the added mass tends toward the solid disk value of [7.27]. The computed damping tends to zero because the damping due to the edge effect of the disk is not accounted for in this theory. Including the edge effect (see Molin, 2011), a better agreement with the experiments is obtained for the damping.

7.2.2.4 Transformation of the Added Mass Matrix to a New Coordinate System

Frequently the added mass matrix is computed in a local coordinate system, for example, as referred to the center axis of a column or pontoon. For further analysis a different platform coordinate system may be preferred. The transformation between the two coordinate systems may be done as follows. Denote coordinates in the original (local) coordinate system by $\mathbf{x_0} = (x_0, y_0, z_0)$ and the new (platform) coordinate system by $\mathbf{x}_1 = (x_1, y_1, z_1)$. Assume the two systems are parallel, so that:

$$\Delta \mathbf{x} = \mathbf{x_1} - \mathbf{x_0} = (\Delta x, \Delta y, \Delta z). \qquad [7.29]$$

The kinetic energy in the fluid while oscillating the body in a certain direction must be independent of the coordinate system used. By considering the kinetic energy using the velocity potentials, it can be shown that the 6 x 6 added mass matrix in the new coordinate system, $\mathbf{A_1}$, is related to the added mass matrix in the original coordinate system, $\mathbf{A}_0$, by:

$$\mathbf{A_1} = \mathbf{K^T A_0 K}, \qquad [7.30]$$

where:

$$\mathbf{K} = \begin{bmatrix} \mathbf{I_{3*3}} & \mathbf{K_1} \\ \mathbf{0_{3*3}} & \mathbf{I_{3*3}} \end{bmatrix}. \qquad [7.31]$$

Here:

$$\mathbf{K_1} = \begin{bmatrix} 0 & -\Delta z & \Delta y \\ \Delta z & 0 & -\Delta x \\ -\Delta y & \Delta x & 0 \end{bmatrix}, \mathbf{I_{3*3}} = \begin{bmatrix} 1 & 0 & 0 \\ 0 & 1 & 0 \\ 0 & 0 & 1 \end{bmatrix}, \mathbf{0_{3*3}} = \begin{bmatrix} 0 & 0 & 0 \\ 0 & 0 & 0 \\ 0 & 0 & 0 \end{bmatrix}. \qquad [7.32]$$

Details of the derivation as well as the more general form valid also when rotations are involved may be found in Korotkin (2008).

7.3 Damping

The damping terms in [7.1] consist of several contributions that may be handled independently. The following terms will be discussed in more detail.

- Linear radiation damping, related to the radiated waves.
- Viscous damping, mainly due to flow separation around the hull.
- Aerodynamic damping, due to the wind turbine, and to some extent the wind forces on the tower.

Most floating structures are lightly damped. This means that the damped natural frequencies are not very different from the undamped natural frequencies. This implies that damping in most cases is important to the responses close to the natural frequencies only. However, the damping is generally both frequency-dependent and amplitude-dependent. This makes it difficult to establish accurate damping estimates. Normally, good physical insight as well as engineering experience is required to come up with realistic damping estimates. Frequently, model testing is applied to study the motion behavior of floating structures. If the tested structure is sensitive to resonant motion, model test results should be interpreted with great care as viscous damping normally is overestimated in model scale as compared to full scale.

7.3.1 Radiation Damping

Radiation damping is considered to be a linear damping contribution. For a general, rigid floating structure the damping matrix will be a full 6×6 matrix with frequency-dependent coefficients. To establish this damping matrix, a 3D radiation-diffraction approach is needed (see Section 6.4). A structure's capability to generate waves is reduced if the structure is deeply submerged. This implies that a surface-piercing vertical column generally contributes more to the wave radiation damping than, e.g., a horizontal pontoon. However, in a strip theory approach, the 2D damping of a pontoon section may be applied to establish an estimate on the damping for the complete pontoon. The horizontal, normal force on the pontoon due to a harmonic motion $\eta_n = \eta_{An} e^{i\omega t}$ normal to a section of the pontoon may be written as:

$$F_{pn}(\omega) = \int_L [A_h^{(2D)}(\omega)(-\omega^2) + B_{rh}^{(2D)}(\omega)(i\omega)] dL \; \eta_n$$

$$= \int_L \left[A_h^{(2D)}(\omega) + \frac{1}{i\omega} B_{rh}^{(2D)}(\omega) \right] dL (-\omega^2 \eta_n) \; . \qquad [7.33]$$

The subscript r indicates radiation damping. In [7.33] it is indicated that both the added mass and damping are frequency-dependent. The radiation effect will only account for waves radiated perpendicular to the pontoon axis. The 6 by 6 damping matrix can now be established similarly as shown for the added mass matrix. A strip theory approach accounts neither for the interaction of the radiated waves from each of the pontoon strips, nor for the interaction between the pontoons. The interaction effects may in some cases be significant for some frequencies and directions of oscillation.

Within the context of ideal fluid flow and linear wave dynamics, there exists a reciprocity relation that relates the wave forces on a fixed body to the forces needed to oscillate the body in otherwise calm water. This is called the Haskind relation (for further discussion, see Newman, 1977; Faltinsen, 1990). The relation is valid for general 3D bodies. Applying the Haskind relation on a vertical column with a rotational symmetry, simple relations between the wave excitation forces and the diagonal of the damping matrix are obtained:

$$B_{rii}(\omega) = \gamma \frac{k}{\rho g c_g} \left| \frac{F_i}{\zeta_A} \right|^2 . \qquad [7.34]$$

Here, F_i is the wave force in direction i, $i = (1, 3, 5)$ when the waves are propagating along the x-axis. $\gamma = 1/4$ for $i = 1$ and 5 and $\gamma = 1/2$ for $i = 3$. In deep water, [7.34] may be written as:

$$B_{rii}(\omega) = \gamma \frac{\omega^3}{\rho g^3} \left| \frac{F_i}{\zeta_A} \right|^2 . \qquad [7.35]$$

The computation of the wave force on a vertical column is addressed in Chapter 6. Note that for a substructure with several columns, there may be significant wave interaction between the columns, modifying the radiated waves and thus the damping. A summation of the damping contribution from each of the columns will thus cause errors. One should rather make a summation of the radiated wave fields, taking phases properly into account, and estimate the damping based upon the radiated energy. This is what is obtained by using 3D potential theory methods.

The Haskind relation may also be invoked to estimate the radiation damping for horizontal pontoons. Having established the wave excitation force on a segment dL of the pontoon, the corresponding contribution to the damping may be obtained. Newman (1962) derived a relation between the 2D wave force and damping for a long horizontal body in deep water and beam seas. For a segment of the pontoon this relation is identical to [7.35] using $\gamma = 1$ and considering three degrees of freedom: the transverse horizontal direction, the vertical direction and rotation about an axis parallel to the body axis.

7.3.2 Viscous Damping

Viscous damping has contributions from all structural elements where flow separation occurs. Pure skin friction is in most cases so small that it may be disregarded. The viscous force is normally expressed as a quadratic quantity with respect to the relative velocity, i.e., on a short, 2D section of a vertical column, the viscous force may be written as:

$$\Delta F_{visc} = \frac{1}{2}\rho C_D D U_{rel}|U_{rel}|\Delta z \ . \quad [7.36]$$

Here, C_D is the drag coefficient, D is the column diameter and $U_{rel} = v_h - \dot{x}(z)$ is the relative horizontal velocity between water and structure at the z-level considered. Δz is the length of the short vertical section considered. It is observed that the viscous force contributes both to excitation via the v_h^2 term and damping via $\dot{x}_h^2$. Further, there is a coupling term between the two that contributes to damping or excitation depending upon the phase between the wave particle velocity and the motion velocity.

7.3.3 Linearization of Viscous Damping

In linear dynamic analysis there is a need for linearization of the viscous effect. This is in particular the case when accounting for viscous damping in frequency domain analyses. Due to the nonlinear nature of the damping and the coupling to the fluid velocity, i.e., wave particle and current velocities, it is in general not possible to perform a consistent linearization of the viscous damping. However, disregarding the fluid velocities and considering a single-degree-of-freedom (SDOF) system, an equivalent linear damping can be derived as follows. Consider a long slender structure, e.g., a cylinder. Denote the 2D damping force acting normal to a short section of length, dz by $F_B dz$. The force is assumed to be composed of a linear and a quadratic contribution, i.e.:

$$F_B dz = (B_1\dot{x} + B_2\dot{x}|\dot{x}|)dz. \quad [7.37]$$

The body velocity normal to the cylinder axis is assumed to be harmonic, i.e., $\dot{x} = -\omega x_A \sin(\omega t)$. To find the equivalent linear damping B_e, the dissipation of energy over one cycle of oscillation, $T = 2\pi/\omega$, is considered. By requiring the dissipated energy to be the same for the equivalent linear system and the quadratic system, B_e is thus found from:

$$\int_T F_B\dot{x}dt = \int_T [B_1\dot{x} + B_2\dot{x}|\dot{x}|]\dot{x}dt = \int_T B_e\dot{x}^2 dt \ . \quad [7.38]$$

Inserting for $\dot{x}$ and working out the integrals, the equivalent damping is obtained as:

$$B_e = B_1 + \frac{8}{3\pi}\omega x_A B_2 \ . \qquad [7.39]$$

It is observed that the equivalent linear damping is proportional to the velocity amplitude, ωx_A. That implies that an iteration procedure usually must be implemented to establish a proper damping estimate. As the damping is of key importance to the resonant response, one will have to guess a resonant response amplitude, estimate the equivalent damping, then compute the response and correct the damping according to the computed response.

Viscous Damping

Consider the following simple 1D example. A small body is exposed to an oscillating flow given by $v = v_A \exp(i\omega t)$*. The body is moving harmonically in the same direction with a velocity* $\dot{x} = \dot{x}_A \exp\left(i(\omega t + \theta)\right)$*. The relative velocity is thus given by* $U_{rel} = \mathrm{Re}\{v - \dot{x}\}$ *. The viscous force is given from [7.36]. Considering one cycle of oscillation, the average dissipated power becomes:*

$$P = -\frac{1}{T}\int_0^T F_{visc}\dot{x}dt,$$

where $T = 2\pi/\omega$. *In Figure 7.4 the dissipated power is plotted as a function of phasing between the fluid velocity and the body velocity. It is observed that for cases with*

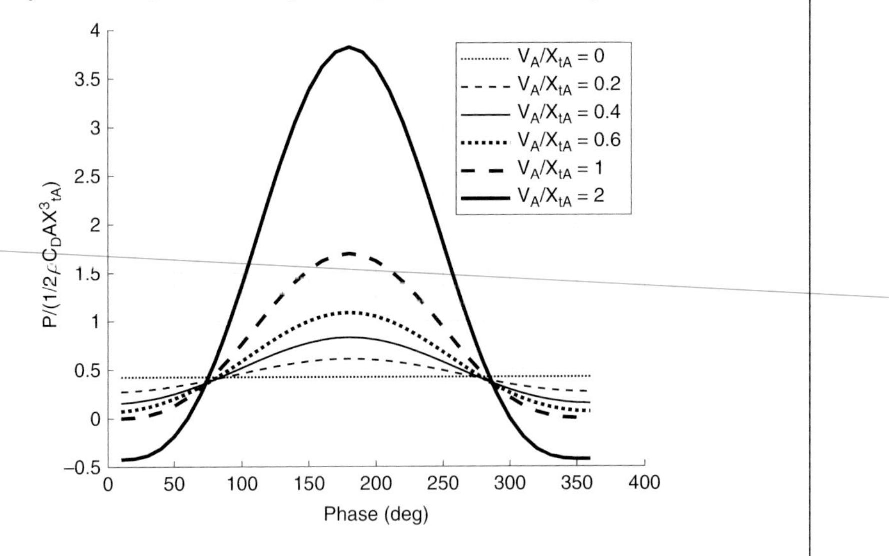

Figure 7.4 Average dissipated power as a function of phase between fluid velocity and body velocity. Amplitude ratio $v_A/\dot{x}_A$ ranging from 0 to 2.

(cont.)

$v_A/\dot{x}_A < 1$, the damping (dissipated power) is positive independent of phasing between the fluid motion and the body motion. However, for $v_A/\dot{x}_A > 1$, the damping may become negative for certain phases, implying an excitation effect. For zero fluid velocity the average dissipated power amounts to $\frac{4}{3\pi}\left[\frac{1}{2}\rho C_D A \dot{x}_A^3\right]$.

The above procedure works fine for a SDOF and in cases where the various modes of motion are uncoupled or close to uncoupled. For most substructures the heave mode has little coupling to other modes, while, for example, the surge and pitch modes may have significant coupling. Frequently the surge motion is referred to the waterline level, while the eigenmode for pitch may have a center of rotation far below the waterline. This causes a significant coupling between the surge and pitch motion when viscous drag forces are accounted for.

To illustrate this point, consider a spar platform designed as a vertical cylinder with constant diameter and a pure surge motion. The drag forces in surge and pitch may then be written as:

$$F_1(t) = C\int_{z_b}^{z_t} \dot{x}_1|\dot{x}_1|dz = C\dot{x}_1|\dot{x}_1|(z_t - z_b) = C\dot{x}_1|\dot{x}_1|L$$

$$F_5(t) = C\int_{z_b}^{z_t} z\,\dot{x}_1|\dot{x}_1|dz = C\dot{x}_1|\dot{x}_1|\frac{\left(z_t^2 - z_b^2\right)}{2}\ . \qquad [7.40]$$

Here, $z_t = 0$ and $z_b = -L$ are the top and bottom coordinates of the cylinder. $C = 1/2\rho C_D D$, with D being the diameter of the cylinder. Computing the dissipated energy as above, the linearized damping in surge is obtained as:

$$B_{11lin} = \frac{8}{3\pi}\dot{x}_{1A}CL\ . \qquad [7.41]$$

Similarly, integrating the pitch moment over one cycle of oscillation and comparing the quadratic and the linear process, a linearized coupling term between the surge motion and pitch moment is obtained as:

$$B_{51lin} = \frac{8}{6\pi}C\dot{x}_{1A}\left(z_t^2 - z_b^2\right)\ . \qquad [7.42]$$

The above approach may be repeated for a pure pitch motion, with the pitch motion referred to $z = 0$. The surge and pitch forces corresponding to [7.40] now become:

$$F_1(t) = C\int_{z_b}^{z_t} z\dot{x}_5|z\dot{x}_5|dz = C\dot{x}_5|\dot{x}_5|\frac{\left(z_t^3 - z_b^3\right)}{3}$$
$$F_5(t) = C\int_{z_b}^{z_t} z\, z\dot{x}_5|\dot{x}_5|dz = C\dot{x}_5|\dot{x}_5|\frac{\left(z_t^4 - z_b^4\right)}{4} \quad . \qquad [7.43]$$

The linearized damping coefficients for the pure pitch motion are obtained as:

$$B_{15lin} = \frac{8}{9\pi}C\dot{x}_{5A}\left(z_t^3 - z_b^3\right)$$
$$B_{55lin} = \frac{2}{3\pi}C\dot{x}_{5A}\left(z_t^4 - z_b^4\right) \quad . \qquad [7.44]$$

From the above relations it is observed that the linearized damping depends upon the choice of surge and pitch velocity amplitude used as basis for the linearization. If one focuses on a good linearization of the pitch damping at the pitch natural period, the coupling effect will cause damping also in surge that may be unrealistic. To succeed in linearization of the damping, one should aim at reducing the coupling terms in the damping matrix as much as possible. This is normally obtained by using a coordinate system in which the modes of motions are close to the eigen-modes of the system.

Viscous Damping in Coupled Motion

Consider a vertical cylinder with length equal to draft 100 m and diameter 10 m. Center of gravity is at -70 m. The 2D added mass and drag coefficients are both set to 1.0. A horizontal mooring system with stiffness 50 kN/m is attached at the waterline level. The natural periods in surge and pitch are 118.6 and 17.70 s. The pitch eigenmode has a center of rotation at z = -61.5 m. The linearized coupled damping matrix has been established by assuming a surge amplitude of 0.7 m and a pitch amplitude of 0.5 deg. The system is set into free oscillations in calm water. The initial surge amplitude is 1.0 m, while the initial pitch and all initial velocities are set to zero. Two cases are considered, one using the quadratic damping and one using the linearized damping matrix. Figure 7.5 shows the results for the two cases.

(cont.)

It observed that the surge motion is well reproduced using the linearized damping (upper-left), even if the surge damping force contains large contributions from the pitch motion (lower-left). Initially, the pitch motion obtained by the linearized equations follows the motions obtained by using quadratic damping well (upper-right). This is because the inertia effects dominate initially. After a while, however, the pitch motion is more and more dominated by the surge natural period in the linearized case. Large differences are also observed in the pitch drag moment (lower right).

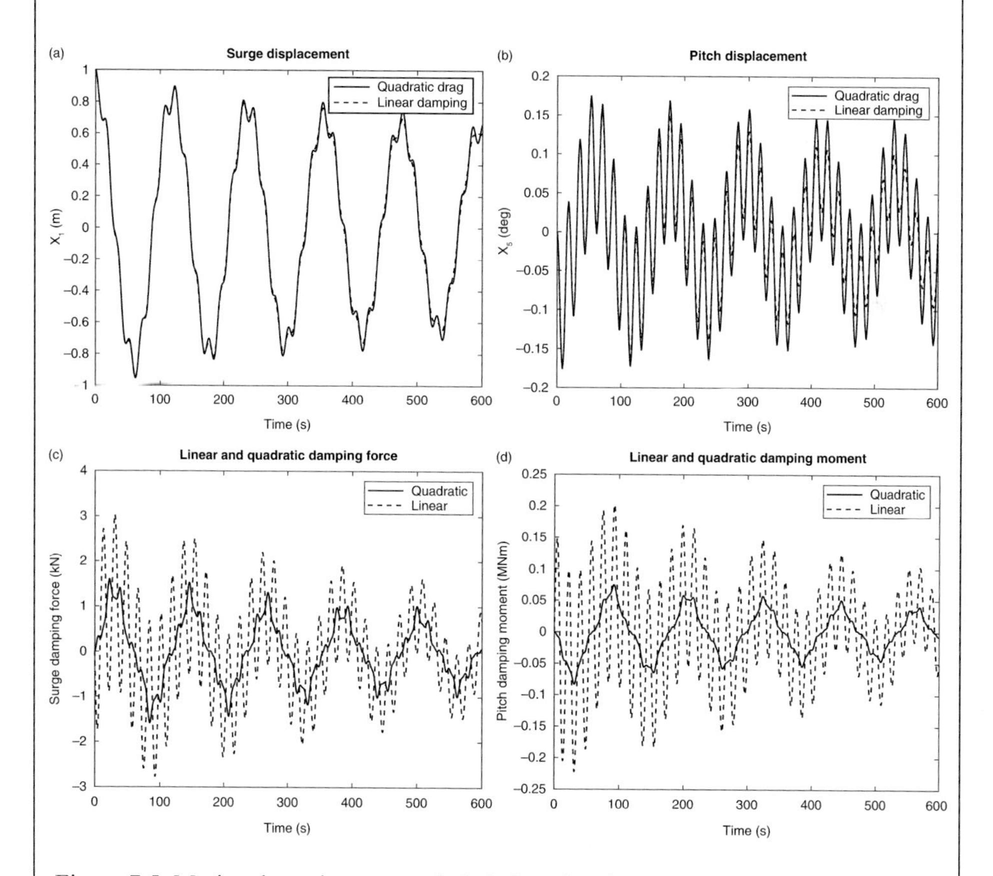

Figure 7.5 Motion decay in surge and pitch for a floating vertical circular cylinder using quadratic and linear damping. Upper figures: displacements after an initial surge of 1.0 m and zero pitch; lower figures: damping force in surge and moment in pitch.

7.3.4 The Drag Coefficient

In most practical cases, the viscous forces are related to the pressure distribution over the structure due to flow separation. That implies that the drag coefficient, C_D, depends upon the body geometry, including surface roughness as well as flow conditions. The flow conditions are expressed via three nondimensional numbers: the Reynolds number, $\mathrm{Re} = \frac{UD}{\nu}$; the Keulegan–Carpenter number, $KC = \frac{U_A T}{D}$; and the relative current number, $= U_c/U_A$. Here, U is a characteristic flow velocity; U_A is the amplitude of the oscillatory velocity, either of the body or the flow; U_c is a steady current velocity; D is a characteristic cross-sectional dimension of the body; ν is the kinematic viscosity of the fluid; and T is the period of oscillation. Thorough discussions of the relations between these parameters and the drag coefficient are given in, e.g., Sarpkaya and Isacsson (1981) and Faltinsen (1990). Recommended values to be used are found in, e.g., DNV (2021c).

For circular cylinders the drag coefficient is sensitive to where flow separation takes place, which again is sensitive to all the above parameters. For cross-sections with a rectangular shape, the drag coefficient is less dependent upon the flow conditions as flow separation occurs at the sharp corners. Classical results for the drag coefficient for a 2D circular cylinder in steady flow as a function of the Reynolds number are shown in Figure 7.6. A drop in the drag coefficient for

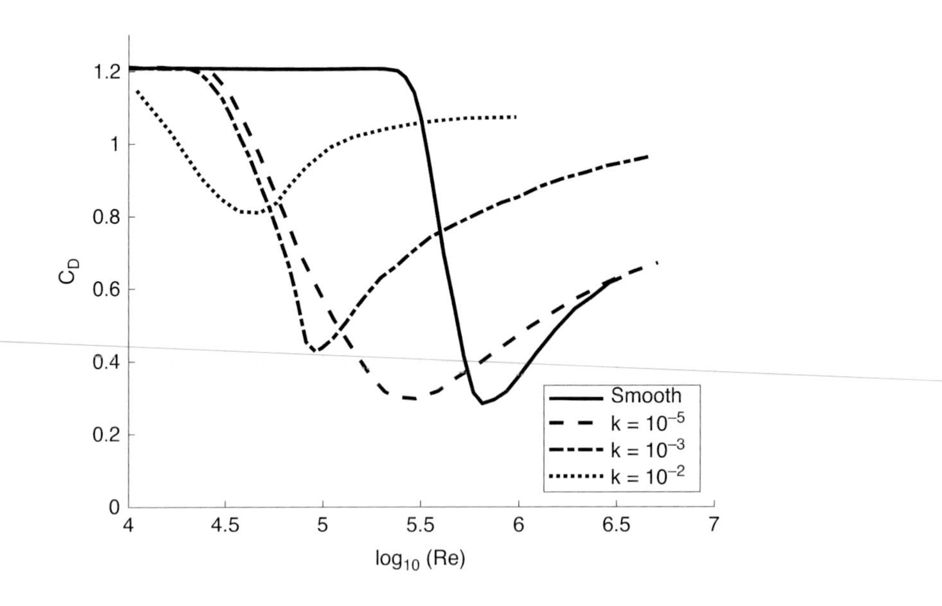

Figure 7.6 Drag coefficient for a 2D circular cylinder in steady flow as a function of the Reynolds number and surface roughness k. Reproduced from DNV (2021c).

the Reynolds number in the order of 10^5 is observed. As the surface roughness of the cylinder increases, the drop occurs at a lower Reynolds number, and is less than for a smooth cylinder.

7.4 Wave Excitation Forces

7.4.1 *Slender Bodies of General Shape*

The estimation of wave excitation forces on floating substructures is now to be addressed. As for the discussion on the added mass coefficients above, structures composed of slender vertical cylinders and a horizontal pontoon using strip theory will be addressed. One of the advantages with this approach is that it is straightforward to use in a finite element analysis of the structure based upon beam elements. However, the global forces are focused upon here as these are needed for estimating the rigid-body motions. Some floating substructures may have a barge-like shape (see Section 4.4.4). To estimate the wave forces on such structures, 3D methods as discussed in Chapter 6 should be used.

As for the added mass, the forces need to be referred to a common point of reference. Further, by using the strip theory approach, it is assumed that the flow over any cross-section of the columns or pontoons may be considered to be 2D, even if the cross-sectional dimensions are changing. No hydrodynamic interaction is assumed between the various structural components.

In computing the six degrees of freedom of rigid-body wave forces, it may be convenient to refer to a coordinate system located at the mean sea surface, with $z = 0$ at the surface level and positive upward.

7.4.2 *Wave Forces on a Vertical Column*

Consider regular waves propagating in direction β relative to the x-axis. The complex wave potential may, see Chapter 2, be written as:

$$\phi = \frac{ig\zeta_A}{\omega}\frac{\cosh[k(z+d)]}{\cosh(kd)}e^{i(\omega t - kx\cos\beta - ky\sin\beta)}. \qquad [7.45]$$

There are two options to estimate the wave force on a vertical circular column. One may either assume a very slender column, with no diffraction effects, and apply the Morison equation or one may include diffraction effects and apply the MacCamy and Fuchs theory. Both these approaches are discussed in Chapter 6. However, the expressions need to be modified to account for the fact that the column does not extend to the sea floor. Using a strip theory approach, this implies that the sectional force is integrated from the bottom to the top of the column, i.e., from $z = z_b$ to $z = z_t$. $(z_b < z_t < 0)$. It is assumed that the column axis is located in (x_c, y_c). Similarly as for the monopile, the surge and sway forces are now obtained as:

$$F_1 = \pi\rho g R^2 \zeta_A C_m \left\{ \frac{\sinh(ks_t) - \sinh(ks_b)}{\cosh(kd)} \right\} e^{i(\omega t + \delta - kx_c \cos\beta - ky_c \sin\beta)} \cos\beta.$$

$$F_2 = \pi\rho g R^2 \zeta_A C_m \left\{ \frac{\sinh(ks_t) - \sinh(ks_b)}{\cosh(kd)} \right\} e^{i(\omega t + \delta - kx_c \cos\beta - ky_c \sin\beta)} \sin\beta. \qquad [7.46]$$

Here, C_m and δ are given in [6.15], $s_t = z_t + d$ and $s_b = z_b + d$. It is observed that the forces have an extra phase shift as the column is offset from $x = y = 0$. The vertical force may be estimated using the pressures from the undisturbed wave, the Froude-Krylov pressure at the bottom and top surfaces of the column, i.e.:

$$F_3 = -\pi\rho R^2 \left\{ \gamma_b \frac{\partial\phi(z_b)}{\partial t} - \gamma_t \frac{\partial\phi(z_t)}{\partial t} \right\}$$

$$= \pi\rho g R^2 \zeta_A \left\{ \frac{\gamma_b \cosh(ks_b) - \gamma_t \cosh(ks_t)}{\cosh(kd)} \right\} e^{i(\omega t - kx_c \cos\beta - ky_c \sin\beta)} . \qquad [7.47]$$

If the column is surface-piercing, $z_t = 0$, there is no wave pressure on the top end and $\gamma_t = 0$. Similarly, if the column is sitting on the bottom, $\gamma_b = 0$. For wetted end surfaces, $\gamma = 1$. Note that a bottom-fixed vertical cylinder piercing the free surface is not exposed to vertical wave forces.

The moments about the x- and y-axes are obtained similarly as in [6.15] and [6.16]; accounting for the horizontal offset, the direction of the waves and that the moment axis is now at the free surface level, the roll and pitch moments are obtained as:

$$F_4 = \pi\rho g R^2 \zeta_A C_m \frac{1}{k} \left\{ \frac{-kz_t \sinh(ks_t) + kz_b \sinh(ks_b) + \cosh(ks_t) - \cosh(ks_b)}{\cosh(kd)} \right\} \cdot.$$

$$e^{i(\omega t + \delta - kx_c \cos\beta - ky_c \sin\beta)} \sin\beta + F_3 y_c$$

$$F_5 = -\pi\rho g R^2 \zeta_A C_m \frac{1}{k} \left\{ \frac{-kz_t \sinh(ks_t) + kz_b \sinh(ks_b) + \cosh(ks_t) - \cosh(ks_b)}{\cosh(kd)} \right\} \cdot$$

$$e^{i(\omega t + \delta - kx_c \cos\beta - ky_c \sin\beta)} \cos\beta - F_3 x_c \qquad [7.48]$$

The last term in the above expressions is due to the moment contribution from the vertical wave force on the column. Note that in the deep-water case, $d \rightarrow \infty$, $\sinh(ks)/\cosh(kd) \rightarrow \cosh(ks)/\cosh(kd) \rightarrow e^{kz}$.

The moment around the z-axis, the yaw moment, is obtained from the horizontal forces:

$$F_6 = -F_1 y_c + F_2 x_c \ . \qquad [7.49]$$

All the above expressions are valid for one single column. If several columns are present, the total force is obtained by summation over all the columns. If a column diameter is changing over the length of the column, a pragmatic approach is to split the column into, e.g., two parts and compute the force on each of the parts separately. This is illustrated in Figure 7.7. The split may be done into two or more parts. To obtain a realistic model, the body volume should be conserved. The vertical wave force at the conical part of the column may be modeled by the wave pressure at the area representing the difference between the cross-sectional area of the cylinders. The modeling of this force may be improved by representing the conical section by more cylinders.

If the distance between the columns is not large compared to the diameter of the columns, the interaction effect may be important. In such cases, a full 3D analysis should be performed to obtain accurate estimates on the wave forces.

7.4.3 Wave Forces on a Horizontal Pontoon

Horizontal pontoons in most cases either have a circular or a rectangular cross-section. In the case of a rectangular cross-section the added mass coefficient in horizontal and vertical directions differs. Consider the horizontal

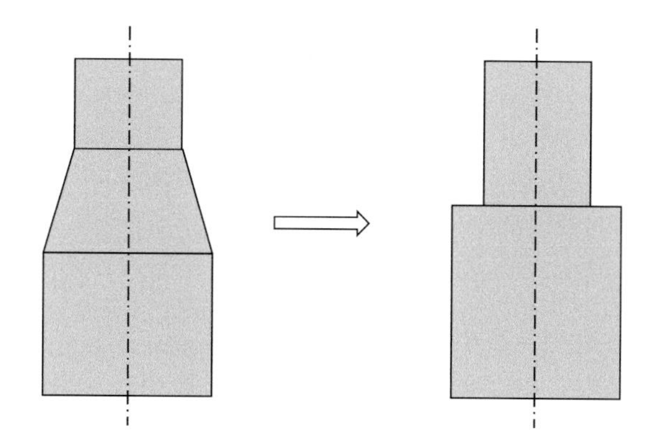

Figure 7.7 Vertical column with conical section modeled by two cylindrical sections.

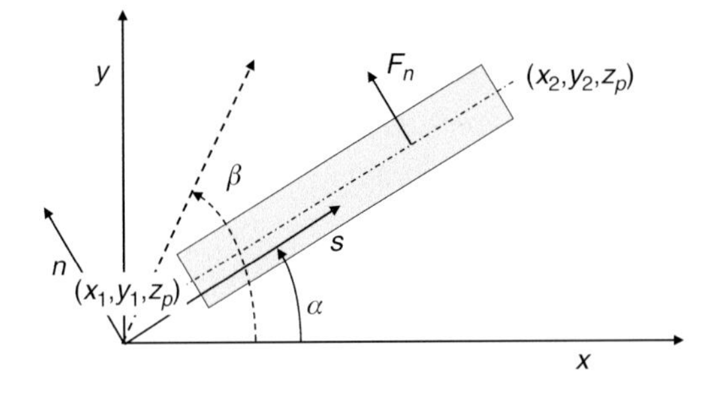

Figure 7.8 A horizontal pontoon. Notations used in deriving the wave forces. α is the direction of the pontoon axis relative to the coordinate system used for the body. (x_1, y_1, z_p) and (x_2, y_2, z_p) are the coordinates of the end points. β is the direction of wave propagation. (s, n, z) are the local pontoon coordinates, parallel and perpendicular to the pontoon axis. The (x, y) and (s, n) planes coincide.

pontoon illustrated in Figure 7.8. A slender body is assumed, implying that the length of the pontoon is much longer than the characteristic cross-sectional dimension. Further, long wavelength theory is used, implying that the wavelength is much longer than the characteristic width of the pontoon. Following the principles outlined in Faltinsen (1990), the vertical and horizontal forces on a 2D section of length ΔL may be written as:

$$
\begin{aligned}
\Delta F_n &= \left[\rho A_p + A_n^{(2D)}\right] a_n \Delta L \\
\Delta F_v &= \left[\rho A_p + A_v^{(2D)}\right] a_v \Delta L.
\end{aligned} \qquad [7.50]
$$

Here, A_p is the cross-sectional area of the pontoon; $A_n^{(2D)}$ is the 2D added mass in horizontal direction, normal to the pontoon axis; $A_v^{(2D)}$ is the 2D added mass in vertical direction; a_n and a_v are the acceleration in the water horizontally, normal to the pontoon axis and in vertical direction respectively.

To obtain the total forces on the pontoon, the forces in [7.50] have to be integrated over the length of the pontoon. To perform this integration, it is convenient to introduce the local (s, n) coordinates, as illustrated in Figure 7.8. The relations between the two coordinate systems are:

$$
\begin{aligned}
s &= x \cos\alpha + y \sin\alpha \\
n &= -x \sin\alpha + y \cos\alpha.
\end{aligned} \qquad [7.51]
$$

The coordinates of the end points of the pontoon axis are thus:

$$
\begin{aligned}
s_2 &= x_2 \cos\alpha + y_2 \sin\alpha \\
s_1 &= x_1 \cos\alpha + y_1 \sin\alpha \\
n_2 &= n_1 = -x_1 \sin\alpha + y_1 \cos\alpha.
\end{aligned} \qquad [7.52]
$$

Considering a pontoon of constant cross-sectional shape, it is only the normal component of the horizontal acceleration, a_n, and the vertical acceleration, a_v in [7.50], that vary along the pontoon length. The horizontal acceleration perpendicular to the pontoon axis may be written as:

$$a_n = -a_x \sin\alpha + a_y \cos\alpha = ia_{nA}[-\cos\beta \sin\alpha + \sin\beta \cos\alpha]e^{i(\omega t - kx\cos\beta - ky\sin\beta)}$$

$$= ia_{nA} \sin(\beta - \alpha)e^{i(\omega t - kx\cos\beta - ky\sin\beta)}$$

$$\text{with } a_{nA} = kg\zeta_A \frac{\cosh\left(k\left(z_p + d\right)\right)}{\cosh(kd)}. \qquad [7.53]$$

Integrating along the pontoon, the following result is obtained for the horizontal force on the pontoon:

$$F_n = \left(\rho A_p + A_n^{(2D)}\right)\int_L a_n dl = \left(\rho A_p + A_n^{(2D)}\right) \sin(\beta - \alpha)ia_{nA}e^{i\omega t}\int_L e^{-i(kx\cos\beta + ky\sin\beta)}dl$$

$$= \left(\rho A_p + A_n^{(2D)}\right) \sin(\beta - \alpha)ia_{nA}e^{i\omega t}\int_L e^{-ik\left(n\sin(\beta-a)+s\cos(\beta-a)\right)}dl$$

$$= \left(\rho A_p + A_n^{(2D)}\right) \sin(\beta - \alpha)ia_{nA}e^{i\omega t}e^{-ik\left(n_1\sin(\beta-a)\right)}\int_{s_1}^{s_2} e^{-ik\left(s\cos(\beta-a)\right)}dl$$

$$= \left(\rho A_p + A_n^{(2D)}\right) \sin(\beta - \alpha)ia_{nA}e^{i\left(\omega t - kn_1\sin(\beta-a)\right)}$$

$$\frac{-1}{ik\cos(\beta - \alpha)}\left[e^{-iks_2\cos(\beta-a)} - e^{-iks_1\cos(\beta-a)}\right]. \qquad [7.54]$$

In the limit $\cos(\beta - \alpha) \to 0$, i.e., the waves are propagating perpendicular to the pontoon axis, the limiting value of the integral is obtained as:

$$\int_{s_1}^{s_2} e^{-ik(s\cos(\beta-a))}dl \to (s_2 - s_1) = L. \qquad [7.55]$$

If the pontoon ends are wetted, a reasonable approximation is to assume that the pressure in the undisturbed wave (the Froude–Krylov pressure) is acting on the surfaces, i.e., the force in axial direction becomes:

$$F_s = A_p[p(s_1) - p(s_2)] = -A_p\rho\left[\frac{\partial\phi(s_1)}{\partial t} - \frac{\partial\phi(s_2)}{\partial t}\right].$$
$$= A_p\rho\frac{1}{k}a_{nA}e^{i(\omega t - kn_1\sin(\beta-\alpha))}[\gamma_1 e^{-iks_1\cos(\beta-\alpha)} - \gamma_2 e^{-iks_2\cos(\beta-\alpha)}]. \qquad [7.56]$$

Here, $\gamma = 1$ for a wetted surface and zero for a dry surface. Frequently, a pontoon is attached to column of larger diameter. The end of the pontoon is then dry. On the other hand, part of the column surface is also dry. It is thus convenient to model both surfaces as wetted. This will almost cancel the global force contribution from the intersection. If local forces are required, this approach will not work.

The vertical force on the pontoon is obtained in a similar way as the horizontal force, i.e., using:

$$F_v = \left(\rho A_p + A_v^{(2D)}\right)\int_{s_1}^{s_2} a_v dl$$
$$= \left(\rho A_p + A_v^{(2D)}\right)a_{vA}e^{i(\omega t - kn_1\sin(\beta-\alpha))}\frac{1}{ik\cos(\beta-\alpha)}[e^{-iks_2\cos(\beta-\alpha)} - e^{-iks_1\cos(\beta-\alpha)}]$$
$$\text{with} \quad a_{vA} = kg\zeta_A\frac{\sinh\left(k\left(z_p + d\right)\right)}{\cosh(kd)}. \qquad [7.57]$$

The forces in the support structure's coordinate system (x, y, z) are obtained as:

$$F_1 = -F_n\sin\alpha + F_s\cos\alpha$$
$$F_2 = F_n\cos\alpha + F_s\sin\alpha \qquad [7.58]$$
$$F_3 = F_v\,.$$

Horizontal Wave Force on Pontoon

An example of the computed horizontal force on a pontoon of length 30 m in a wave of length 15 m is shown in Figure 7.9. The force perpendicular to the pontoon axis is shown. The force is given as a function of the angle between wave propagation and the pontoon axis. By presenting the result in the format $Abs[F_n/\sin(\beta - a)]\ \sin(\beta - a)$*, the sign of the force relative to the pontoon normal axis is retained. It is observed that the extreme forces are obtained for* $(\beta - \alpha) = \pm 90^o$*. Further, zero force is obtained for waves propagating along the pontoon axis. For* $(\beta - \alpha) = \pm 60^o$ *additional zero values appear. For these angles one wavelength will cover the full pontoon length, i.e.,*

(cont.)

$L\cos(\beta-\alpha)=\lambda.$

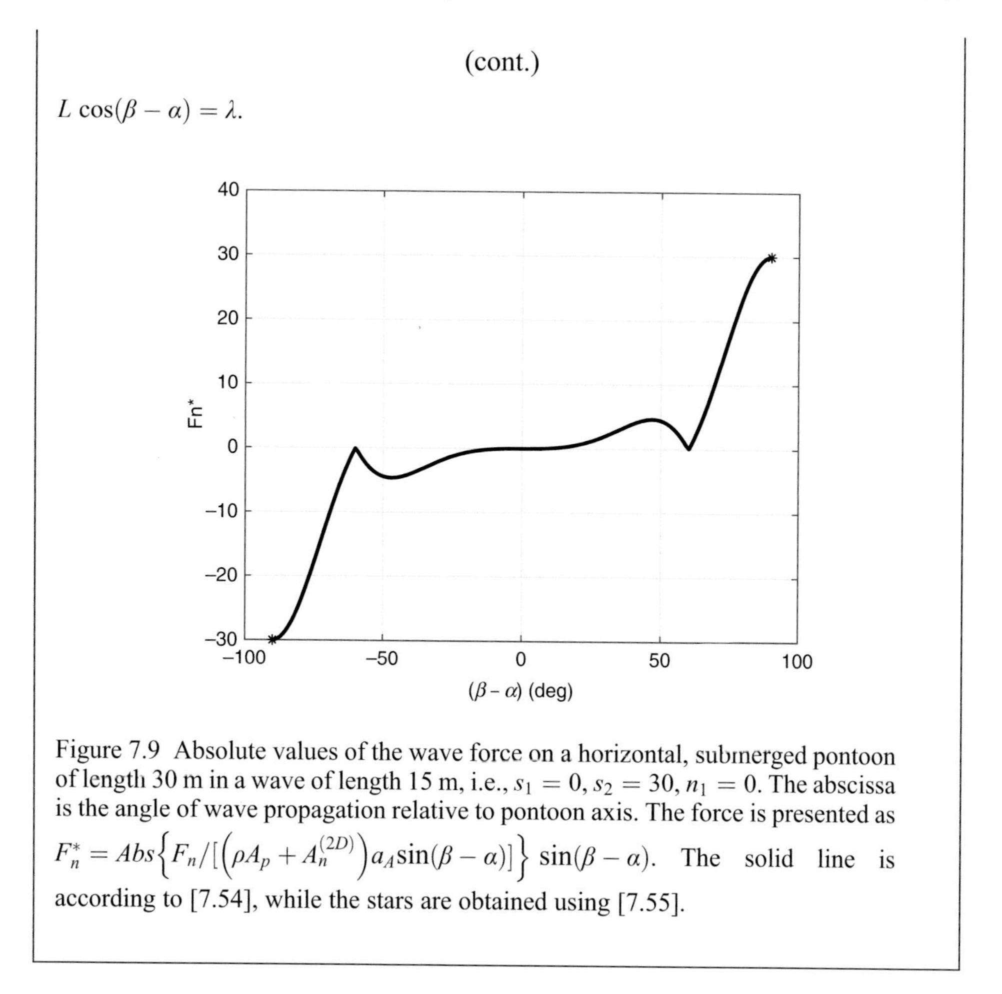

Figure 7.9 Absolute values of the wave force on a horizontal, submerged pontoon of length 30 m in a wave of length 15 m, i.e., $s_1=0, s_2=30, n_1=0$. The abscissa is the angle of wave propagation relative to pontoon axis. The force is presented as $F_n^*=Abs\left\{F_n/\left[\left(\rho A_p+A_n^{(2D)}\right)a_A\sin(\beta-\alpha)\right]\right\}\ \sin(\beta-\alpha)$. The solid line is according to [7.54], while the stars are obtained using [7.55].

7.4.4 Moments Acting on a Horizontal Pontoon

Recall that the (x,y) and (s,n) planes coincide. Similar as for the pontoon forces, the moments about the (s,n,z) axes may be written as:

$$M_n=\left(\rho A_p+A_v^{(2D)}\right)\int_{s_1}^{s_2}(-s)a_v ds+F_s z_p$$

$$M_s=\left(\rho A_p+A_h^{(2D)}\right)\int_{s_1}^{s_2}(-z_p)a_n ds+\left(\rho A_p+A_v^{(2D)}\right)\int_{s_1}^{s_2}n_1 a_v ds. \qquad [7.59]$$

$$M_z=\left(\rho A_p+A_h^{(2D)}\right)\int_{s_1}^{s_2}s a_n ds-F_s n_1$$

It is observed that these expressions resemble those of the forces, with one important difference: the factor s in the integral terms for M_n and M_z. Working out these integrals and relating them to the integrals involved in the force expressions, the moments can be written as:

$$\begin{aligned} M_n &= -KF_v + z_p F_s \\ M_s &= -z_p F_n + n_1 F_v \\ M_z &= KF_n - n_1 F_s. \end{aligned} \quad [7.60]$$

Here, K is given by:

$$K = \frac{s_2 e^{-iks_2 \cos(\beta-\alpha)} - s_1 e^{-iks_1 \cos(\beta-\alpha)}}{e^{-iks_2 \cos(\beta-\alpha)} - e^{-iks_1 \cos(\beta-\alpha)}} + \frac{1}{ik \cos(\beta-\alpha)}. \quad [7.61]$$

Note that K is complex and thus contains phase information. In the coordinate system of the support structure, the moments become:

$$\begin{aligned} F_4 &\equiv M_x = M_s \cos\alpha - M_n \sin\alpha \\ F_5 &\equiv M_y = M_n \cos\alpha + M_s \sin\alpha. \\ F_6 &\equiv M_z \end{aligned} \quad [7.62]$$

7.4.5 Viscous Drag Effects

The viscous forces, as written in [7.36], contain the relative velocity between water and structure. For a slender vertical structure, this reads $U_{rel} = v_h - \dot{x}$. Here, v_h is the horizontal component of the fluid velocity and $\dot{x}$ is the horizontal velocity of the structure. The viscous drag forces are frequently estimated using a strip theory approach, assuming the length of the structure is much larger than the characteristic cross-sectional dimension. The drag force on a strip of a vertical structural member thus becomes, assuming the fluid velocity is larger than the structural velocity:

$$\Delta F_D = \frac{1}{2}\rho D C_D |v - \dot{x}|(v - \dot{x})\Delta z = \frac{1}{2}\rho D C_D [v^2 - 2v\dot{x} + \dot{x}^2]\Delta z \quad \text{for } (v - \dot{x}) > 0. \quad [7.63]$$

v^2 represents an excitation term, while the two remaining terms may represent damping, i.e., a force opposing the motion or an excitation, depending upon the phasing between the velocity components and the relative magnitude between them. In waves, the largest velocities are present close to the free surface, and the largest viscous excitation effects are thus present in this region. At greater depth, the viscous damping effect may be more important. In the above expression, the horizontal relative velocity is used to estimate the normal force. For a slender structural member of general orientation, one should use the relative velocity component normal to the axis of the member in estimating the force. This "cross-flow principle" is normally assumed to hold if the flow direction is between 45 and 90 deg relative to the member axis (DNV, 2021c). In DNV (2021c) additional recommendations on how to handle the viscous drag forces are also given. In Section 6.5.1, the viscous wave forces in the splash zone are discussed. The same effects are experienced on columns of floating structures, with the additional effect of the motion velocity of the structure.

Due to the nonlinearity of the viscous forces, time domain simulations are normally required in cases where the viscous effects play an important role in the forcing.

7.4.6 Cancellation Effects

In the design of floating support structures, the geometric layout can efficiently be utilized to minimize the wave excitation loads at certain frequencies. Consider the simple half of a semisubmersible in Figure 7.10. The half semisubmersible consists of two columns and one pontoon. It is assumed that the columns are sitting on top of the columns. Assume the waves' direction of propagation is perpendicular to the paper plane. The undisturbed pressure in the water, the Froude–Krylov term in the wave excitation pressure, is then constant along the length of the pontoon. The vertical force acting on the semisubmersible is approximately given from the Froude–Krylov pressures acting on the top and bottom of the pontoon multiplied by corresponding areas:

$$F_3 = p_B A_B - p_T A_T. \tag{7.64}$$

Here, A_B and A_T are the wetted area of the bottom and the top of the pontoon respectively. In deep water the pressure is given from $p = \rho g \zeta_A e^{kz+i\omega t}$. Thus, the force becomes zero for a wave number k given by:

$$k = \frac{\ln(A_B/A_T)}{z_T - z_B}. \tag{7.65}$$

The difference between the top and bottom areas is given from the cross-sectional area of the columns. By choosing a suitable column cross-sectional area, pontoon dimensions and submergence, a wanted wave period for cancellation may be obtained. It is observed that this expression also holds if the platform consists of two parallel pontoons. In the case of two parallel pontoons, there will also be a close-to-zero vertical excitation force if the distance between the pontoons is half a wavelength. However, as the zero vertical force corresponds to a wavelength about half the distance between the pontoons, this wavelength will cause a maximum in the roll motion of the structure.

Consider waves propagating in the paper plane (Figure 7.10). If the wavelength is approximately twice the distance between the columns, the horizontal acceleration in the wave acting on the two columns will have opposite phase. Thus, a close-to-zero horizontal excitation force is acting on the platform. It should be noted that wavelengths that correspond to close-to-zero wave excitation forces on the complete structure in many cases correspond to the wavelengths giving the largest internal forces in the structure. This is easily understood by considering the case of opposite phase of the forces on the two columns.

For the spar platform, the lower part of the hull is normally designed with larger diameter than the diameter at the water line (see Figure 7.7). This difference in diameter is required to ensure a sufficient buoyancy while at the same time keeping the natural frequency in heave below the range of wave frequencies. As for the pontoon, the vertical excitation force may be approximated by the Froude–Krylov force on the bottom of the spar minus the vertical component of the Froude–Krylov force acting on the conical part, simplified as illustrated in Figure 7.7 (right). Thus, a cancellation effect of the vertical wave force is obtained for a certain wave frequency. In principle it is possible to design both a semisubmersible and a spar to have a cancellation frequency at the heave natural frequency. Theoretically, this could significantly

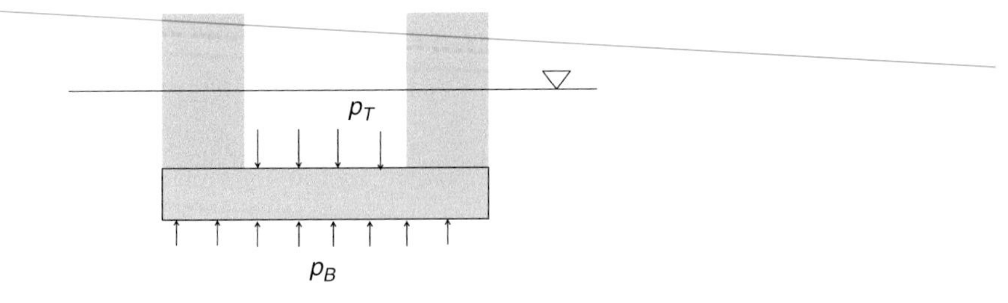

Figure 7.10 Half of a semisubmersible consisting of two columns and one pontoon.

reduce the resonant motions. However, due to other design requirements, this option is not used in practical design.

7.4.7 Wave Forces on Large-Volume Structures: Boundary Element Method

The basic principles for the 3D boundary element method are outlined in Section 6.4. In Table 7.1, the added mass and damping for a horizontal pontoon as computed by strip theory and a 3D boundary element method are compared. In the below example the corresponding wave excitation forces are compared.

One may question why strip theory approaches should be used when full 3D tools are available. There are several reasons for this. Strip theory is much faster, both in establishing the numerical model and performing the computations. This feature makes the method well suited for use in optimization tools. Further, it is easy to identify the added mass and excitation force components related to the various structural components. Further, strip theory is ideal for implementing hydrodynamic forces into a program for global structural analysis of the foundation as the sectional forces are readily available. However, the 3D boundary element technique is superior in computing the hydrodynamic loads for complex structures accounting for interaction phenomena between the various structural components.

Wave Forces on a Horizontal Pontoon

The horizontal pontoon used in the example in Section 7.2.2.2 is considered. The wave forces are computed both using strip theory, using the added mass coefficients from the previous example, and using the 3D boundary element method.

The draft and orientation of the pontoon is as before. Water depth of 100 m is assumed. The waves are propagating in positive x-direction. The real and imaginary part of the wave forces in the six degrees of freedom as a function of frequency is obtained as displayed in Figure 7.11. The solid lines are the real part of the forces as computed by strip theory; the dashed lines are the corresponding imaginary part. The dots and crosses are the results from the 3D boundary element method. The forces are scaled by a factor $\rho g \zeta_A B^2$ *for the linear forces.* ζ_A *is the wave amplitude. The moments, computed around origin, are scaled by* $\rho g \zeta_A B^3$.

A clear cancellation effect is observed for modes 1–3 around 0.25 Hz, corresponding to a wavelength of about 26 m, which is the projected length of the pontoon in the direction of wave propagation.

(cont.)

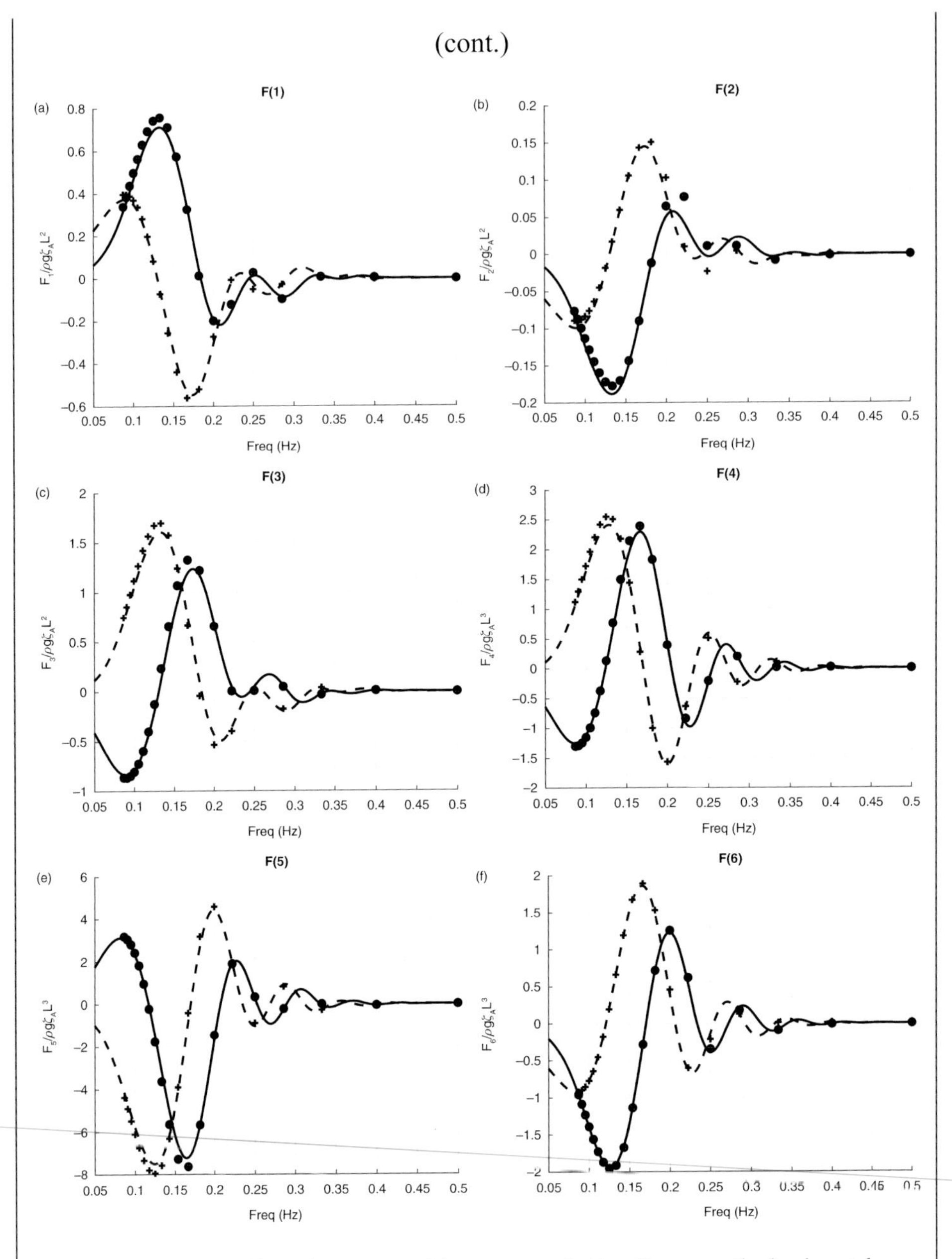

Figure 7.11 Real and imaginary part of the wave excitation forces on the horizontal pontoon shown in Figure 7.2.

7.4.8 Time Domain Simulations with Frequency-Dependent Coefficients

As briefly mentioned in Section 7.1, the hydrodynamic added mass and damping coefficients are frequency-dependent. The frequency dependency of the added mass is frequently ignored if the structure is slender or deeply submerged (see discussion of the Morison equation versus the MacCamy and Fuchs solution in Chapter 6). The frequency dependence of the hydrodynamic coefficients is related to body's capability to generate waves when oscillating. Thus, there exists a relation between the frequency-dependent part of the added mass and the wave radiation damping.

One of the attractive properties of the linear formulation of the hydrodynamic coefficients and excitation forces is the option of solving the equations of motion in the frequency domain. However, even if it may be justified to linearize the hydrodynamic problem, that may not be the case for other parts of the problem such as the aerodynamic loads. The equations of motion for the complete floating wind turbine must thus be solved in time domain. This requires special attention to the frequency-dependent added mass and damping. The problem was addressed by Cummins (1962) and Ogilvie (1964). Falnes (2002) and Naess and Moan (2013) also discuss how the frequency-dependent hydrodynamic coefficients may be transferred to time domain. In time domain, the linear equations of motion may be written (the "Cummins equation") as:

$$(\mathbf{M} + \mathbf{A}_\infty)\ddot{\boldsymbol{\eta}}(t) + \int_0^t \mathbf{K}(t-\tau)\dot{\boldsymbol{\eta}}(\tau)d\tau + C\boldsymbol{\eta}(t) = \mathbf{F}(t). \qquad [7.66]$$

$\mathbf{K}$ is known as the retardation function or the impulse response function. The equation is obtained by a Fourier transform of the linear equations of motion in frequency domain:

$$\{-\omega^2[\mathbf{M} + \mathbf{A}(\omega)] + i\omega\mathbf{B}(\omega) + \mathbf{C}\}\boldsymbol{\eta}(\omega) = \mathbf{F}(\omega). \qquad [7.67]$$

The added mass and damping coefficients are spit into a constant and a frequency-dependent term, $\mathbf{A}(\omega) = \mathbf{A}_\infty + \mathbf{A}'(\omega)$ and $\mathbf{B}(\omega) = \mathbf{B}_\infty + \mathbf{B}'(\omega)$. Here, the index $_\infty$ denotes the asymptotic value as the frequency tends to infinity. For a stationary body, i.e., a body with zero mean forward speed, $\mathbf{B}_\infty = 0$, no waves are created as the frequency of oscillation tends to infinity. The integral term in [7.66] may be regarded as a memory effect, as it contains information of all past time. It may be assumed that the body is as rest for $t < 0$. Further, the causality condition is invoked, i.e., the system cannot react upon future forces. Utilizing symmetry properties of

$\mathbf{A}(\omega)$ and $\mathbf{B}(\omega)$, and the requirement that $\mathbf{K}$ must be real, it can be shown that the retardation function can be written on two different forms, using either the radiation damping or the frequency-dependent part of the added mass:

$$\begin{aligned}\mathbf{K}(t) &= \frac{2}{\pi}\int_0^\infty [\mathbf{B}(\omega)] \cos(\omega t) d\omega \\ &= -\frac{2}{\pi}\int_0^\infty \omega[\mathbf{A}(\omega) - \mathbf{A}_\infty] \sin(\omega t) d\omega.\end{aligned} \quad [7.68]$$

These expressions also show that the damping and the frequency-dependent part of the added mass are both related to the body's ability to radiate waves when oscillating. More details upon these issues are found in Falnes (2002). In principle, it is thus straightforward to obtain the retardation function if the frequency-dependent added mass or damping are known, e. g., from a boundary element panel code analysis. However, such codes have issues related to so-called "irregular frequencies" (see Section 6.4) and low accuracy as the wavelength approaches the size of the panels. Thus, to establish the high-frequency limit of the added mass may involve some challenges. For further discussion of these issues, see Faltinsen (2005).

The convolution integral in [7.66] may be costly to evaluate, in particular for long simulations and thus large t. In practical simulations the integration is truncated. The memory effect is assumed to negligible after some finite time. Various ways to speed up the evaluation of the convolution integral for implementation in state-space simulation models have been suggested. Duarte et al. (2013) compare several methods for approximating the retardation functions and discuss accuracy as well as computational speed.

Figure 7.12 gives examples of the frequency-dependent radiation damping and the corresponding retardation functions. The structure considered is a spar platform with shape as given in Figure 6.10. For such a geometry, it is straightforward to obtain an accurate estimate of the radiation damping. For a semisubmersible with a more complex geometry, this is more demanding. It is observed that the retardation functions as shown in Figure 7.12 tend to zero after a few oscillations. After approximately 30 s, the retardation functions are approximately zero and the memory effect has vanished.

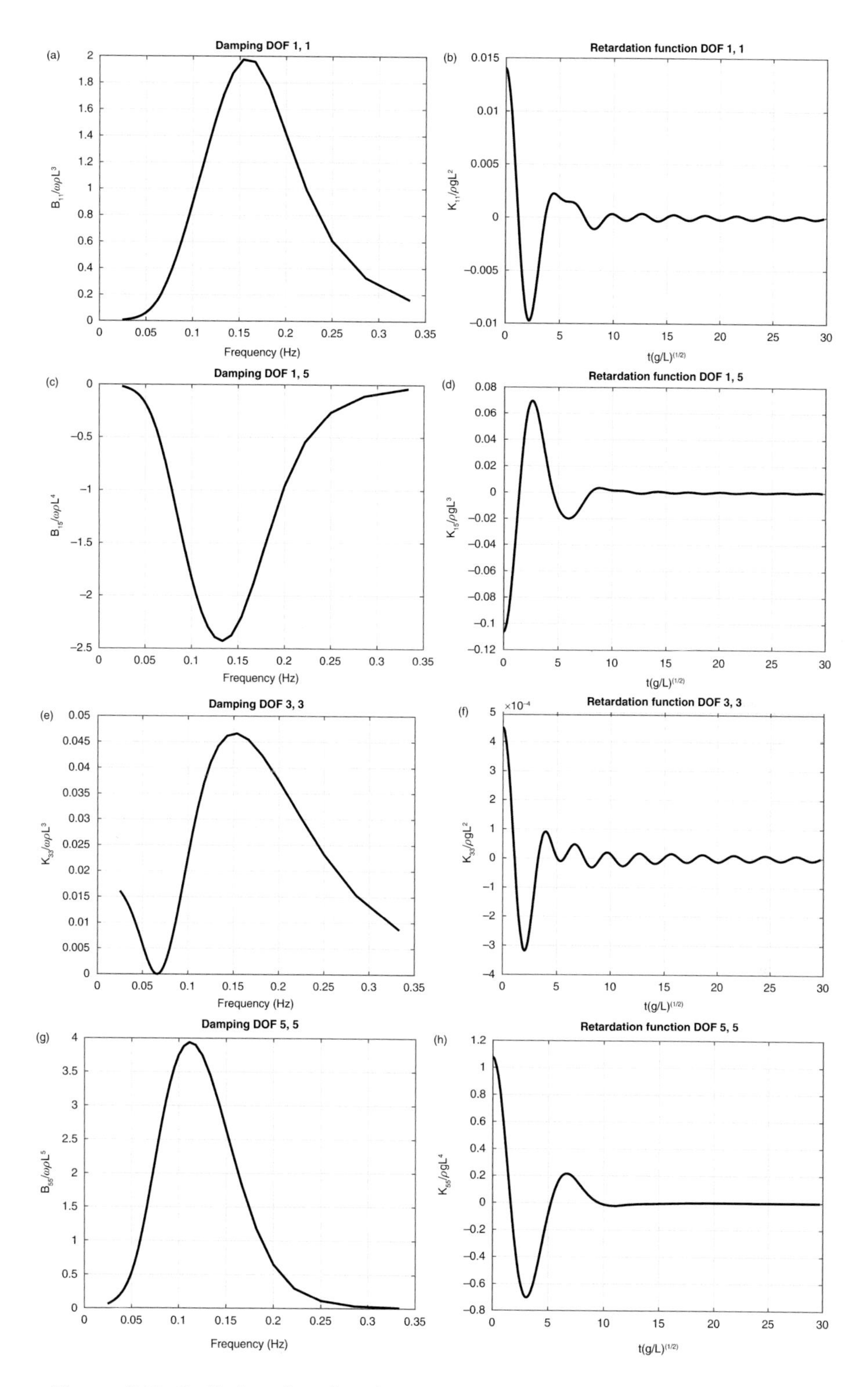

Figure 7.12 Radiation damping for a spar platform with draft of 76 m and maximum diameter of 14.4 m (see Figure 6.10). Left: damping in surge, heave, pitch and coupled surge-pitch; right: corresponding retardation functions. Length parameter used for scaling, $L = 10\ m$.

7.5 Restoring Forces

7.5.1 Hydrostatic Effects

The restoring forces acting on a floating wind turbine substructure are due to the hydrostatic effects and the mooring lines. The hydrostatic forces are, for normal motions, assumed to be linear. The 6 x 6 hydrostatic stiffness matrix may be written as:

$$\mathbf{C}_h = \begin{bmatrix} 0 & 0 & 0 & 0 & 0 & 0 \\ & 0 & 0 & 0 & 0 & 0 \\ & & \rho g S & \rho g S_2 & -\rho g S_1 & 0 \\ & & & \rho g(S_{22} + V z_B) - M g z_G & -\rho g S_{12} & -\rho g V x_B + M g x_G \\ & Sym & & & \rho g(S_{11} + V z_B) - M g z_G & -\rho g V y_B + M g y_G \\ & & & & & 0 \end{bmatrix}. \qquad [7.69]$$

Here, M is the mass of the body; V is the submerged volume of the body; (x_G, y_G, z_G) is the CG of the body; (x_B, y_B, z_B) is the volume center of the submerged body (center of buoyancy, CB); S is the water plane area; $S_i = \int_S x_i dS$ are the first moments of the water plane area, $S_{ij} = \int_S x_i x_j dS$ are the second moments of the water plane area; $x_1 = x$ and $x_2 = y$. The hydrostatic stiffness matrix is symmetric. For a freely floating body $\rho V = M$ and $(x_G, y_G) = (x_B, y_B)$, several of the off-diagonal terms in [7.69] thus become zero. This is, however, not the case if the static mooring forces are significant as compared to the buoyancy force. It is assumed that M is a rigid mass, i.e., there is no fluid that may move inside the body.

7.5.2 Effect of Catenary Mooring Lines

The geometry and loads in catenary mooring lines are discussed in Section 7.6. The restoring forces are generally very dependent upon the pretension and offset of the top end of the mooring line. However, given a certain position of the top end, the linearized contribution to the stiffness matrix from each of the mooring lines may be computed as follows.

Initially the line is assumed to give restoring effects resulting from motion in the plane of the catenary only. The catenary line is assumed to be located in a local (x, z) plane with origin in the upper end of the mooring line. The 3 x 3 restoring matrix for each line $\mathbf{C}^{(l)}$ has thus only the following non-zero elements: $C_{11}^{(l)}, C_{13}^{(l)}, C_{31}^{(l)}, C_{33}^{(l)}$. The local plane is rotated an angle θ about a vertical axis relative to the global coordinate system. The restoring force matrix in a coordinate system parallel to the substructure's coordinates then becomes:

$$\mathbf{C}^{(0)} = \boldsymbol{\gamma}^T \mathbf{C}^{(l)} \boldsymbol{\gamma}. \qquad [7.70]$$

Here, the transformation matrix due to the rotation about the z-axis is given by:

$$\gamma = \begin{pmatrix} \cos(\theta) & \sin(\theta) & 0 \\ -\sin(\theta) & \cos(\theta) & 0 \\ 0 & 0 & 1 \end{pmatrix}. \quad [7.71]$$

The line is supposed to be attached to the support structure at (x_t, y_t, z_t) in the substructure's coordinate system. The 6 x 6 stiffness matrix referred to the substructure becomes then:

$$\mathbf{C}^{(m)} = \begin{pmatrix} \mathbf{C}^{(0)} & \mathbf{C}^{(0)}\boldsymbol{\alpha} \\ \boldsymbol{\alpha}^T\mathbf{C}^{(0)} & \boldsymbol{\alpha}^T\mathbf{C}^{(0)}\boldsymbol{\alpha} \end{pmatrix}, \quad [7.72]$$

with the $\boldsymbol{\alpha}$ given by:

$$\boldsymbol{\alpha} = \begin{pmatrix} 0 & z_t & -y_t \\ -z_t & 0 & x_t \\ y_t & -x_t & 0 \end{pmatrix}. \quad [7.73]$$

Mooring System Stiffness

As an example, we may consider the stiffnesses in surge, heave and pitch for a mooring system consisting of three symmetrically spaced mooring lines ($N_l = 3$). We then obtain the following contributions to the restoring matrix for the platform:

$$\begin{aligned} C_{11}^{(m)} &= \sum_{j=1}^{N_l} C_{11}^{(l)} \cos^2\theta_j = \frac{3}{2} C_{11}^{(l)} \\ C_{33}^{(m)} &= \sum_{j=1}^{N_l} C_{33}^{(l)} = 3C_{33}^{(l)} \\ C_{15}^{(m)} &= \sum_{j=1}^{N_l} C_{11}^{(l)} z_t \cos^2\theta_j = \frac{3}{2} C_{11}^{(l)} z_t \\ C_{55}^{(m)} &= \sum_{j=1}^{N_l} \left[C_{11}^{(l)} z_t^2 + C_{33}^{(l)} r_m^2 \right] \cos^2\theta_j = \frac{3}{2} \left[C_{11}^{(l)} z_t^2 + C_{33}^{(l)} r_m^2 \right]. \end{aligned} \quad [7.74]$$

The last expression *at each line corresponds to the result for the symmetrical three-point mooring. θ_j is the azimuth angle of the mooring line attachments as referred to the platform coordinate system, z_t is the vertical coordinate of the mooring line attachments and r_m is the radius of mooring line attachments, i.e., $r_m = \sqrt{x_{ti}^2 + y_{ti}^2}$. The contribution from $C_{13}^{(l)}$ and $C_{31}^{(l)}$ becomes zero in this symmetrical case.*

7.5.3 Effect of Tether Mooring

Tether mooring systems normally consist of one or more vertical lines with pretension (see Chapter 4). The pretension level is governed by the design requirement that the line should never go slack. Tether mooring has for a long time been applied on offshore oil and gas platforms. The advantage of using tether mooring is that the heave, roll and pitch motions are all very small, i.e., restrained modes. In the oil and gas industry, this has opened up the option of having dry well-heads on a deck of floating platforms. Tether mooring is normally combined with a hull design that minimizes the dynamic wave loads in the tethers. In contrast to catenary mooring lines, tether mooring implies permanent vertical loads on the anchors. This calls for special anchor designs, for example, bucket or gravity anchors with a submerged weight at least equal to the pretension in the tether.

A challenge using multiple tethers (three or more) for floating wind turbines is the large overturning moment due to the wind thrust on the turbine. This overturning moment is $M_w = T_w H$, where T_w is the wind thrust and H is the vertical distance between the rotor axis and the point of attachment of the tethers. Using four tethers in a square layout as an example, the force in each of the tethers to compensate for the overturning moment amounts to $F_t = M_w/2D_t = T_w H/2D_t$. Here D_t is the distance between the tethers. Thus, the dynamic loads in the tethers increase with the H/D_t ratio. For most wind turbines $H >> D_t$, implying large dynamic load variations in the tethers as compared to the dynamic wind thrust. This put requirements to the pretension in the tethers which again may be a driver for the buoyancy of the substructure and the size of the anchors.

The 3 x 3 (surge, sway, heave) restoring matrix for one single tether, referred to the top end of the tether, is:

$$C_t = \begin{bmatrix} T/L & 0 & 0 \\ 0 & T/L & 0 \\ 0 & 0 & AE/L \end{bmatrix}. \qquad [7.75]$$

Here, T is the tether tension, L is the tether length and AE is the axial stiffness per unit length of the tether. Using [7.72] and [7.73] to transfer this stiffness matrix to the platform origin, the symmetrical 6 x 6 restoring matrix becomes:

$$C^{(m)} = \frac{1}{L}\begin{bmatrix} T & 0 & 0 & 0 & Tz_t & -Ty_t \\ & T & 0 & -Tz_t & 0 & Tx_t \\ & & AE & AEy_t & -AEx_t & 0 \\ & & & \left(AEy_t^2 + Tz_t^2\right) & -AEx_ty_t & -Tx_tz_t \\ & Sym & & & \left(AEx_t^2 + Tz_t^2\right) & -Ty_tz_t \\ & & & & & T\left(x_t^2 + y_t^2\right) \end{bmatrix}. \qquad [7.76]$$

7.6 Mooring Lines

The main purpose of a conventional mooring system is to keep the floating structure at location, i.e., to avoid drift-off due to mean forces from wind, waves and current. Except for tether systems, used for tension leg platforms, mooring systems are normally not designed to restrict the dynamic wind and wave forces. A mooring system should thus be sufficiently strong to take the maximum average plus slowly varying forces, and at the same time sufficiently compliant to avoid extreme loads due to dynamic offset. The force-displacement characteristics of mooring lines are normally very nonlinear, as illustrated in Figure 7.13, which shows how a certain horizontal mean force (in this case approximately 2400 kN) corresponds to a certain static horizontal offset of the top end of the mooring line (fair lead). In this specific example the fair lead position is offset by approximately 620 m from the anchor position. Adding wind- and wave-induced horizontal motions on top of the mean offset, the corresponding mooring line tension will vary according to the nonlinear force-displacement characteristic. It is observed that if the mean force increases, the tension amplitude corresponding to a certain motion amplitude will increase.

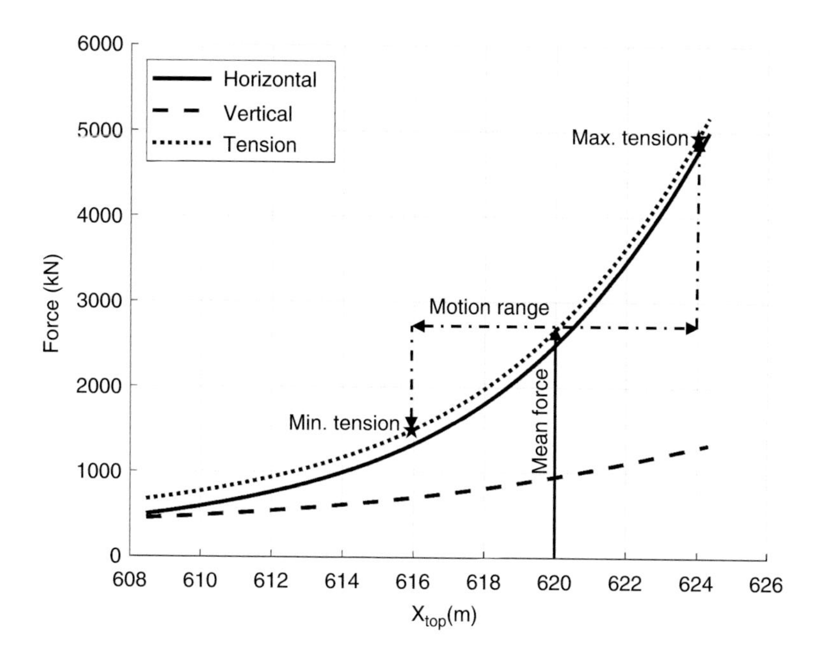

Figure 7.13 Force-displacement characteristic for a mooring line of length 627.0 m in unstretched condition. The top end is located 71.2 m above the sea floor. Submerged weight per unit length is 2.46 kN/m. The axial stiffness EA = 892.6 MN. The x-axis gives the horizontal distance from the anchor location to the top end (fair lead). The vertical arrow indicates a mean load in the line and the corresponding mean distance from the anchor. The dot-dashed horizontal line indicates a double amplitude of wave- and wind-induced motions, causing the tension in the line to vary between the values given by the stars.

To avoid excessively large extreme load amplitudes and to ensure that the anchor can withstand the loads, it is normally required that the length of the mooring line shall be sufficient to ensure that the mooring line force at the anchor position acts as a purely horizontal force, even during extreme load cases.

The pretension is important to ensure a proper stiffness in yaw for a floater. In the case of a vertical-axis wind turbine (VAWT) the mooring system must also resist the generator torque. This puts special design requirements on the mooring system. If a VAWT is placed on top of, e.g., a spar platform, and the mooring lines initially have a radial pattern, the torque will cause a rotation of the foundation until the mooring line tension component in the circumferential direction balances the generator torque. At the same time the turbine mean thrust load must be carried by the lines.

7.6.1 The Concept of Effective Tension

Before discussing the geometric and restoring characteristics of mooring lines, the concept of "effective tension" will be explained.

Hydrostatic forces arise due to the pressure acting on the surface of a body. The total hydrostatic force vector is given by integrating the pressure over the wetted surface:

$$\mathbf{F}_{hs} = \int_{S_w} p\mathbf{n}dS. \tag{7.77}$$

Here, S_w is the wetted area of the body and $\mathbf{n}$ is the unit surface normal vector. If the body is fully submerged (or surface-piercing, with zero pressure at the free surface), the surface fully encloses the volume of the body. In that case the surface integral can be rewritten to a volume integral by observing that the hydrostatic pressure is given from a potential field, and invoking the Gauss theorem, i.e.:

$$\oiint_{S_w} p\mathbf{n}dS = \iiint_V \nabla\, pdV = \rho g V n_3. \tag{7.78}$$

Thus, a purely vertical buoyancy force is obtained. In the case of the cable and considering a short segment, the segment is not wetted at the end surfaces. Thus, the volume integral does not describe the hydrostatic effect and must be corrected for missing pressure at the end surfaces. This correction is illustrated in Figure 7.14. The buoyancy force is applied as if the end surfaces are wetted, i.e., [7.78] is used. Next, the forces on the end surfaces are corrected to compensate for the missing hydrostatic pressure. This introduces an additional axial tension amounting to $p_e A$. This is not a physical tension creating stress in the line, but a correction term entering the equations for the equilibrium of the line. Thus, in computing the axial tension in computing the line geometry, the effective tension is to be used:

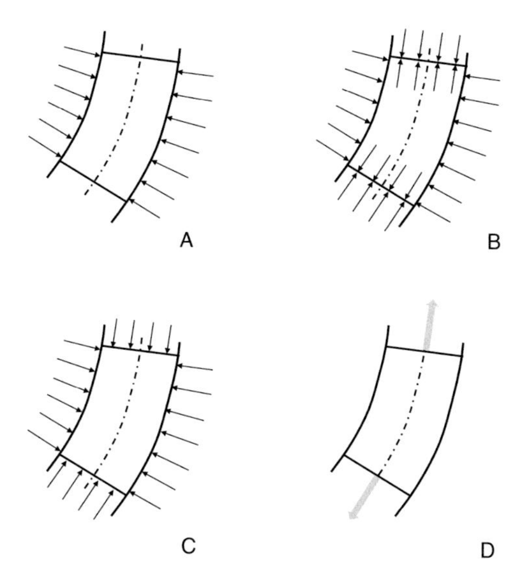

Figure 7.14 Illustration of the effective tension concept for a short segment of length Δs. Case A shows the real pressure distribution. Case B gives the same net forces as Case A. Case B may be replaced by Case C plus Case D. Case C represents a fully wetted segment with vertical force $\rho g A \Delta s$. The end forces in Case D are the needed correction forces, $p_e A$. A is the cross-sectional area of the line and p_e is the external hydrostatic pressure.

$$T_E = T + p_e A \ . \tag{7.79}$$

T is the tension as obtained using the forces acting at the ends of the line and a submerged weight per unit length as given by $w_0 = mg - \rho g A$. In [7.79] the hydrostatic pressure is denoted p_e to show that this is an external pressure. For pipelines, a similar correction is obtained considering the effect of the internal pressure.

Assume the line segment has a length Δs_0 at zero tension. Under tension the length is Δs. The strain is then given by $\varepsilon = (\Delta s - \Delta s_0)/\Delta s_0$. However, also the external pressure influences the axial strain of the segment. For a cylindrical body exposed to external pressure p_e on the sides and an axial tension T, the linear axial strain is obtained as:

$$\varepsilon = \frac{1}{E}\left[\frac{T}{A} + 2\nu p_e\right] . \tag{7.80}$$

Here, E is the Young's modulus of elasticity and ν is the Poisson's ratio,[1] which are both material-dependent. Small strains are assumed, and thus a linear stress-strain

[1] The Poisson's ratio, after the French mathematician and physicist Siméon Poisson, is a measure of the deformation of a material perpendicular to the direction of loading. If a rod has an axial load causing an axial strain of ε_a, then the change in transverse dimension is given by $\varepsilon_n = -\nu\varepsilon_a$, where ν is the Poisson's ratio. Thus, if the Poisson's ratio is positive, a stretching of the rod will cause a transverse contraction.

relation may be assumed. For steel within the elastic range of deformation the Poisson's ratio is approximately 0.3, while for synthetic ropes it is close to 0.5. A Poisson's ratio of 0.5 implies that the line, within a first-order approximation, conserves the volume when tensioned. If the Poisson's ratio is less than 0.5, the volume increases when the line is tensioned. If it is assumed that $v = 0.5$ is a reasonable value for the mooring line, it is observed that the axial elongation may be computed using the effective tension and disregarding the Poisson effect:

$$\varepsilon = \frac{1}{E}\left(\frac{T}{A} + 2vp_e\right) \simeq \frac{T_E}{EA} \ . \tag{7.81}$$

For mooring lines, the effective tension concept does not change the forces significantly. For larger-diameter structures, however, the effect is very important. E.g., deep-water pipelines may experience axial compression forces in the wall. However, they will not buckle as the effective tension is positive.

7.6.2 Inelastic Catenary Line

The static equilibrium equations for mooring lines are derived under the assumption of no bending stiffness in the line and small axial elongations. In the following, the axial elongation is ignored. Consider a line suspended between two points, A and B, as shown in Figure 7.15. The line has a length L and it is assumed that the line is submerged in water. The vertical forces acting are the weight and the buoyancy, $w_0 = (w - \rho g A)$. Here, $w = mg$ is the weight in air per unit length of the line and m is the mass per unit length. ρ is the density of water and A the cross-sectional area. No horizontal forces are acting along the line; the horizontal component of the tension in the line H is thus constant along the line. At each end of the line, vertical and horizontal point forces are acting.

Consider a small section, Δs, of the line, as illustrated in Figure 7.16, and consider equilibrium in the tangential and normal direction of the line:

$$\begin{aligned} &-(T + p_e A) - w_0 \sin\phi \cdot \Delta s + (T + p_e A + \Delta T)\cos(\Delta\phi) = 0. \\ &-w_0 \cos\phi \cdot \Delta s + (T + p_e A + \Delta T)\sin(\Delta\phi) = 0 \ . \end{aligned} \tag{7.82}$$

Here, the variation of the external pressure over the segment has been ignored. Using $T + p_e A = T_E$, $\cos(\Delta\phi) \rightarrow 1$ and $\sin(\Delta\phi) \rightarrow \Delta\phi$ and letting the segment length tend to zero, [7.82] is rewritten as:

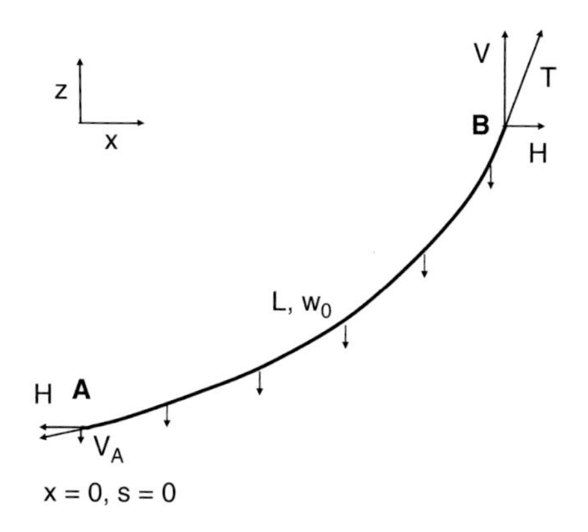

Figure 7.15 Forces acting on a line with constant weight per unit length suspended in water between Points A and B.

$$\frac{dT_E}{ds} = w_0 \sin\phi$$
$$T_E \frac{d\phi}{ds} = w_0 \cos\phi \ . \qquad [7.83]$$

The x and z derivatives with respect to the line coordinate are obtained as:

$$\frac{dx}{ds} = \cos\phi = \frac{1}{\sqrt{1+\tan^2\phi}} = \frac{H}{\sqrt{H^2 + [V - w_0(L-s)]^2}}$$
$$\frac{dz}{ds} = \sin\phi = \frac{\tan\phi}{\sqrt{1+\tan^2\phi}} = \frac{V - w_0(L-s)}{\sqrt{H^2 + [V - w_0(L-s)]^2}} . \qquad [7.84]$$

By integrating these equations, the so-called "inelastic" catenary equations are obtained:

$$x = \frac{H}{w_0}\left\{\sinh^{-1}\left[\frac{V - w_0(L-s)}{H}\right] - \sinh^{-1}\left[\frac{V - w_0 L}{H}\right]\right\}$$

$$z = \frac{H}{w_0}\left\{\sqrt{1 + \left[\frac{V - w_0(L-s)}{H}\right]^2} - \sqrt{1 + \left[\frac{V}{H}\right]^2}\right\} . \qquad [7.85]$$

Here, (x, z) are the coordinates of the line with $z = 0$ at the upper end (B) where $s = L$. $z = -D$ and $s = 0$ at the lower end (A). Further details of the derivation of the catenary equation can be found in Faltinsen (1990) or Triantafyllou (1990).

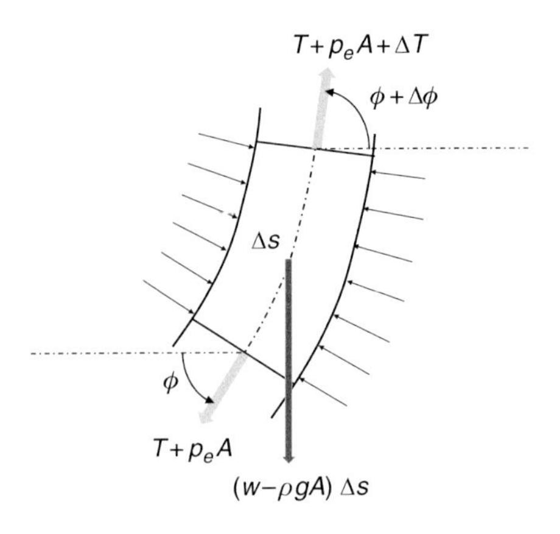

Figure 7.16 Forces acting upon a short segment of the line.

Faltinsen (1990) also shows how one may account for the fraction of the line resting upon the sea floor.

7.6.3 Elastic Catenary Line

For mooring line applications, the elastic elongation of the line may be of importance, even if it is assumed to be small. A first-order approximation of the elastic effects is thus acceptable. The stretched length of the segment in Figure 7.16 is $\Delta s_e = (1+\varepsilon)\Delta s$, with $\varepsilon \ll 1$. Assuming the Poisson's ratio $v = 0.5$, the volume of the segment will not change under the action of tension or external pressure. Thus, the buoyancy force acting upon the tensioned segment is equal to that of the unstretched segment, $w_1 \Delta s_1 = w_0 \Delta s$. The equations for the balance between horizontal and vertical forces on a line segment are thus similar as for the inelastic line, accounting for the new segment length:

$$\begin{aligned} \frac{dT_E}{(1+\varepsilon)ds} &= w_1 \sin\phi = \frac{w_0}{(1+\varepsilon)} \sin\phi \\ T_E \frac{d\phi}{(1+\varepsilon)ds} &= w_1 \cos\phi = \frac{w_0}{(1+\varepsilon)} \cos\phi \, . \end{aligned} \qquad [7.86]$$

The x and z coordinates of the stretched line are obtained from:

$$\begin{aligned} \frac{dx}{ds} &= (1+\varepsilon)\cos\phi \simeq \left(1+\frac{T_E}{EA}\right)\cos\phi = \cos\phi + \frac{H}{EA} \\ \frac{dz}{ds} &= (1+\varepsilon)\sin\phi \simeq \left(1+\frac{T_E}{EA}\right)\sin\phi = \sin\phi + \frac{V - w_0(L-s)}{EA} . \end{aligned} \qquad [7.87]$$

Integrating as for the inelastic equations, the elastic catenary equations are obtained, assuming small linear strain and conserved volume:

$$x = \frac{H}{w_0}\left\{\sinh^{-1}\left[\frac{V - w_0(L-s)}{H}\right] - \sinh^{-1}\left[\frac{V - w_0 L}{H}\right]\right\} + \frac{H}{EA}s$$

$$z = \frac{H}{w_0}\left\{\sqrt{1 + \left[\frac{V - w_0(L-s)}{H}\right]^2} - \sqrt{1 + \left[\frac{V - w_0 L}{H}\right]^2}\right\} + \frac{s}{EA}\left[V - w_0 L + \frac{1}{2}w_0 s\right]$$

$$\phi = \arctan\left(\frac{dz}{dx}\right) = \arctan\left(\frac{V - w_0(L-s)}{H}\right). \qquad [7.88]$$

For further discussion and details, see Triantafyllou (1990).

7.6.4 Restoring Characteristics

As an example, the mooring line described in Figure 7.13 is used. Static force-displacement characteristics in the plane of the catenary are considered. This technique is valid for static loads and slow motions only. As the speed of the motion increases, or the frequency of oscillation increases, viscous forces acting upon the line become important and modify the restoring characteristics. This effect is discussed in more detail below.

The line configurations for various horizontal tension levels are shown in Figure 7.17. Note the large changes in touch-down position even for small changes in position of the fair lead (top end). Changing the horizontal force from 50 kN to 4500 kN moves the fair lead by 41.4 m. The corresponding change in touch-down position is 425.1 m. The elastic elongation of the line is only 0.51% at the largest tension level, i.e., the line length increases from 627.0 m to 630.2 m.

The force-displacement characteristic of a mooring line is in most cases related to the change in geometry of the line, not the elastic elongation. As seen from Figure 7.17, even small horizontal displacements at the top end of the line cause large transverse motions of the line. This is a very important effect in the case of dynamic excitation of the mooring line.

Figure 7.18 shows the restoring coefficients as a function of the horizontal force level. The restoring coefficients are defined as $C_{xx} = \partial F_x / \partial x$, $C_{xz} = \partial F_x / \partial z$, $C_{zz} = \partial F_z / \partial z$. The stiffness is largest in the horizontal direction, C_{xx}, and increases in this case almost linearly with the horizontal force. The vertical stiffness, C_{zz}, is

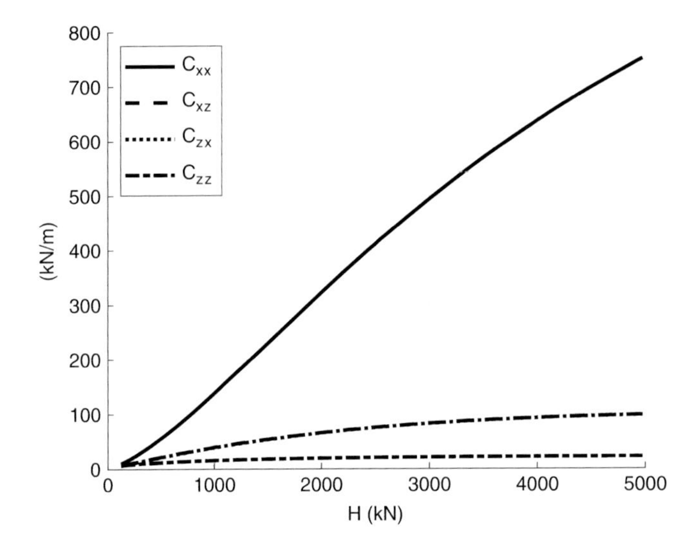

Figure 7.18 Restoring coefficients for the mooring line described in Figure 7.13 as a function of horizontal force (HF) level.

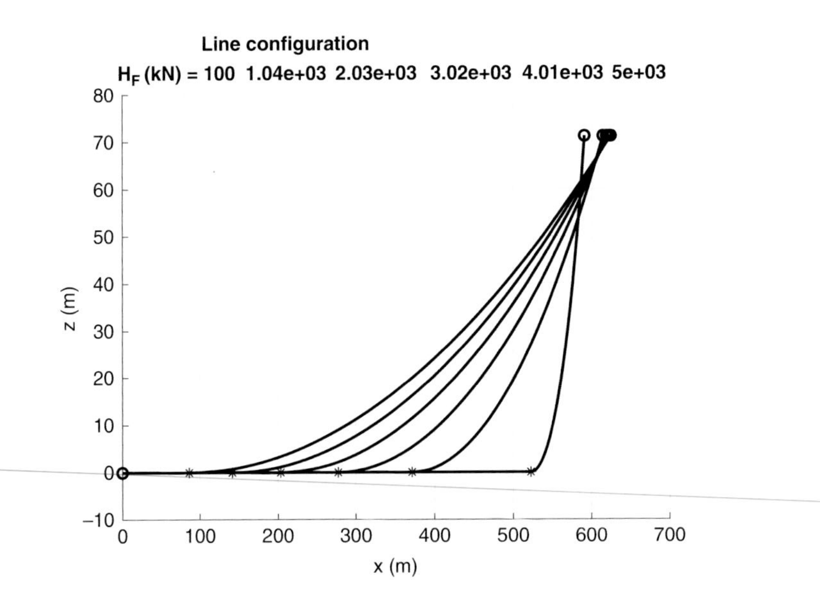

Figure 7.17 Static configuration of the line described in Figure 7.13 for various horizontal force (HF) levels. The stars denote the points of bottom touch-down, while the circles show the anchor and fair lead positions. For the case with H = 3020 kN, the touch-down is at 203.4 m. The horizontal distance from anchor to fair lead is 621.1 m, and the length of the line between touch-down and fair lead is 424.3 m (both lengths referred to in unstretched condition. For a horizontal tension of 3020 kN, the line stretches by 0.34%).

only a small fraction of the horizontal stiffness. The coupling stiffness, $C_{xz} = C_{zx}$, is also significantly less than the horizontal stiffness, but it is worth noting that a horizontal displacement may cause a significant vertical force.

For further discussion it is useful to consider the mooring line stiffness composed of two contributions in a series coupling. The two contributions are the *geometric stiffness* effect and the *elastic stiffness*. The elastic stiffness, C_E, is due to the elasticity of the line, i.e., EA/L; the geometric stiffness, C_G, is the stiffness due to the force-displacement relations found from the inelastic catenary equations, [7.84]. The total stiffness may be written as $C = (1/C_E + 1/C_G)^{-1}$.

For low tension levels, the geometrical change is relatively large even for a small change in force, as illustrated in Figure 7.17. In this force range the geometric stiffness thus dominates the total stiffness. As the tension level increases, the geometric changes are reduced. This is in particular the case if the touch-down point has moved all the way to the anchor. The geometry then approaches a straight line and the stiffness asymptotically approaches the elastic stiffness. This is illustrated in Figure 7.19.

7.6.5 Dynamic Effects

From the above discussion it is observed that a small displacement of the top end of the line may cause large transverse displacements along the line. If the top-end displacement takes place at a finite velocity, as in an oscillatory

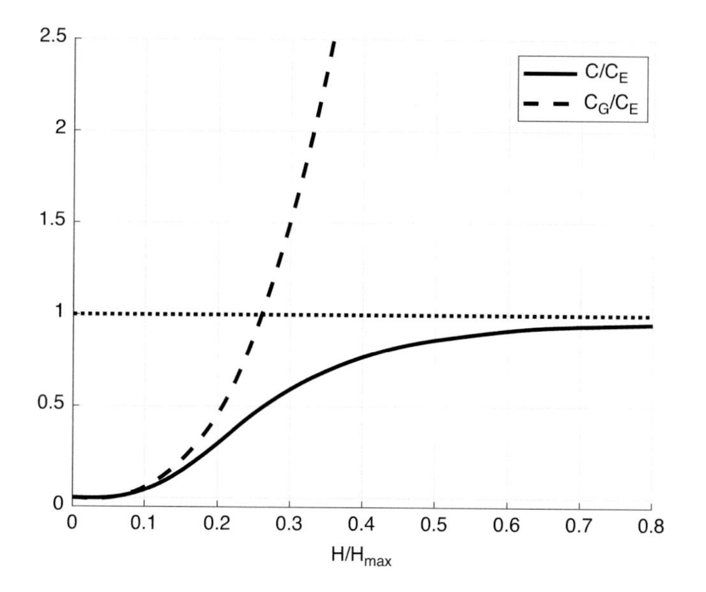

Figure 7.19 Illustration of the relative importance of the elastic and geometric stiffness as a function of the horizontal force level.

motion, drag forces will act along the line. Considering a small segment of the line of length Δs, the local transverse and tangential drag force along the segment may be written as:

$$f_n(s)\Delta s = -\frac{1}{2}\rho C_n d\dot{u}_n|\dot{u}_n|\Delta s$$
$$f_t(s)\Delta s = -\frac{1}{2}\rho C_t d\dot{u}_t|\dot{u}_t|\Delta s\ . \qquad [7.89]$$

These forces will oppose the change of line geometry. The top-end force will thus be larger than in the static case. To solve this force-displacement relation in the general case, including the above viscous drag forces as well as inertia effects, a time-domain finite element approach is required. However, it is possible to establish some estimates to illustrate the effect of the viscous forces. For this purpose, it is assumed that the line may be modeled as an elastic catenary with homogeneous properties, and the elastic deformations are assumed to be much smaller than the geometric deformations.

The line configuration, together with the notations used, are shown in Figure 7.20. The motion of the line, relative to the initial static equilibrium configuration, is denoted $u(s)$, with normal and tangential components u_n and u_t. It is assumed that the motion tangential to the top end is dominating the deformation of the line, while top-end motion normal to the line is of minor importance.

At the initial configuration, the line has a top tension T. By adding a top tension ΔT, the elastic additional deformation of the top end, u_{ET}, is obtained by integrating the strain along the full length of the line:

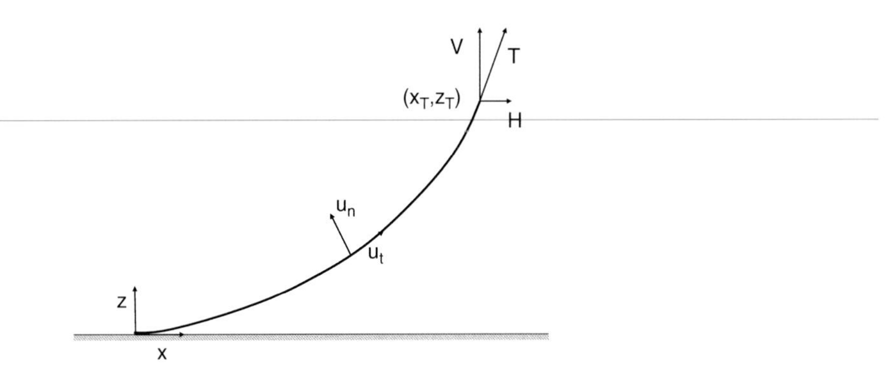

Figure 7.20 Line configuration used for the simplified dynamic analysis.

$$u_{Et}(s=L)=u_{ET}=\int_0^L \frac{\Delta T(s)}{EA(s)}ds. \qquad [7.90]$$

Here, the index T refers to the value at the top, $s=L$. The geometric contribution to the top-end tangential displacement is thus $u_{GT}=u_T-u_{ET}$, where u_T is the total top-end displacement. The corresponding static displacements along the line are denoted $u_{0t}(s)$ and $u_{0n}(s)$. It is now assumed that the dynamic displacement along the line is proportional to the static displacement, i.e., the geometric dynamic displacement can be written as:

$$\begin{aligned} u_n(s) &= \alpha u_{0n}(s). \\ u_t(s) &= \alpha u_{0t}(s)\ . \end{aligned} \qquad [7.91]$$

If the top-end motion is assumed to be harmonic and the drag forces are linearized, α becomes a complex coefficient relating the quasi-static geometric displacements and the dynamic displacements. The velocities and accelerations along the line are thus given by:

$$\begin{aligned} \dot{u}(s) &= i\omega\alpha u_0(s). \\ \ddot{u}(s) &= -\omega^2\alpha u_0(s)\ . \end{aligned} \qquad [7.92]$$

The normal and tangential dynamic loads acting along the line may be written as:

$$\begin{aligned} f_n(s) &= -\frac{1}{2}\rho C_{Dn} d\dot{u}_n|\dot{u}_n| - \left(m+\frac{\pi}{4}\rho d^2 C_m\right)\ddot{u}_n = f_{nd}+f_{nm}. \\ f_t(s) &= -\frac{1}{2}\rho C_{Dt}\pi d\dot{u}_t|\dot{u}_t| - m\ddot{u}_t = f_{td}+f_{tm}\ . \end{aligned} \qquad [7.93]$$

Here, ρ is the density of the water; C_{Dn} is the drag coefficient for flow normal to the line; d is the diameter of the line; m is the mass per unit length; C_m is the added mass coefficient normal to the line; and C_{Dt} is the longitudinal drag (skin friction) coefficient. No added mass is included in the tangential direction. There must be a balance between the dynamic forces along the line, as expressed by [7.93], and the end forces acting on the line. It is convenient to consider the moment of the forces about the touch-down point. The moment of the distributed dynamic forces may be written as:

$$M_{LD} = \int_0^L f_n(s)[\cos\phi(s)x(s) + \sin\phi(s)z(s)]ds.$$
$$+\int_0^L f_t(s)[-\cos\phi(s)z(s) + \sin\phi(s)x(s)]ds. \quad [7.94]$$

Here, ϕ is the angle between the tangent of the line and the horizontal plane. Similarly, the moment due to the top-end forces minus the quasi-static contribution may be written as:

$$M_{TD} = -(\Delta T_T - \Delta T_{T0})[-\cos\phi_T z_T + \sin\phi_T x_T] = -\Delta T_{TD}K. \quad [7.95]$$

ΔT_{T0} is the quasi-static top-end force. The difference between the dynamic and the quasi-static top-end force is given by the differences in the top-end strain, ε_T, i.e.:

$$\Delta T_{TD} = \Delta T_T - \Delta T_{T0} = (\varepsilon_T - \varepsilon_{T0})EA. \quad [7.96]$$

As it is assumed that the dynamic displacement is following the quasi-static mode shape, there will be proportionality between the strain at the upper end and the elastic displacement along the line. Thus, the following relation is obtained between the top-end strain and the top-end elastic displacement:

$$\frac{(\varepsilon_T - \varepsilon_{T0})}{\varepsilon_{T0}} = \frac{(u_{TE} - u_{TE0})}{u_{TE0}}. \quad [7.97]$$

The above equations may now be solved by requiring $M_{LD} = M_{TD}$. α is obtained as:

$$\alpha \simeq 1 - \frac{u_{TE0}}{u_{TG0}} \frac{\Delta T_{TD}}{\Delta T_{T0}}. \quad [7.98]$$

As ΔT_{TD} depends upon α, an iteration procedure must be used to solve for α. The total dynamic top tension, including the quasi-static contribution, may be written as:

$$\Delta T_T = \Delta T_{T0} \frac{u_T - \alpha u_{TG0}}{u_{TE0}}. \quad [7.99]$$

As both drag and inertia forces are acting on the line, there will be a phase shift between maximum force and maximum displacement. An “apparent stiffness” and damping force for the line may be derived. In very slow

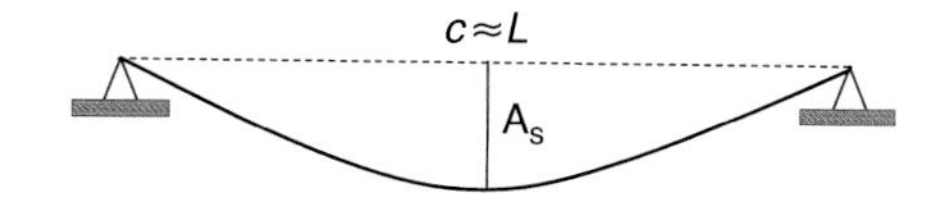

Figure 7.21 Horizontal elastic line with sag, and axial stiffness AE/L.

oscillations, α will be close to unity and the dynamic tension variation will be close to the quasistatic value. As the drag forces on the line increase, either due to a higher frequency of oscillation or due to a larger oscillation amplitude, α will decrease and the dynamic tension will increase. This effect is called "drag-locking" as the drag forces restrict the transverse motion of the line and force shifts the deformation from a geometric to an elastic deformation. The apparent line stiffness thus increases.

If the assumption of a quasi-static deformation pattern of the line is to be reasonable, the frequency of oscillation should be significantly lower than the lowest natural frequency of the line. A rough estimate of the natural frequencies for transverse oscillation may be obtained by using the natural frequencies of an elastic horizontal line with sag and fixed at both ends, as illustrated in Figure 7.21. The horizontal force is T and the sag at the mid-span A_s. The sag is assumed to be much less than the length of the line, causing the chord-length, c, to be approximately equal to the line length and the horizontal force approximately equal to the line tension. Under these assumptions, the two first natural frequencies are obtained as:

$$\omega_1 \simeq \frac{\pi}{L}\sqrt{\frac{T}{(m+A)}\left(1+\frac{\pi^2}{8}\frac{AE}{T}\frac{A_s^2}{L^2}\right)}.$$
$$\omega_2 \simeq \frac{2\pi}{L}\sqrt{\frac{T}{(m+A)}}. \qquad [7.100]$$

Here, A is the added mass per unit length for transverse oscillations. The first natural frequency corresponds to a half-sine mode along the span. It therefore involves axial deformations. The second natural frequency corresponds to a full sine mode and does not involve axial deformations.

Dynamic Amplification of Line Tension

Consider a catenary line of length 600 m and with homogeneous mass and stiffness properties. Further details are specified below.

(cont.)

Line weight	*290 N/m*
Vertical distance from sea floor to fair lead	*60 m*
Axial stiffness (EA)	*610 MN*
Line mass	*34.0275 kg/m*
Line diameter	*0.07 m*
Added mass coefficient (normal)	*1.0*
Drag coefficient (normal)	*1.0*
Drag coefficient (tangential)	*0.05*
Horizontal force	*777.55 kN*
Vertical distance sea floor–fair lead	*100 m*
Results from static analysis:	
Vertical force, top of line	*165.30 kN*
Total top force	*794.93 kN*
Stretched length of line	*600.77 m*
Top-end angle	*12.00 deg*
Horizontal position of touch-down	*29.04 m*
Horizontal stiffness, Cxx	*272.81 kN/m*
Vertical stiffness, Czz	*1.395 kN/m*
Coupled stiffness, Czx	*28.69 kN/m*
Elastic stiffness, EA/L	*1016.67 kN/m*
Length of secant (touch-down–fair lead)	*570.68 m*
Angle of secant	*6.00 deg*
Line sag	*14.86 m*

The two lowest transverse natural frequencies are estimated at 1.01 and 1.58 rad/s respectively. Figure 7.22 shows the dynamic amplification of the top tension as a function of oscillation frequency for two top-end tangential motion amplitudes, u_{Tt} = *0.1 m and 0.45 m. The results are compared to results from a nonlinear time-domain finite element program (FEM) using the Morison equation for the hydrodynamic forces* (Ormberg and Bachynski, 2012).

Based upon the above estimates of the natural frequencies, one may expect the results to be valid for frequencies well below 1 rad/s only. The importance of the drag forces diminishes as the amplitude of oscillation is reduced. For the 0.1 m amplitude it is observed that the dynamic load amplitude is lower than the quasi-static amplitude in the low-frequency range. This is explained by the effect of the inertia loads, acting in opposite phase to the restoring force.

For the largest amplitude, the drag forces are more important and a more rapid increase in the dynamic tension is observed as the frequency increases. As the frequency of oscillation increases, the difference between the present simplified approach and the FEM results increases. An important reason for this is the assumption of a motion

(cont.)

pattern similar to the quasi-static deformations. The somewhat irregular look of the FEM results is due to irregularities in the time-domain results causing difficulties in identifying the relevant amplitudes of the dynamic loads.

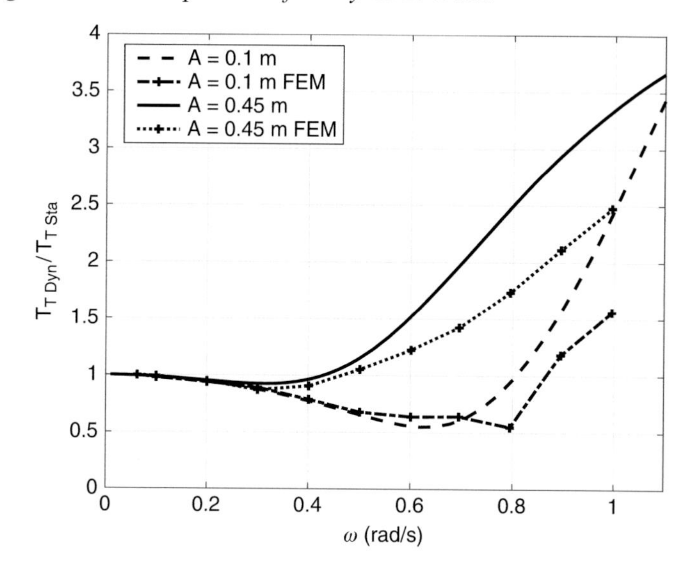

Figure 7.22 Dynamic amplification of top tension as a function of motion frequency. Two motion amplitudes of top end: 0.1 and 0.45 m. Solid and dashed lines: present approximation; lines with cross markers: FEM analysis using the Morison equation for the hydrodynamic forces.

7.7 Low-Frequency Wind-Induced Motions

As discussed in Chapter 2, several methods exist for generating the point spectra of the wind velocity and the coherence of the wind field. Few measurements are available to validate the various methods, in particular for heights beyond 100 m above sea level and for the horizontal coherence. Frequently, the models applied assume neutral stability of the atmosphere. As discussed in Chapter 2, atmospheric conditions with almost neutral stability are not the most common situation. The uncertainty about relevant spectral models and coherence functions is significant in the low-frequency range. Therefore, great care should be exercised in the choice of wind field model when studying the dynamic response of floating wind turbines.

Nybø, Nielsen and Godvik (2022) used various models to generate the wind field and studied the low-frequency dynamic response of a 15 MW offshore wind turbine on a spar floater. The rotor diameter is 240 m and the height of the rotor center is 135 m above sea level. The natural periods for the rigid-body motions (surge, sway, heave, roll and pitch) for this kind of floaters may be in the range of 25–200 s. This range of periods is normally not considered important for bottom-fixed wind turbines. Depending upon the technique used to generate the wind field, large differences in energy and coherence may appear in this low-frequency range. This is particularly the case in non-neutral atmospheric stability conditions. For the lowest frequencies it may be questioned whether the frequency content is part of a stationary process or due to nonstationarity of the wind conditions. Independent of the cause, the result may be excitation of low-frequency motion modes.

Nybø et al. (2022) used four different wind field formulations for the response analyses: the standard Kaimal spectrum with an exponential coherence model; the Mann spectral tensor model; a wind field generated by Large Eddy Simulation (LES) code; and a model using a wind spectrum based upon measured wind speed time series combined with a Davenport coherence model fitted to wind speed measurements at two different vertical levels. This wind field is called TIMESR. The LES wind field and the TIMESR both account for atmospheric stability in the coherence, while the Kaimal and the Mann implementations assume neutral stability. However, the turbulence intensity and mean wind shear are in these two models fitted to measured data. Thereby, the mean wind speed, turbulence intensity and wind shear are similar in all the applied models. In particular in the low-frequency part, below 0.1 Hz of the spectra, significant difference in the energy content is observed between the four methods. This is illustrated in Figure 7.23. It is observed that the power spectra density (PSD) of the LES spectrum is significantly lower than what is obtained by the other three methods. The coherence computed from the wind fields also differs greatly. These differences are important for the excitation of the various rigid modes of motion.

In Figure 7.24 illustrations of the importance of coherence to the excitation loads are given. Case A illustrates large coherence in vertical direction over a length scale similar to the rotor diameter, i.e., the dynamic wind speed is in phase over this length scale. In the horizontal direction, the coherence is assumed to be lower, causing opposite phase of the wind speed between the two sides of the rotor disk. In Case B, the length scales of the vertical and horizontal coherences have switched, i.e., large coherence in horizontal direction is assumed, while in the vertical direction the wind speed has opposite phase in the lower versus the upper part of the rotor. In Case C, the flow is assumed to be coherent over most of the rotor disk, while in Case D, the turbulent structures are assumed to have very short extent giving a very low coherence over the rotor disk area both in the vertical and the horizontal direction.

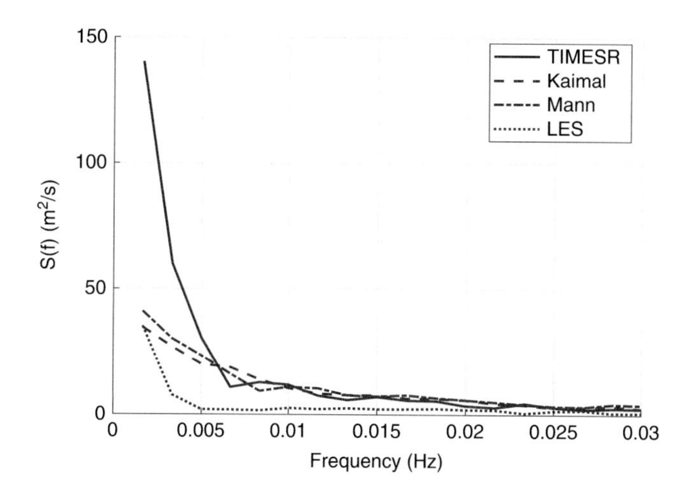

Figure 7.23 Power spectral density (PSD) in the low-frequency range of the wind spectrum calculated by four different methods. Neutral atmospheric stability and 13 m/s mean wind speed are assumed. From Nybø et al. (2022) by permission of John Wiley and Sons, license No. 5460741249347.

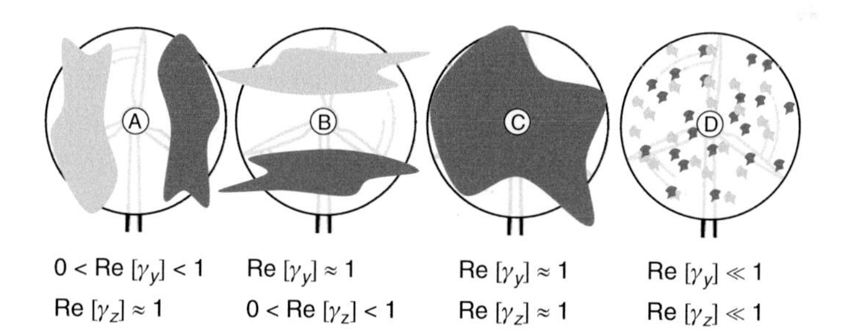

Figure 7.24 Illustration of different coherent structures that give different excitation of the various modes of motion. Dark- and light-gray tones indicate areas where the dynamic wind speed is out of phase. Reproduced from Nybø et al. (2022) by permission of John Wiley and Sons, license No. 5460741249347.

As indicated in the table in Figure 7.24, the three cases will excite the different modes of motion differently. Case C will cause the largest excitation in surge, while Cases A and B will cause lower loads as the total load on the rotor disk is partly canceled out due to the phase differences. The same argument can be used for the platform pitch motion, referred to the water line and the mooring loads. For the yaw motion, however, it is observed that Case A will excite this mode more severely than Cases B and C. Which case is most relevant is obviously dependent upon frequency and rotor diameter considered. To which extent the atmospheric stability

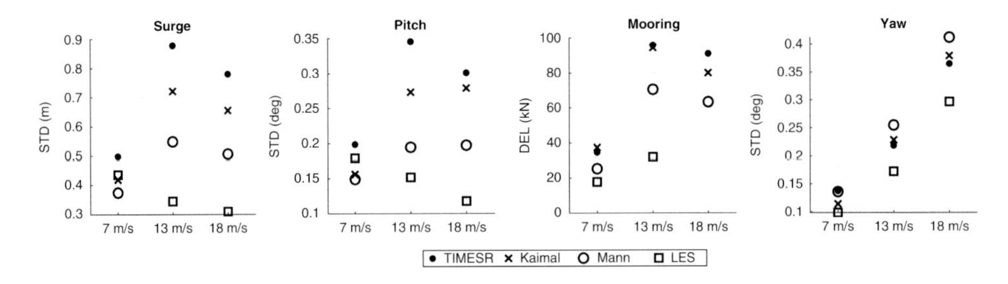

Figure 7.25 Standard deviation of the platform surge, pitch and yaw motion due to low frequency loads of the wind field only. For the mooring line loads a damage equivalent load (DEL) is given. Neutral atmospheric stability. 7, 13 and 18 m/s mean wind speed. Reproduced from Nybø et al. (2022) by permission of John Wiley and Sons, license No. 5460741249347.

and height over sea level change the structure of the coherence is a topic for ongoing research.

Figure 7.25 shows examples of the results obtained by Nybø et al. (2022). The standard deviations of the surge, pitch and yaw motions in 7, 13 and 18 m/s mean wind speed are given. Further, the damage equivalent load (DEM) for the most heavily loaded mooring line is given. The DEM is an expression for the fatigue loading in the line. The four wind spectra discussed above are used and neutral atmospheric stability is assumed. The turbulence intensity is in the range of 2–6.6%. From the figure it is observed that there is a large scatter between the results as obtained by the different methods for generating the wind fields. The spectra obtained by TIMESR contain much energy at the lowest frequencies, which is reflected in the platform surge and pitch motions and is also related to a high coherence level over the rotor disk. On the other hand, it is observed that the Mann model, causing low surge and pitch motions, results in the largest yaw motions. This is closely related to the lower coherence obtained by using the Mann model. Investigating various atmospheric stability conditions, additional differences between the results are observed.

7.8 Control Issues for Floating Wind Turbines

7.8.1 Introduction

In discussing the control issues for bottom-fixed wind turbines (see Section 3.9), the main control objectives were to maximize power production below rated wind speed and to keep constant power above rated wind speed. It was also demonstrated that the controller should be designed with a slow response relative to the eigenfrequencies of the structure. The controller thereby reacts properly to variations in the incoming wind speed but does not interfere with the structural dynamics. An exception is when the controller is tuned to provide damping to, for example, the first fore-aft elastic bending mode of the tower.

For floaters, an additional perspective appears. Due to the rigid-body motions of the floater, the rotor may have a significant motion perpendicular to the rotor plane. The floater is also designed so that the rigid-body natural periods are outside the range of wave periods. This implies natural periods above 20 s for slack (catenary) moored floaters or below about 4 s for tension leg floaters. In this section the slack-moored floater is considered. The pitch natural period may be in the range of 25–50 s, while the surge natural period may be more than 1 min. When the rotor is moving toward the wind, the relative velocity between air and rotor increases. The control system, as outlined in Section 3.9, will interpret this as an increased wind velocity and as the period of motion period is so slow, the control system will adjust rotor speed, torque and blade pitch accordingly. As will be discussed, the effect may be a reduced or an increased damping of the floater motion. Other damping components such as wave radiation damping and viscous damping are low for such low-frequency motions. The floating wind turbine may thus require a modified control system to ensure proper damping and motion behavior.

7.8.2 *Action of a Conventional Controller*

Consider the turbine outlined in Figure 7.26. The wind is supposed to blow in positive x-direction and the surge and platform pitch motions are considered only. The relative wind velocity at nacelle level can be written as:

$$U_r = U_w - U_{nac} = U_w - \dot{\eta}_1 - z_n \dot{\eta}_5. \qquad [7.101]$$

Here, U_w is the wind velocity at nacelle level and z_n is the vertical position of the nacelle. η_1 and η_5 are the surge and pitch motions respectively. It is assumed that $U_w > U_{nac}$. It is further assumed that the motions are so slow that the control system adjusts the turbine to the stationary power and thrust. The thrust on the turbine is thus given by:

$$T = \frac{1}{2} C_T \rho_a A U_r^2, \qquad [7.102]$$

where A is the rotor area and C_T is the thrust coefficient.

To illustrate the effect of the rigid-body motion, one frequency of motion is considered only. This frequency may correspond to an eigenfrequency. As the eigenmode does not need to correspond to the modes as defined by the coordinate system, both a surge and pitch component need to be included. Each of the two velocity components are thus written as:

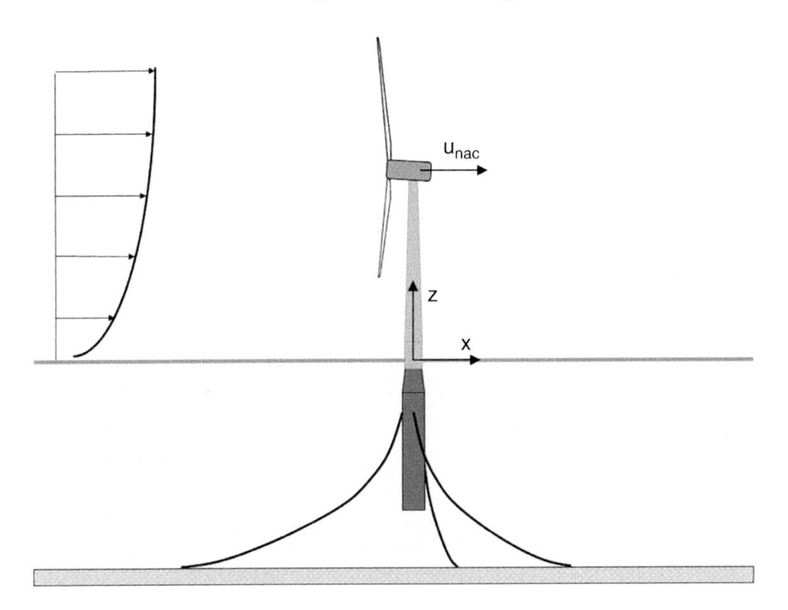

Figure 7.26 Coordinate system located at calm water level and velocity of horizontal nacelle motion.

$$\dot{\eta}_i = \dot{\eta}_{iA} \cos(\omega t + \phi_i), \quad i = 1, 5. \qquad [7.103]$$

Here, ϕ_i is the phase of each component. During one cycle of oscillation the platform motion energy absorbed by the turbine may be written as:

$$E_d = -\int_0^{T_\eta} TU_{nac}dt. \qquad [7.104]$$

Here $T_\eta = 2\pi/\omega$.

7.8.2.1 Below Rated Wind Speed

Below rated wind speed, the controller is tuned for maximum power production. In this case both the power and thrust coefficients are almost constant. C_T may thus be assumed to be constant in estimating the absorbed energy during one cycle of oscillation and [7.104] gives the following result:

$$\begin{aligned} E_d &= -\int_0^{T_\eta} TU_{nac}dt = -\frac{1}{2}C_T\rho_a A\int_0^{T_\eta} U_r^2 U_{nac}dt \\ &= \frac{1}{2}C_T\rho_a AU_{w0}T_\eta[\dot{\eta}_{1A}^2 + 2z_n\dot{\eta}_{1A}\dot{\eta}_{5A}\cos\phi + z_n^2\dot{\eta}_{5A}^2]. \end{aligned} \qquad [7.105]$$

Here, $\phi = \phi_5 - \phi_1$ and U_w is replaced by the mean wind velocity, U_{w0}. Using a linear 2DOF model, the damping force and dissipated energy over one cycle of oscillation can be written as:

$$\mathbf{F_{DL}} = \mathbf{B_L}\dot{\boldsymbol{\eta}}.$$
$$E_{DL} = \int_0^{T_\eta} \dot{\boldsymbol{\eta}}^{\mathbf{T}} \mathbf{B_L} \dot{\boldsymbol{\eta}} dt \ . \quad [7.106]$$

Here, $\dot{\boldsymbol{\eta}} = (\dot{\eta}_1 \ \dot{\eta}_5)^T$ is the vector containing the (complex) surge and pitch velocities. Equating [7.105] and [7.106], the following elements in the damping matrix due to the rotor thrust are found:

$$\begin{aligned} B_{11} &= C_T \rho_a A U_{w0}. \\ B_{15} = B_{51} &= C_T \rho_a A U_{w0} z_n \\ B_{55} &= C_T \rho_a A U_{w0} z_n^2. \end{aligned} \quad [7.107]$$

C_T is in the order of 0.8–0.9 for wind velocities below rated. It is thus observed that the wind turbine contributes with a positive and, as can be shown, very significant damping when operating below rated wind speed and a conventional control system for bottom-fixed turbines is used. This is valid both for pitch and surge motions.

7.8.2.2 Above Rated Wind Speed

Above rated wind speed, the conventional blade pitch controller is set to maintain constant power and rotational speed. Thus, the thrust coefficient varies with the mean wind velocity, as demonstrated in Section 3.8. For the slow platform motions considered here, it may be assumed that the blade pitch controller and the thrust force behave almost as in the stationary case, i.e., there is no time delay between the change in relative velocity and corresponding thrust.

As a first approximation it is assumed that the thrust coefficient varies linearly with the relative wind velocity. This is an acceptable assumption for small velocity variations around the mean velocity U_{w0}. Thus, the thrust coefficient is written as:

$$C_T(U_r) = C_T(U_{w0}) \left[1 + k_{CT} \frac{U_d}{U_{w0}} \right]. \quad [7.108]$$

Here, $U_d = -U_{nac} = -\dot{\eta}_1 - z_a \dot{\eta}_5$ is the dynamic variation in the relative wind velocity due to the combined surge and pitch motion. It is assumed that $U_d \ll U_{w0}$. Under the above assumptions, the instantaneous thrust force becomes:

$$T = \frac{1}{2} C_T \rho_a A U_r^2 = \frac{1}{2} \rho_a A C_T(U_{w0}) \left[1 + k_{CT} \frac{U_d}{U_{w0}} \right] [U_{w0} + U_d]^2. \qquad [7.109]$$

Let U_d be harmonic with period T_η and amplitude U_{dA}, then the absorbed energy, similarly as in [7.105], is obtained as:

$$\begin{aligned} E_d &= -\int_0^{T_\eta} T U_{nac} dt = -\frac{1}{2} \rho_a A \int_0^{T_\eta} C_T U_r^2 U_{nac} dt \\ &= \frac{1}{2} \rho_a A U_{w0}^3 C_{T0} T_\eta \left(\frac{U_{dA}}{U_{w0}} \right)^2 \left[1 + \frac{k_{CT}}{2} + \frac{3k_{CT}}{8} \left(\frac{U_{dA}}{U_{w0}} \right)^2 \right]. \end{aligned} \qquad [7.110]$$

The last term in the bracket may be disregarded as this is small (second-order) compared to the two other terms. Inserting for U_{dA}, the following expression is obtained for the energy absorbed during one cycle of oscillation:

$$E_d = \frac{1}{2} C_{T0} \rho_a A U_{w0} T_\eta [\dot{\eta}_{1A}^2 + 2 z_n \dot{\eta}_{1A} \dot{\eta}_{5A} \cos\phi + z_n^2 \dot{\eta}_{5A}^2] \left[1 + \frac{k_{CT}}{2} \right]. \qquad [7.111]$$

This is the same expression as for the below-rated case, except for the last bracket. The linearized damping coefficients are thus obtained as:

$$\begin{aligned} B_{11} &= C_T \rho_a A U_{w0} \left[1 + \frac{k_{CT}}{2} \right] \\ B_{15} = B_{51} &= C_T \rho_a A U_{w0} z_n \left[1 + \frac{k_{CT}}{2} \right] \\ B_{55} &= C_T \rho_a A U_{w0} z_n^2 \left[1 + \frac{k_{CT}}{2} \right]. \end{aligned} \qquad [7.112]$$

Above rated wind speed, the slope of the thrust coefficient versus wind speed is negative. k_{CT} is thus negative. From [7.112] it is observed that the damping coefficients become negative if $k_{CT} < -2$. In Figure 7.27, $k_{CT}(U_{w0}) = \frac{U_{w0}}{C_T} \frac{dC_T}{dU_w}$ is plotted for the thrust characteristic of the NREL 5 MW reference turbine discussed in Section 3.8.1. It is observed that $k_{CT} < -2$ for velocities above rated wind speed.

The consequences of this negative damping above rated wind speed have been illustrated both in numerical simulations and in full scale. Skaare et al. (2011) demonstrated that a 2.3 MW turbine mounted on a spar foundation (the

Hywind Demo turbine) exhibited negative damping for the platform pitch mode when operated with a conventional control system. Such behavior is also shown in simulations (see Figure 7.28).

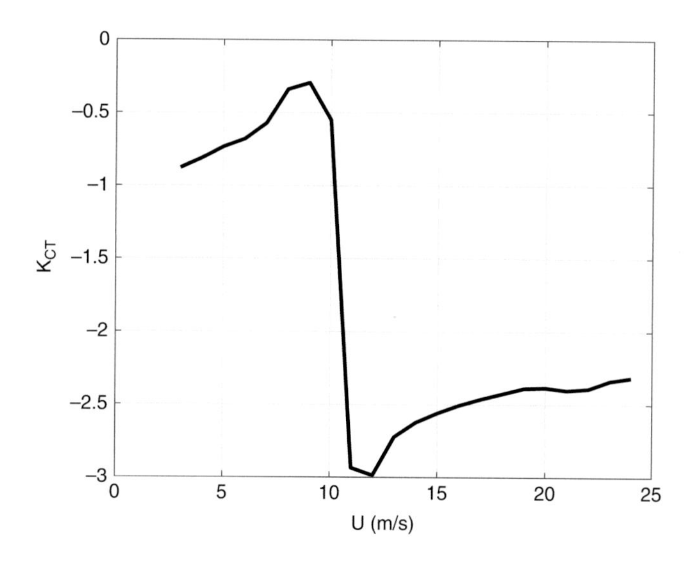

Figure 7.27 k_{CT} as a function of mean wind velocity for the turbine with thrust curve as given in Figure 3.46. It is observed that $k_{CT} < -2$ for velocities above rated wind speed.

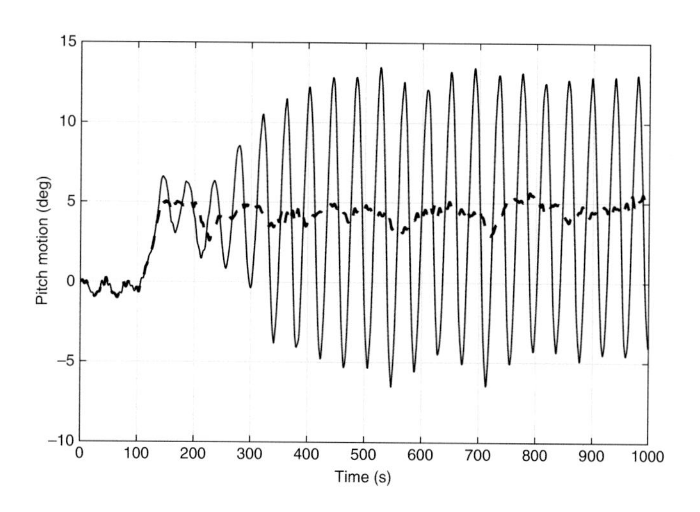

Figure 7.28 Platform pitch for a spar floating wind turbine. Tower pitch angle in degrees versus time. Mean wind speed 14 m/s and turbulent intensity 8.9%. Significant wave height 2.6 m, start-up from rest. Solid line: platform pitch with conventional turbine control; dashed line with motion controller switched on. Data by courtesy of Bjørn Skaare, Equinor.

7.8.3 Control of Low-Frequency Motions

7.8.3.1 Controller for Bottom-Fixed Turbines

Larsen and Hanson (2007) demonstrated by simulation the build-up of excessive motion of a floating wind turbine if a conventional controller[2] was used above rated wind speed. They demonstrated how a bottom-fixed wind turbine with tower natural frequency of 0.5 Hz and a controller natural frequency of 0.1 Hz had a nice and stable response, while unstable behavior occurred after shifting the tower natural frequency to 0.05 Hz. Above rated wind speed they used a PI controller with constant torque and blade pitch control for rotor speed. The reason for using constant torque rather than constant power is to reduce dynamic loads in the drive train. The penalty is somewhat larger variations in output power.

To investigate further if the principles for the bottom-fixed turbine controller could be used for a floating wind turbine and obtain stable response, Larsen and Hanson (2007) kept the PI controller and tested out four natural frequencies for the controller. The frequencies were 0.02, 0.04, 0.05 and 0.1 Hz. For each natural frequency they tuned the controller gains and studied the wind turbine behavior. The natural frequency of the turbine pitch motion was 0.035 Hz. From previous experience they aimed for a relative damping ratio of 0.8. They found that a controller natural frequency of 0.02 Hz was superior to the other choices, confirming the observations in Chapter 3 that the controller natural frequency should be significantly below the structural natural frequencies to avoid destructive interaction.

The controller proposed by Larsen and Hanson (2007) removed the unstable behavior and reduced the platform pitch motion significantly. However, simulations in turbulent wind revealed large dynamic variations in quantities including power, torque and rotational speed. Increased variations in these quantities must be expected as the controller is too slow to adjust for the variation in the incident wind speed. Additional variations are to be expected when wave-induced motions are included.

The following section briefly discusses a couple of other options to avoid excitation of the modes with low natural frequencies, e.g., surge and pitch.

7.8.3.2 Use of a Notch Filter

A notch filter is a filter that is used to remove components of certain frequencies in a stochastic signal. A classic example is removing noise due to the 50 Hz grid frequency in electrical appliances. In the present case a notch filter is used to remove the input to the blade pitch control signal in a narrow frequency range. If this narrow frequency range is centered on the platform pitch natural frequency, the hypothesis is that the resonant excitation is avoided. Some basics about the notch filter and how it can be implemented in time domain are given in Appendix D.

[2] See Section 3.9 for a discussion of conventional controllers.

As shown in this appendix, a notch filter can be applied to a discrete stochastic time series $x(t)$ to obtain a filtered time series $y(t)$ by the following recursive function:

$$y_n = \sum_{k=0}^{M} c_k x_{n-k} + \sum_{j=1}^{N} d_j y_{n-j}. \qquad [7.113]$$

The coefficients c_k and d_j are obtained from the filter frequency, bandwidth and sampling interval.

The notch filter can be used in simulation models where the conventional controller is represented by, for example, a look-up table for the turbine thrust. The notch filter may then be used to simulate the floater motion controller. Used together with a conventional controller, problems may occur due to, e.g., the phase shift introduced by the notch filter. Assuming a linear relation between small variations in blade pitch angle and variation in thrust force, the effect of the notch filter will be illustrated by filtering the relative wind velocity between the wind and the turbine, governing the thrust force.

Transient Response Using a Notch Filter on the Wind Speed

Consider a spar-like floating wind turbine with 5 MW rated power. The main characteristics are given in Table 7.2.

Table 7.2 *Main characteristics of spar substructure and wind turbine.*

Rotor diameter	125.00 m
Height of rotor axis	85.00 m
Rotational speed	12.10 rev/min
Draft of support structure	120.00 m
Center of gravity	-78.15 m
Center of buoyancy	-62.60 m

	Dry Mass	Hydrodynamic Mass
Surge[3], M_{11}	8149 Mg	7797 Mg
Pitch, M_{55}	6.377E7 Mgm^2	3.800E7 Mgm^2
Coupled surge-pitch, M_{15}	-6.199E5 Mgm	-4.842E5 Mgm

[3] To avoid confusion between force and mass, the use of ton is avoided. Therefore, megagram is used for mass: 1 Mg $=10^6$ g = 1000 kg.

(cont.)

	Hydrostatic	Mooring
Surge restoring, C_{11}		47.070 kN/m
Pitch restoring, C_{55}	1123 MNm/rad	239.600 MNm/rad
Coupled surge-pitch, C_{15}		-3.358 MN/rad

The power and thrust coefficients for the turbine considered are shown in Figure 7.29. The coupled surge and pitch platform rigid-body motions are considered. A small linear damping (approximately 2%) is included in the model to account for an approximate hydrodynamic damping and to ensure stability of the numerical integration. The natural periods for the coupled surge-pitch system are obtained as 115.7 s and 29.8 s respectively. The longest natural period corresponds to an almost pure horizontal translation, while the fastest natural period corresponds to a rotational mode with center of rotation 69.1 m below the free surface, i.e., about midway between the center of buoyancy and the center of gravity.

The wind turbine is exposed to a wind field with mean wind speed 15 m/s at nacelle level and a turbulence intensity of 10%. The wind field is assumed to be coherent over the rotor disk area. A forward Euler integration scheme with time step 0.1 s is used for the simulations.

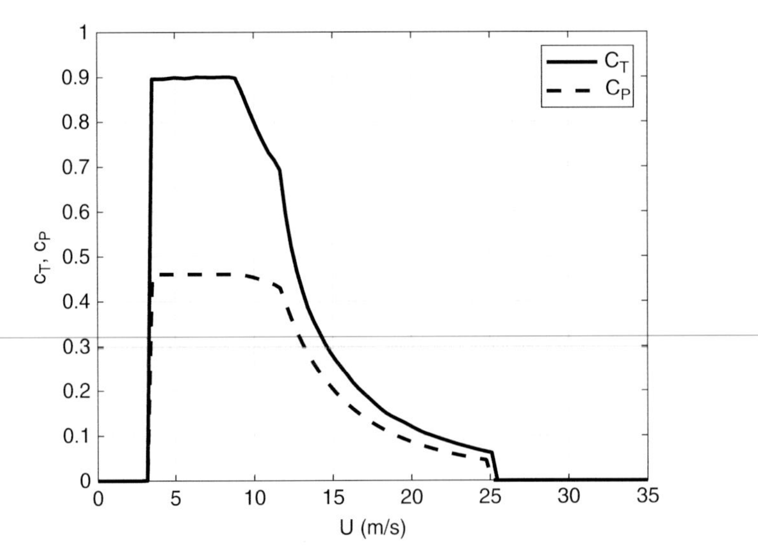

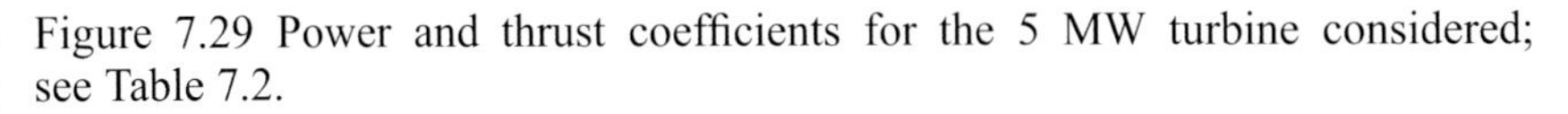
Figure 7.29 Power and thrust coefficients for the 5 MW turbine considered; see Table 7.2.

(cont.)

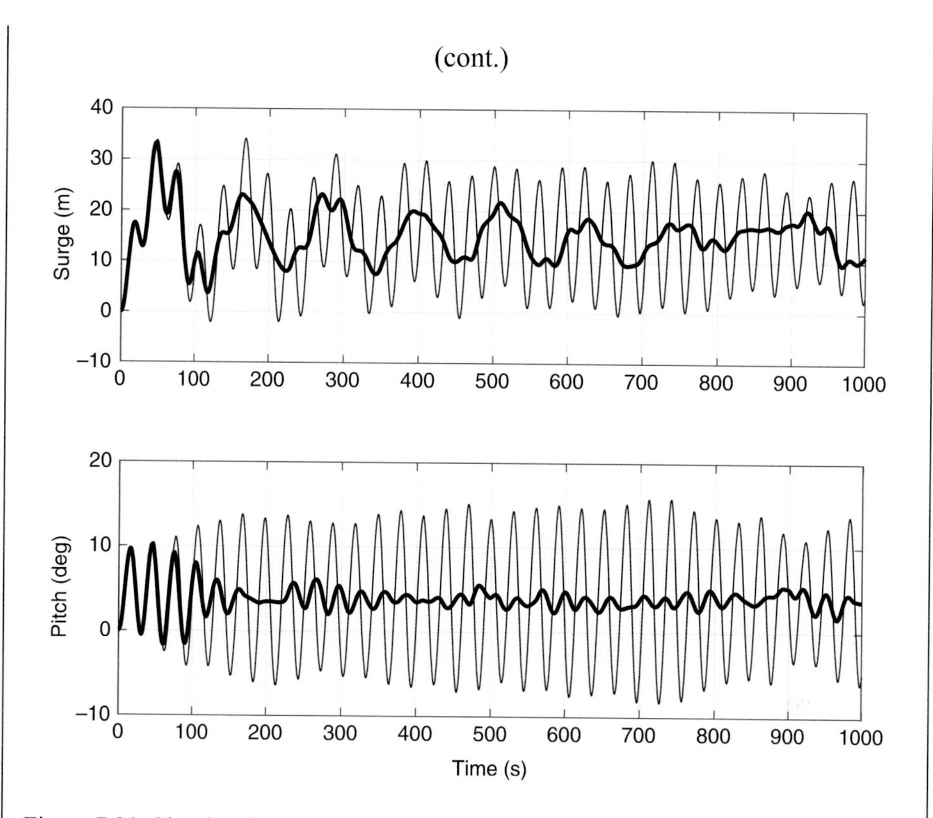

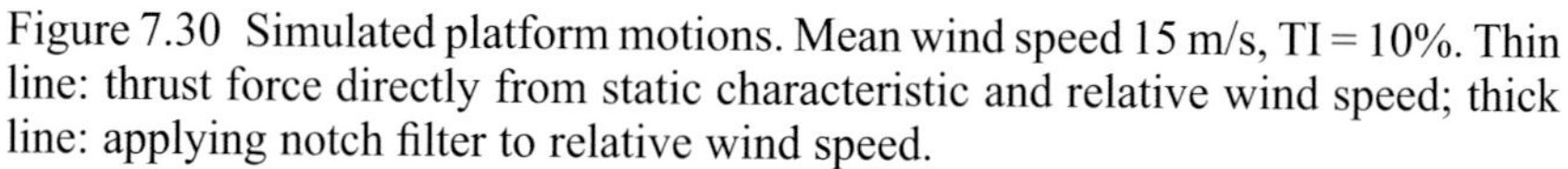

Figure 7.30 Simulated platform motions. Mean wind speed 15 m/s, TI = 10%. Thin line: thrust force directly from static characteristic and relative wind speed; thick line: applying notch filter to relative wind speed.

Two cases are considered in the simulations. First, a simulation is performed using the relative wind speed as input to the thrust force calculations and applying the quasi-static thrust characteristic to obtain the turbine thrust. The second simulation is performed in a similar way, the only difference being the relative wind speed, which now is passed through the notch filter prior to the thrust force calculation. The notch filter has a center frequency of 1/29.8 s, corresponding to the platform pitch natural frequency, and a bandwidth $\varepsilon = 0.1$*. The computed surge and pitch motions for the two cases are displayed in Figure 7.30.*

From Figure 7.30 it is observed that initially the motions are dominated by the transient effects. This is to be expected as the initial position of the platform is at origin with zero platform pitch angle and the wind force is set on at full strength at $t = 0$*. After a while the transient effects dies out. In the case without a notch filter, a large resonant pitch motion builds up. We have here an example of where the eigenmodes differ from the defined modes of motion. In the*

(cont.)

analysis, the pitch is defined as rotation about the y-axis, located at the water plane. However, the eigenmode for rotation has a center of rotation far below the water plane. Thus, the rotational eigenmode ("pitch mode") also dominates the surge motion. The pitch motion is limited by the system damping and because excitation of the natural frequency is avoided by the use of the notch filter. Further, the relative wind speed drops below the rated wind speed during some of the oscillations. Below rated wind speed, the aerodynamic loads contribute to damping.

Applying the notch filter, the platform pitch motion dies out after the transient and has only small amplitude motions caused by the turbulent wind force. The surge motion shows some long periodic resonant motions. These is because the wind forcing at the surge natural period is not filtered out.

7.8.3.3 Motion Control Strategies

Section 3.9.3 discussed the control of a bottom-fixed wind turbine operation above rated wind speed. A PI control of the blade pitch angle was used:

$$\Delta\beta = K_p \Delta\omega_g + K_i \int_0^t \Delta\omega_g dt, \quad [7.114]$$

where K_p and K_i are the proportional and integral gain respectively. If one considers a SDOF system, for example, the pure pitch motion of a floating wind turbine, it is possible to introduce a correction for the nacelle motion to avoid the negative damping discussed above. One may add a term proportional to the nacelle velocity on the right-hand side of [7.114], $K_{fb}\dot{\eta}_n$. This is a feedback term. De Souza (2022) investigated this approach for a spar platform with a 20 MW wind turbine combined with a rotor torque control. De Souza found that the approach may have stability issues, in particular when introducing a low-pass filter on the input to the controller. The filter introduces a phase shift which reduces the stability.

To improve the stability of the above feedback control, a term proportional to the nacelle offset may be introduced. Then [7.114] is modified to:

$$\Delta\beta = K_p \Delta\omega_g + K_i \int_0^t \Delta\omega_g dt - K_p K_{ff} \dot{\eta}_n - K_i K_{ff} \eta_n \; . \quad [7.115]$$

Here, K_{ff} is a feed-forward gain, less than zero. De Souza (2022) showed that this approached improved the stability even with a filter present.

Skaare et al. (2007) used an estimator-based control strategy consisting of a numerical model of the wind turbine together with measurements to estimate the incoming wind. The measurements can be nacelle acceleration, thrust force rotor torque and/or power. Skaare et al. (2007) implemented the estimator-based control system using a tool for computing the dynamics of floating wind turbines and found significant improved fatigue life of the tower due to the reduced pitch motion. A minor reduction in the mean power output was observed as well as some increased variability of power and rotational speed.

7.8.3.4 Use of Energy Shaping Control

Pedersen (2017) demonstrates how the negative damping can be mitigated by varying the rotor speed and thus using the rotor as an intermediate energy storage. He calls the principle "energy shaping control" (ESC). In the following, the main principle of the approach is outlined and demonstrated.

As above, the horizontal velocity of the floater at nacelle level is considered. Considering the combined surge and pitch motion, the velocity at nacelle level becomes $\dot{\eta}_n = \dot{\eta}_1 + z_n\dot{\eta}_5$.

It is assumed that the power output from the rotor follows the relative velocity between the wind and the rotor in a quasi-static manner, i.e., the delivered power from the rotor may be approximated by:

$$P = T(U_{w0} - \dot{\eta}_n). \qquad [7.116]$$

It is assumed that $|\dot{\eta}_n| << U_{w0}$. At above-rated wind speeds, the conventional controller aims at keeping the power as well as the rotor speed constant. The thrust, and thus the power, is thus controlled by pitching the blades.

Now assume that the rotational speed of the rotor may vary. Ignoring losses, the equilibrium in instantaneous power may then be written as:

$$\frac{d}{dt}\left[\frac{1}{2}I\Omega^2\right] + P = T(U_{w0} - \dot{\eta}_n). \qquad [7.117]$$

The first term in [7.117] is the power used for changing the kinetic energy in the rotor-generator assembly. Consider the variation around an equilibrium condition. At equilibrium $\Omega = \Omega_0$, $P = P_0$ and $T = T_0$. I.e., for a small deviation from the equilibrium condition, [7.117] may be written as:

$$I(\Omega_0 + \Delta\Omega)\left(\dot{\Omega}_0 + \Delta\dot{\Omega}\right) + P(\Omega_0) + \Delta P = (T_0 + \Delta T)(U_{w0} - \dot{\eta}_n). \qquad [7.118]$$

Now $\dot{\Omega}_0 = 0$ and $P(\Omega_0) = T_0 U_{w0}$. For small deviations from equilibrium, we thus obtain:

$$\Delta P = \frac{dP(\Omega_0)}{d\Omega}\Delta\Omega$$
$$\Delta T = \frac{dT(U_{w0})}{dU}\Delta U_{rel} = -\frac{dT(U_{w0})}{dU}\dot{\eta}_n, \qquad [7.119]$$

Omitting terms of higher order, and using [7.117], the following linear equation for the variation of the rotational speed is obtained:

$$I\Omega_0\Delta\dot{\Omega} + \frac{dP(\Omega_0)}{d\Omega}\Delta\Omega = \Delta T U_{w0} - T_0\dot{\eta}_n\ . \qquad [7.120]$$

Again, we observe that if the rotor velocity is constant, $\Delta\Omega = \Delta\dot{\Omega} = 0$, the following relation is obtained:

$$\frac{\Delta T}{T_0} = \frac{\dot{\eta}_n}{U_{w0}}. \qquad [7.121]$$

That is, if the axial velocity of the rotor increases (in the direction of the incident wind) the thrust force also increases, and a negative damping effect is obtained.

To avoid this negative damping effect, Pedersen (2017) proposes allowing for variation in the rotational speed. The variation of the rotational speed is controlled by an augmented reference signal for the rotor speed controller. Pedersen thus assumes that the actual rotational speed Ω is observed with high accuracy and that the set-point for the controller is augmented by a correction term $\Delta\Omega$ such that the controller aims for a rotational speed of $\Omega_r \simeq \Omega = \Omega_0 + \Delta\Omega$. The new rotational velocity is obtained by solving a differential equation with left-hand side equal to [7.120] but with a new right-hand side:

$$I\Omega_0\Delta\dot{\Omega} + \frac{dP(\Omega_0)}{d\Omega}\Delta\Omega = -\alpha T_0\dot{\eta}_n. \qquad [7.122]$$

By requiring that the right-hand sides of [7.120] and [7.122] are equal, the following is obtained:

$$\frac{\dot{\eta}_n(1-\alpha)}{U_{w0}} = \frac{\Delta T}{T_0}. \qquad [7.123]$$

It is observed that for $\alpha > 1$, the variation in thrust force will act in the opposite direction of $\dot{\eta}_n$, i.e., a positive damping is obtained. The thrust force on the turbine may now be written as:

$$T(t) = \frac{1}{2}\rho\pi R^2 C_t(U_w)U_w^2(t) + \Delta T = \frac{1}{2}\rho\pi R^2[C_t(U_w)U_w^2(t) + (1-\alpha)C_t(U_{w0})\dot{\eta}_n U_{w0}]. \qquad [7.124]$$

Here, $U_w(t)$ is the instantaneous incident wind velocity.

Now, $|\dot{\eta}_n|/U_{w0} \ll 1$, $C_t(U_w)/C_t(U_{w0})$ is close to unity and α is of order 1 but larger than 1. Thus, it is clear from [7.124] that the damping force is positive but small compared to the average wind thrust.

Use of Energy Shaping Control

The same 5 MW turbine as in the previous example is considered. The rotational inertia of the rotor-generator assembly is set to $I = 5.026E06$ kgm^2. *The power is given by* $P = Q\Omega$, *Q being the rotor torque. The derivative of the power with respect to the rotor velocity is obtained as:*

$$\frac{dP(\Omega_0)}{d\Omega} = Q_0 + \Omega_0 \frac{dQ}{d\Omega}. \qquad [7.125]$$

For simplicity, it is assumed that the second term may be ignored. In lack of a proper controller, a perfect control of the rotational speed is assumed. I.e., the rotational speed obeys [7.122], where the nacelle velocity is obtained by solving for the combined platform pitch and surge motions. The instantaneous wind thrust is then obtained from [7.124]. Simulations with and without ESC are performed. In the ESC case the augmentation factor α *is set to 2.*

First, a steady case with steady wind speed of 15 m/s is considered. The wind is started instantaneously at $t = 0$. *In Figure 7.31 the computed surge and pitch responses are shown and compared for the two cases. A significant transient response is observed in both cases. In the initial phase the responses are similar for the case with and without ESC, but after a while the ESC case responses are attenuated and decay toward the steady-state value, about 15 m for surge and 3 deg for pitch. The case without ESC continues with large oscillations dominated by the pitch natural period. Figure 7.32 shows the corresponding variations in rotor thrust and rotational velocity.*

(cont.)

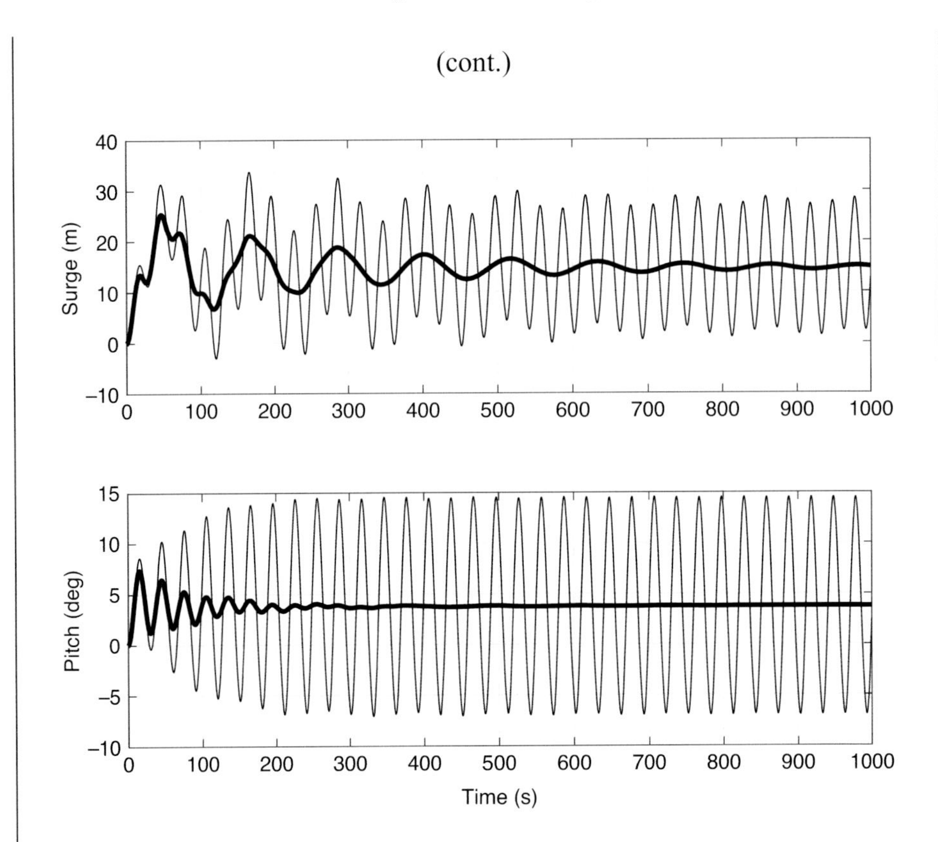

Figure 7.31 Steady 15 m/s wind. Initial surge and pitch = 0. Thick line: ESC with $\alpha = 2$; thin line: without ESC.

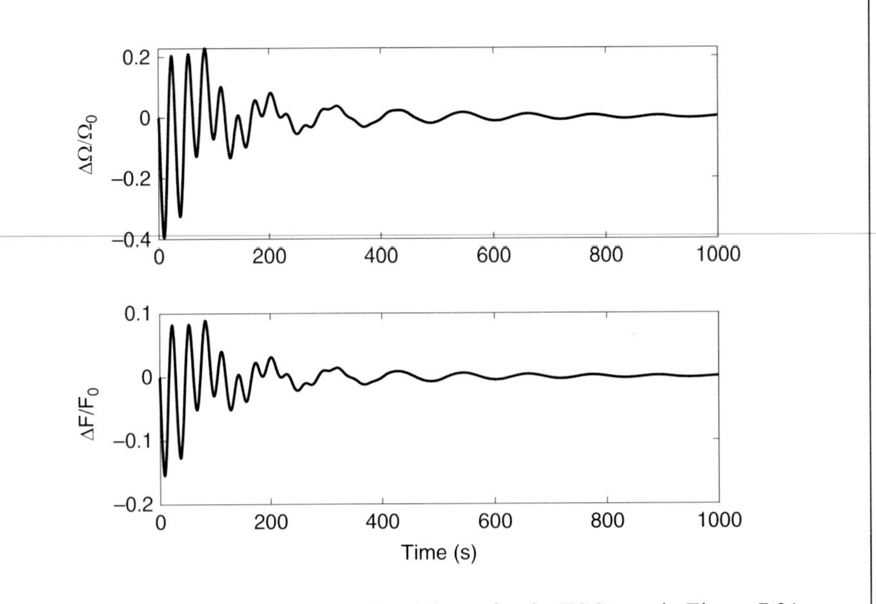

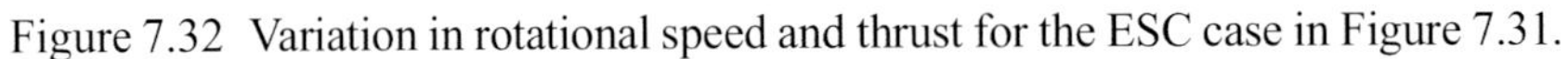

Figure 7.32 Variation in rotational speed and thrust for the ESC case in Figure 7.31.

(cont.)

Second, the same case as above is considered but including 10% turbulence intensity. A Kaimal wind spectrum is applied with coherence of 1 over the rotor area. The computed motions are shown in Figure 7.33. Large, almost steady-state pitch motions are obtained without the ESC, while in the ESC case the motions are reduced significantly after the end of the initial transient motions.

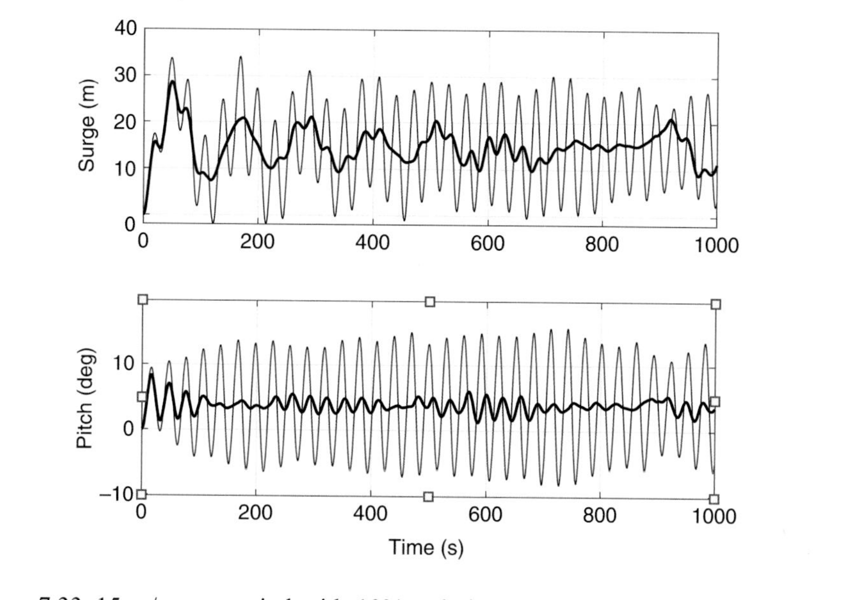

Figure 7.33 15 m/s mean wind with 10% turbulence intensity. Thick line: ESC with $\alpha = 2$; thin line: without ESC.

7.8.3.5 Examples from the Hywind Demo Development

In the development of the Hywind demo project, several new controllers were developed (see, e.g., Nielsen, Hanson and Skaare, 2006; Skaare et al., 2011; Skaare, Hanson and Nielsen, 2007; and Skaare et al., 2014). The detailed specifications of the controllers are proprietary, but results from the implementation of two of them are given in Skaare et al. (2011). The key idea is that the standard controller should work as before, keeping constant power production above rated wind speed and collectively adjusting the blade pitch angle to compensate for the variation in wind speed. The additional new motion controller "sits on top" of the standard controller, using the turbine motions as an extra input. By using the information of the turbine pitch motion an extra blade pitch signal can be given to provide the damping required to avoid excessive resonant motion, like what is demonstrated by the ESC in Section 7.8.3.4.

Similarly, as the blade pitch angle can be used to provide damping of the turbine pitch motion, one may use variation in generator torque to provide damping of the roll motion. In designing the controllers and determining the gains, one must evaluate issues such as: How much damping is needed? How do the motions influence fatigue of the structure? Does the controller impose excessive blade pitch activity and thus reduced lifetime of the actuators? Does the controller interact with nonresonant motions that are difficult to control, for example, wave-induced motions?

By a proper setting of the controller, energy from the wave-induced motions may be extracted by the wind turbine. This may have some relevance at below-rated wind speeds. The basic principles are given in Nielsen, Hanson and Skaare (2006). The key challenge is that large motions will be required to extract a significant amount of energy.

7.8.4 Some Possible Dynamic Instabilities

7.8.4.1 Heave-Pitch (Roll) Coupling

Section 8.4 discusses the Mathieu instability in the context of a lifting operation from a moving crane. The vertical motion of the crane tip causes a variation in the tension in the lifting line. The line tension is entering into the restoring force for the pendulum motion of the load, and thus the equation of motion of the pendulum motion has a time-varying stiffness. The differential equation for the pendulum motion is an example of the Mathieu equation, which, in a more general form, may be written as:

$$\ddot{\eta} + 2\zeta\omega_0\dot{\eta} + \omega_0^2[1 + \varepsilon\cos(\omega t)]\eta = 0. \qquad [7.126]$$

Here, it has been assumed that the time variation in the stiffness term is harmonic. ω_0 is the undamped natural frequency and ζ is the damping ratio. It can be shown that the time-varying stiffness term may cause instability of the system. Even without any external excitation, η may increase from an initial small value. For small ε, the instability occurs for $\omega/\omega_0 = 2/n,\ n = 1,\ 2,\ 3,\ \ldots$. For larger ε, the range of instability increases, while increased damping reduces the range of instability (see Figure 8.8).

For a floating platform, pitch or roll motions may be excited by large heave motions. Consider the pitch motion of a spar substructure. The pitch-restoring term is given in [7.69] as:

$$C_{55} = \rho g(S_{11} + Vz_B) - Mgz_G. \quad [7.127]$$

Within linear theory, C_{55} is a constant.[4] However, if the floating turbine performs large heave motions, and the pitch restoring is evaluated at the actual vertical position of the substructure, it is observed that both the displaced volume V and the centers of buoyancy and gravity will vary with the heave motion (Haslum and Faltinsen, 1999). If the substructure has a heave motion $\eta_3(t)$, the change in submerged volume becomes $-S\eta_3$, with S being the water plane area of the substructure. The center of buoyancy will move $\frac{1}{2}\eta_3$ upward. Thus, the pitch-restoring term in the heave-displaced position is obtained as:

$$\widetilde{C}_{55} = \rho g(V - S\eta_3)\left(z_B + \frac{1}{2}\eta_3 - z_G - \eta_3\right). \quad [7.128]$$

For a spar substructure, the water plane area in most cases will contribute little to the pitch restoring. Thus, by accounting for the heave motion, the pitch restoring can be approximated by:

$$\widetilde{C}_{55} \simeq C_{55}\left(1 - \frac{1}{2}\frac{\eta_3}{(z_B - z_G)}\right). \quad [7.129]$$

If $\eta_3(t)$ is harmonic and η in [7.126] is replaced with η_5, the time-varying pitch-restoring term is obtained as:

$$\varepsilon \cos(\omega t) = -\frac{1}{2}\frac{\eta_{3A}}{(z_B - z_G)} \cos(\omega t). \quad [7.130]$$

Considering the regions of instability, large pitch (or roll) motions may occur if the ratio between the period of the heave motion T and the pitch natural period T_0 is 1/2, 1, ... From Figure 8.8, it is observed that the range of instability is large in particular in the region around $T/T_0 = 1/2$. Large heave motions may occur if the heave natural period is in the range of the wave periods. Therefore, a normal design requirement is for the heave natural period to be larger than the wave periods and the pitch natural period to be more than twice the wave periods.

For a lightly damped substructure, resonant heave motions may also be excited by second-order wave loads and other nonlinear effects. Haslum and Faltinsen (1999) demonstrate how a heave motion consisting of one component at a wave frequency and a slower component at the heave resonant period may contribute to

[4] For floating bodies such as ships, we have $\rho V = m$. C_{55} is then frequently written as $\rho g V GM_0$, where GM_0 is the initial metacentric height. For floating bodies with large waterplane area, S_{11} is the dominating term, while for spars the term $V(z_B - z_G)$ dominates GM_0.

Mathieu instability. The two motion components will create a heave envelope process with frequency $\Delta\omega = \omega_{wave} - \omega_{03}$. Thus, if the pitch natural frequency coincides with $\Delta\omega$, large pitch motions may occur via the Mathieu instability. The nonlinear effect causing heave at the heave natural period is a coupling effect between the heave and pitch motion. When the spar substructure performs a pitch motion, forces acting perpendicular to the axis of the spar cause a vertical force. The heave motion at resonance is thereby amplified, which, via the Mathieu instability, again amplifies the pitch motion. Increased heave and pitch damping will mitigate the resonant responses.

7.8.4.2 Roll-Yaw Coupling

Haslum et al. (2022) point out an aerodynamic coupling effect that may cause large roll and yaw motions of a floating wind turbine. They illustrate the effect by considering a spar platform. The effect may be explained as follows. Consider a spar substructure, as illustrated in Figure 7.1. However, to minimize the coupling effects in the inertia matrix, the origin of the coordinate system is moved downward to the CG of the floating turbine. For simplicity only the roll and yaw motions are considered. Under the assumption that the thrust T on the rotor acts perpendicular to the rotor plane, the thrust in x and y-direction becomes $T_x = T\cos\eta_6$ and $T_y = T\sin\eta_6$. The thrust acts a distance h above origin, thus the roll and yaw moments due to the thrust may be written as:

$$\begin{aligned} F_4 &= -T_y h\cos\eta_4 \simeq -Th\eta_6 = -C_{46}\eta_6. \\ F_6 &= T_x h\sin\eta_4 \simeq Th\eta_4 = -C_{64}\eta_4 \end{aligned} \quad [7.131]$$

Here, small rotations have been assumed and the aerodynamic stiffness coupling terms $C_{46} = -C_{64}$ introduced. Assuming a diagonal mass matrix and that the linearized hydrostatic and mooring stiffnesses are also diagonal, the inertia and stiffness matrices for the coupled roll-yaw motion may be written as:

$$M = \begin{bmatrix} m_{44} & 0 \\ 0 & m_{66} \end{bmatrix}, \qquad C = \begin{bmatrix} C_{44} & C_{46} \\ -C_{46} & C_{66} \end{bmatrix}. \quad [7.132]$$

The eigenvalues for the undamped system are obtained using [5.21]:

$$\lambda^2 = \frac{1}{2}\left[\frac{C_{44}}{m_{44}} + \frac{C_{66}}{m_{66}} \pm \sqrt{\left(\frac{C_{44}}{m_{44}} - \frac{C_{66}}{m_{66}}\right)^2 - 4\frac{C_{46}^2}{m_{44}m_{66}}}\right]. \quad [7.133]$$

Without the last term, representing the aerodynamic coupling, the two solutions for λ^2 represent the two uncoupled natural frequencies, $\omega_{i0} = \sqrt{\frac{C_{ii}}{m_{ii}}}, i = 4, 6$. Including the coupling term, but with a weak coupling so the term inside the square root sign still is positive, two real natural frequencies are still obtained. If the coupling term increases so that the term inside the square root becomes negative, the natural frequencies become complex, i.e., $\lambda_i = \alpha_i \pm i\beta_i$. Thus, the undamped response will behave as:

$$\eta_i(t) = Ae^{i\lambda_i t} = Ae^{i\alpha_i t}e^{\pm\beta_i t}. \qquad [7.134]$$

The first term represents a harmonic oscillation while the second term represents an exponential decaying term and an exponential growing term. Thus, to avoid instability in the undamped case, the term under the square root sign in [7.133] must be positive. This can be expressed as a requirement to the frequency difference between the uncoupled roll and yaw frequencies:

$$|\omega_{40}^2 - \omega_{60}^2| \geq 2\frac{C_{46}}{\sqrt{m_{44}m_{66}}}. \qquad [7.135]$$

As $C_{46} = Th$, the requirement to the difference between the roll and yaw natural frequencies increases as the thrust force increases and as the distance between the CG and the nacelle increases.

The above considerations illustrate the main causes of the instability of the coupled roll-yaw motion. Haslum et al. (2022) have studied the phenomenon using a state-of-the-art computer program for floating wind turbines. Due to several coupling effects, nonlinearities and damping, an exponential increase of the motions is not observed, but rather limit-cycle motion responses. An important observation is that damping may reduce the limiting thrust force for stability as given by [7.135]. This is explained by the fact that a phase shift between the two motions will cause the thrust force to partly act in phase with the motion velocity. Damping causes such a phase shift. A force acting in phase with the motion velocity will transfer or extract energy into that mode of motion.

Exercises Chapter 7

1. Write down the (linear) contribution to the 6DOF restoring matrix for a tension leg platform with four vertical tethers in a quadratic pattern. Define the quantities involved.

2. Write down the (linear) contribution to the 6DOF restoring matrix from a tension leg platform with three vertical tethers in symmetric triangular pattern. Define the quantities involved.
3. Consider a four-legged tension leg platform with a quadratic layout and consider loads in the x-direction only (see Figure 4.7).
   a. Use quasistatic considerations and derive the relation between the wind thrust at nacelle level and the tether loads.
   b. Find a criterion for the minimum pretension in the tethers to avoid slack.
   c. Make a similar consideration as in Exercises 3a and 3b, but assume the loads are wave loads acting at the water line level.
4. The stiffness of a catenary mooring line consists of an elastic and a geometric contribution. Show how the total stiffness is obtained by combining the two.
5. Assume a wind turbine tower can be modeled as a vertical beam with uniform mass distribution. The tower is 120 m tall and has a mass of 5000 kg/m. Compute the 6DOF mass matrix for the tower when:
   a. the center of the coordinate system is at the bottom of the tower and the z-axis coincides with the tower axis.
   b. the center of the coordinate system is at the bottom of the tower and the z-axis is displaced 20 m in the negative x-direction and 10 m in the negative y-direction. The z-axis is parallel to the tower axis.
   c. the tower is located as in Exercise 5b and a nacelle modeled as a point mass of 450 Mg is located on top of the tower.
6. Consider the vertical wave forces on a spar platform (see Figure 7.7). The diameter of the lower part is D_L and of the upper part is D_U. The lower part of the cone starts at draft z_L and ends at z_U. The draft of the spar is L.
   a At which draft will you replace the cone with an abrupt change in diameter?
   b Assume deep-water waves. At which wave frequency will we have zero vertical wave force? Assume you may estimate the wave force by the Froude-Krylov contribution.
   c Write down the expression for the heave natural frequency.
7. Consider a semisubmersible as outlined in Figure 7.10. We will consider the two columns and the single pontoon only. The waves are assumed to propagate in the direction of the pontoon axis. The pontoon is 30 m long, 5 m wide and 3 m high. The bottom of the pontoon is located 20 m below the water surface. The columns have a square cross-section with the length of each side equal to 5 m. Deep-water waves may be assumed.
   a. For which wave period do we obtain cancellation of the heave wave force? Use the Froude-Krylov approximation.

   b. For which periods do we expect cancellation of the wave forces in surge?
   c. For which wave period do we obtain cancellation of wave forces in pitch? Use the Froude-Krylov approximation and make an estimate both by ignoring and including the horizontal forces of the columns.

8. In [7.63] the drag force on an element is given by considering the relative motion between the fluid and the motion. Assume a sinusoidal motion velocity of both fluid and body. Consider four different phases between the motion velocity and the fluid velocity, 0, $\pi/4$, $\pi/2$ and π, and assume the amplitude of the fluid velocity to be twice the body velocity.
   a. What is the minimum and maximum drag force on the body during one cycle of oscillation in the four cases?
   b. What is the dissipated damping energy during one period of oscillation in the four cases?
9. Assume a synthetic mooring line has a diameter of 0.12 m and a minimum breaking strength (MBS) of 10 MN. Assume that the mean tension in the line is 20% of the MBS. How large is the difference between the effective tension and the real tension in the line as a function of water depth?
10. Consider the thrust coefficient displayed in Figure 7.29 and consider a floating wind turbine with the nacelle 120 m above sea level and a natural period in pitch of 30 s. Consider platform pitch only. The static platform pitch at 7 m/s wind speed is 2 deg. Ignore damping and make a SDOF simulation routine and consider the following cases.
    a. Start with an initial platform pitch angle of 2 deg and use a steady wind speed of 7 m/s, 10 m/s, 12 m/s and 20 m/s (you may use k_{CT} from Figure 7.27.) Discuss the motion behavior in the various cases.
    b. Use one of the wind time histories in "WindTimeSeries.txt" and scale the velocities so that mean wind speeds of 7 m/s, 10 m/s, 12 m/s and 20 m/s are obtained. Check if the obtained turbulence level is reasonable. Simulate the dynamic response.
11. Show how the linear dynamic equation for the rotational speed in [7.120] is obtained.
12. Consider a spar-like floating wind turbine with the following main particulars. The substructure has only one diameter. Assume the vertical mass distribution of the tower and substructure to be uniform. Three mooring lines are used, placed symmetrically around the substructure. Main data is as follows.

Height of rotor center:	70 m
Rotor diameter:	82 m
Thrust coefficient:	0.8
Wind speed:	10 m/s
Tower:	from z = 0 to z = 68 m
Substructure:	from z = 0 to z = −95 m
RNA mass:	140 Mg
Center of gravity RNA, z_G:	70 m
Tower mass:	170 Mg
Center of gravity tower, z_G:	32 m
Substructure mass:	1300 Mg
Center of gravity substructure, z_G:	−39 m
Radius substructure:	4.1 m
Ballast:	from −95 to −79 m
Center of gravity ballast, z_G:	−87 m
Horizontal stiffness single mooring line:	18 kN/m
Vertical stiffness single mooring line:	2 kN/m
Vertical connection of mooring lines:	−20 m

Consider the surge, heave and pitch motions only. The origin of the coordinate system is at the water line level. Compute:

a. mass of ballast (ignore the vertical load from the mooring and power cable)
b. mass matrix
c. added mass matrix
d. hydrostatic restoring matrix
e. mooring line restoring matrix
f. natural frequencies with and without mooring attached
g. static pitch due to wind thrust
h. wave excited motions in surge, pitch and heave due to a wave with an amplitude of 1 m propagating in positive x-direction. Use strip theory. Use a range of wave periods from 3 to 40 s. Assume infinite water depth

13. Show how the linearized damping in [7.39] is obtained.

8
Marine Operations

Marine operations play an important role in the installation and maintenance of offshore wind farms. For bottom-fixed wind turbines, many lifting operations take place offshore, including installation of the substructure, transition piece, tower, nacelle, rotor blades etc. Most of these operations are weather-sensitive, in the sense that wind speeds and wave height must stay below certain threshold values during the operation. Most floating wind turbine concepts are designed to be assembled in sheltered waters, thus the number of marine operations in open water, at the wind farm site, may be reduced; however, installation of mooring systems and power cables still involves operations that are subject to weather restrictions.

An example of the order of magnitudes involved is the marine operations involved in the installation of the Dudgeon wind farm.[1] To complete the installation work, about 30,000 lifting operations took place, of which about 500 are categorized as heavy-lift operations. These include the lifting of tower segments, nacelles etc. More than 4000 vessel days were used.

During the operation of a wind farm, regular access is required to the turbines for inspection and minor repairs. However, "heavy maintenance" must also be planned for. This involves replacing major components such as turbine blades, gearboxes etc.

A short description of various marine operations will be given here, after which the concept of "weather windows" will be discussed. A weather window is a period

[1] Dudgeon wind farm is located about 30 km off the shore of North Norfolk. It started operation in 2017. The wind farm consists of 67 turbines, each with rated power of 6 MW. The turbines are installed in water depth of 18–25 m using monopiles (see http://dudgeonoffshorewind.co.uk/ (accessed March 12, 2018)).

in which, for example, the significant wave height stays below a certain value representing the maximum allowable for the operation at hand.

Lifting frequently involves crane operations where a crane sits on a vessel moving in the waves. This situation is modeled in a simplistic manner to illustrate some of the dynamic issues involved in such operations.

Impact between bodies and snatches in lifting gear should be avoided. Therefore, a method for analyzing the probability of such events is discussed in some detail.

8.1 Installation Operations

In this section, some typical marine operations performed during the installation of an offshore wind turbine are briefly discussed. The operations depend upon the support structure involved as well as available installation vessels. Methods and equipment used are rapidly developing. This is partly due to the development of equipment to reduce installation costs, but also due to the steady increase in turbine sizes, the number of offshore wind farm projects and new knowledge.

8.1.1 Bottom-Fixed Wind Turbines

Presently, monopiles are the most frequently used substructure for offshore wind turbines in shallow water depths. The dimensions and mass of monopiles increase with turbine size and water depth. The length of monopiles also depends upon the characteristics of the bottom sediments. The dimensions of monopiles were studied at the University of Strathclyde (Kiełkiewicz et al., 2015) and are summarized in Table 8.1. The monopiles are normally transported to the site by a transport barge, then lifted into vertical position by a crane vessel and hammered into the soil. Strict tolerances are set for the verticality of the piles.

On top of the monopile a transition piece is mounted. This component may have a weight of several hundred tons.

The tower may be installed as a single unit or in two to three segments. The tower has a considerable weight; for the two turbines listed in Table 8.1, the tower weights are 247 t and 654 t, with tower lengths of approximately 60 m and 85 m.

The nacelle and hub are normally lifted as one unit, while the blades may either be installed one by one, or two of the blades may be installed together with the nacelle and the third blade installed afterward.

Table 8.1 *Typical weights for bottom-fixed wind turbine components (Kiełkiewicz et al., 2015).*

Wind Turbine	Weight of Rotor, Nacelle and Hub (t)	Weight of Tower (t)	Diameter of Monopile (m)	Water Depth (m) Min/Max	Length of Pile (m) Min/Max	Weight of Monopiles (t)		Weight of Transition Piece (t)	
						min	max	min	max
Vestas V112 – 3.3 MW	155.0	247.0	6.0	15/30	38/75	426.9	1156.2	250.5	342.1
			7.0	15/40	35/78	488.1	1252.2	338.7	392.9
Vestas V164 – 8 MW	480.0	653.9	8.0	15/35	43/81	775.6	1838.2	467.8	590.5
			9.0	15/45	40/87	841.0	2155.4	623.4	715.6

Figure 8.1 Illustration of marine operations related to offshore wind turbines. From left: heavy-lift vessel installing monopiles; cable-laying operation; installation of nacelle and rotor blades; rock installation around the monopile; access during service and maintenance. Reproduced with permission by the Ulstein Group, Norway.

As can be understood from the weights and physical dimensions involved, the lifting operations are very sensitive to dynamic effects excited by wave and wind loads. To minimize the motions of the crane vessel (known as a heavy-lift vessel, HLV), these are in most cases jack-up vessels, resting at four to six legs on the sea floor, as illustrated by the rotor blade installation vessel in Figure 8.1. The jack-up and jack-down phases are the most weather-sensitive phases for the vessel. During these phases, wave-induced motions may cause the legs to impact on the sea floor. In lifted position, the vessel is not very sensitive to wave loads, but the lifting operations are sensitive to wind loads (see example in Table 8.2).

As discussed in Chapter 4, jackets and gravity substructures may also be used. Jackets are normally installed by a lifting operation similar to that for monopiles. However, after setting the jacket at the sea floor, each leg must be secured to the sea floor by piles.

Gravity foundations are mostly used at locations not suited for piling. Most gravity foundations are designed to be floated out to site and then lowered to the sea floor and filled by, for example, heavy gravel to secure sufficient stability. If the foundation is to be towed a long distance, this may represent a critical marine operation.

8.1.2 Floating Wind Turbines

Floating wind turbines are installed in waters too deep for the use of jack-up installation vessels. However, most floating wind turbines are designed to be assembled in sheltered waters preferably along a quay. After finalizing commissioning work, the complete assembly is towed to the installation site, where only the hook-up of mooring lines and power cables remains. However, in case of heavy maintenance work, one must choose between disconnecting the floating wind turbine and towing it to a quay or sheltered waters, or using a floating crane vessel to perform the necessary replacements on-site.

Installation of the anchors is a key marine operation for floating wind turbines. The anchors used depend upon the bottom sediments as well as the mooring principle used. For a catenary moored platform located at a site with soft sediments, either drag-in fluke anchors or suction anchors may be used. The suction anchor is an upside-down bucket driven into the sea floor by applying internal under-pressure. The fluke anchor needs to be dragged into the sea floor by a load similar to the maximum mooring load expected during operation. This may require an anchor installation vessel with very large towing capacity.

For tension leg platforms (TLPs), vertical mooring loads act on the anchors. A normal requirement is that the mean vertical force on the anchor should be carried by the weight of the anchor, while at least fractions of the dynamic load may be taken by friction or suction effects. Thus, the anchors for TLPs must be heavy, requiring additional mass to be added in the form of, for example, gravel after installation. Several TLP concepts are not stable prior to mooring and de-ballasting. Very specialized installation procedures may thus be required for such concepts. The procedures may involve the use of additional floating and stabilizing aids for the installation phase.

The completed floating wind turbine is to be towed to the site for installation. For this purpose, a tug is applied. In addition to the towing tug, frequently two additional tugs are used. These are connected at the rear of the towed turbine and are used to control the trajectory of the towed turbine in restricted waters, as well as to control its directional stability. The directional stability of the towed turbine is controlled by hydrodynamic forces, in particular the yaw moment acting on the towed structure during forward speed; the length of the towing line; its point of attachment on the towed structure and its tension. Both static and dynamic instabilities may occur. In the case of static instability, a steadily increasing yaw angle of the towed structure will take place until nonlinear effects limit the yaw angle. In the case of dynamic instability, yaw oscillations of increasing amplitude will occur. Details of the directional stability of towed structures are discussed in more detail in Faltinsen (1990).

While towing in restricted waters, a short towing line is preferred to keep control the trajectory of the towed structure. In open waters, however, a long towing line is preferred. This reduces the effect of the induced velocities from the tug's propeller. The induced velocities in the propeller wake increase the resistance of the towed structure. Further, in open waters, both the tug and the towed structure exhibit wave-induced motion. If the towing line is short, the stiffness will be large, and thus the motions will induce large dynamic forces in the towing line. Using a long towing line and possibly inserting a segment of synthetic rope with low elastic stiffness, e.g., nylon, the elastic stiffness of the towing line will be significantly reduced and thus the dynamic loads also reduced. See the discussion of mooring lines in Section 7.5.

Many floating offshore wind substructures have vertical circular cylinders as part of the structure (see Figures 4.3 and 4.4). Towing a vertical circular cylinder through water implies that vortices are shed from the cylinder, creating a wake behind the cylinder. The vortex shedding is due to the separation of the flow from the cylinder and occurs on both sides of the cylinder. The shedding may be symmetric, occurring simultaneously on both sides of the cylinder, or asymmetric, such that the vortices are shed alternately from the two sides of the cylinder. The vortex shedding will induce forces on the cylinder. The symmetric vortex shedding induces forces in the direction of towing, denoted as inline forces. The asymmetric vortex shedding induces both inline forces and forces perpendicular to the direction of towing, denoted as crossflow forces. In most cases the crossflow forces are much larger than the inline forces.

The vortex shedding frequency can be written as $f_s = S_t\, U/D$, where S_t is the Strouhal number, U is the velocity of the cylinder through the water, and D is the cylinder diameter. The Strouhal number depends upon the Reynolds number for the flow and roughness of the cylinder. The Reynolds number is defined by $\mathrm{Re} = UD/\nu$, where ν is the kinematic viscosity. For a smooth cylinder in subcritical flow, $300 \lesssim \mathrm{Re} \lesssim 3 \cdot 10^5$, the Strouhal number $S_t \simeq 0.2$. In the transitional range, $3 \cdot 10^5 \lesssim \mathrm{Re} \lesssim 3.5 \cdot 10^6$, the separation pattern is chaotic and the Strouhal number may vary over a wide range, from 0.2 to above 0.4. In the supercritical range, $3.5 \cdot 10^6 \lesssim \mathrm{Re}$, the vortex shedding pattern becomes more regular and asymmetric with a slightly increasing Strouhal number, from about 0.22 towards 0.3 at $\mathrm{Re} = 10^7$. In the case of asymmetric vortex shedding, the inline forces will occur at twice the vortex shedding frequency, while the crossflow forces will occur at the vortex shedding frequency. For a more thorough discussion of the vortex shedding patterns, see Blevins (1977).

The above considerations are based upon 2D experiments and considerations. For a long cylinder, i.e., where the length-to-diameter ratio is much larger than one, the vortex shedding will have limited correlation along the length of the cylinder (Blevins, 1977). However, if the vortex shedding frequency is close to a natural frequency, resonant motions may be triggered. The motion will contribute to synchronization of the vortex shedding along the cylinder and thus increase the forcing. The crossflow motion amplitude for a lightly damped cylinder may exceed one diameter. However, at amplitudes above approximately half a diameter, the vortex shedding pattern is modified and the excitation forces modified; accordingly, the process is "self-limiting." The vortex-induced motion of a rigid structure is frequently abbreviated VIM.

The vortex shedding may also trigger elastic resonant modes. This phenomenon is denoted vortex-induced vibrations (VIV). In this case the resonant oscillations will also contribute to synchronization of the vortex shedding pattern along the structure and thus increase the loading.

In determining the natural frequency of a structure, the added mass must be estimated. However, a phenomenon called "lock-in" may occur. This means that the vortex shedding frequency becomes controlled by the oscillation frequency and may thus differ from the vortex shedding frequency in the case of a fixed cylinder (Blevins, 1977; Larsen and Koushan, 2005). This can be interpreted as a change in timing of the vortex shedding and that the vortices thus modify the added mass of the cylinder. The consequence is that VIM and VIV may not only happen in a narrow range of towing speeds but may lock-in in a wider range of speeds when first triggered. The effect of lock-in is more important in water than in air due to the higher density.

Frequency of Vortex Shedding, VIM and VIV

Consider a vertical cylinder with diameter 10 *m towed at a speed of* 1 *m/s. In water with temperature* 15^0C, *the kinematic viscosity is about* $1.14 \cdot 10^{-6}$ m^2/s. *The Reynolds*

(cont.)

number becomes approximately 10^7 *and the flow is thus supercritical. Assuming a Strouhal number of* 0.25 *gives a vortex shedding frequency* $f_s = 0.025$ *Hz. This may be close to, for example, the roll natural frequency of the towed structure.*

Consider a tower of a wind turbine exposed to a wind velocity of $10\ m/s$, *and assume a constant tower diameter of* 6 *m. The kinematic viscosity of air at* 15^0C *is approximately* $1.48 \cdot 10^{-5}$. *Thus, the flow is supercritical in this case as well. The vortex shedding frequency becomes* $f_s = 3.27$ *Hz. This frequency may excite an elastic bending mode of the tower.*

VIV and VIM are both very sensitive to damping. Increased damping reduces both the inline and crossflow motion amplitudes significantly. Also measures to avoid synchronization of the vortex shedding along the cylinder will reduce the amplitude of oscillation. A classical measure to break up the synchronization along the length of the cylinder is by attaching helical strakes, as seen on most large chimneys. Wake effect from other nearby structures will also disturb the synchronization. Examples on such nearby structures are additional columns on the floater, pontoons, and wind turbine rotor in front of tower.

8.2 Access

Offshore wind turbines must be accessible by people at regular intervals. In the earliest days of offshore wind, a very simple access system was used, in which the turbine was equipped with a vertical ladder from the sea level to the access platform, located above the level of the wave crests. A small crew boat with a bumper in the bow was used for the transfer of people. The boat used thrusters to push toward the ladder while the crew jumped from the boat to the ladder and climbed up. This operation required very calm wave conditions and could involve a significant hazard to the crew.

In recent years, several new systems have been developed to make access less weather-sensitive and safer. The systems normally involve larger vessels and a motion-compensated platform or gangway. The vessel may either push toward the wind turbine during the access operation, or it may stay a few meters away from the turbine, using dynamic positioning (DP) and a motion-compensated gangway directly to the access platform, as illustrated in Figure 8.1. In some cases, helicopters may be used for direct access to the nacelle.

8.3 Weather Windows

8.3.1 Introduction

A weather window is a weather situation, normally a wave condition or a wind condition, that allows certain marine operations to be performed. I.e., the significant wave height and/or wind speed should be below certain thresholds and the duration of this condition should be sufficient to complete the operation. Using wave statistics, for example, one may estimate the number and duration of events with significant wave height below a certain threshold during a year or season (winter, spring, summer, autumn). However, in a real operational decision situation, one must rely upon weather forecasts to determine if the weather window is sufficient to complete the operation at hand. Depending upon the duration of the operation and the specific weather situation, the weather forecast can be more or less reliable. The limiting weather parameter is usually either wind or waves. In some cases current can be a limiting factor as well. However, the current speed is much more predictable than wind and waves and is thus easier to account for. In the following, weather windows related to waves are considered, but weather windows related to wind can be handled in a similar manner.

In Figure 8.2 (left), the hourly significant wave height is plotted for one year. The crossings of a threshold level (3 m is used as an example) are marked by crosses (up-crossing) and stars (down-crossings). A clearly seasonal variation of the significant wave height is observed, with low wave heights during the summer and larger wave heights during the autumn and winter seasons. Thus, to establish useful statistics for the duration of weather windows, the figures should be worked out on a monthly or seasonal basis.

In Figure 8.2 (right), data for approximately one month of the time history is extracted. It is easily observed how sensitive the data are to the threshold level. If the operations considered have an operational threshold of $H_s < 5\,m$, the operations could be performed almost uninterrupted during this month. On the other hand, if the threshold is as low as $H_s < 2\,m$, only a minor fraction of the time during this month is available, and the duration of each period is short. It is thus of great importance to design the marine operations for as large waves as possible and to be able to interrupt the operations if the wave height should exceed the operational limit for a period of time. The figure also illustrates how missing data may destroy the duration statistics. For the period between 1830 h and 1980 h, wave data are missing, and the missing values are replaced by linearly interpolated values. One or more crossings of the 3 m level may have occurred during this period.

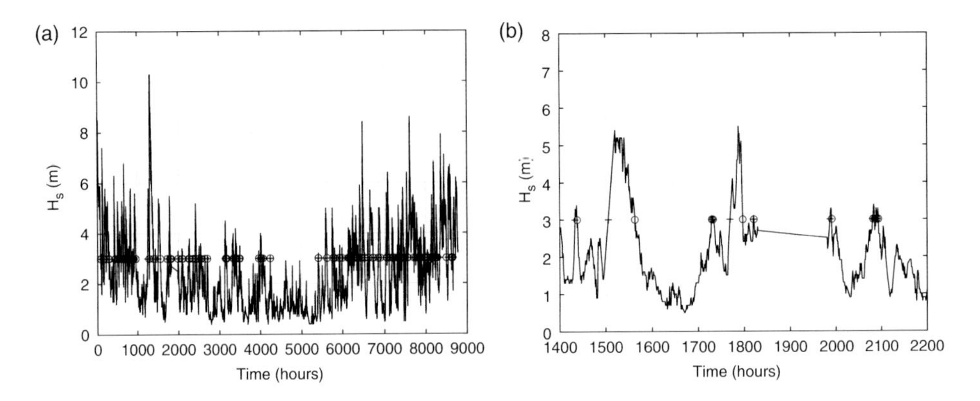

Figure 8.2 Example of significant wave height recorded every hour for one year. The crosses indicate an up-crossing of the threshold level, in this example 3 m, and the circles indicate a down-crossing of the threshold level. Right: illustration of missing data filled in using linear interpolation.

8.3.2 Duration Statistics for Calm Wave Conditions

If long time series of wind and/or wave data are available, statistics of the weather windows may be established. Graham (1982) investigated such time series and established some statistical models useful for planning purposes.

Consider a total period T_{tot}. During this period the significant wave height is below a chosen threshold N_c number of times and the total duration of the significant wave height staying below the threshold is denoted T_c. The average duration of the calm periods is thus $\overline{\tau}_c = T_c/N_c$. Graham (1982) considered the relation between the cumulative probability of the significant wave height $P(H_s)$ and the average duration of periods with wave heights below this. He found that the relation could be fitted to a Weibull distribution and that the average duration of periods with significant wave height below H_s can be approximated by:

$$\overline{\tau}_c = A[-\ln(P(H_s))]^{-\frac{1}{B}} . \quad [8.1]$$

A and B are fitted to historical data and Graham (1982) found that $A = 20$ hours and $B = 1.3$ fitted wave data well for the North Sea. The important observation from [8.1] is that *sea states with the same probability of exceedance have the same average duration*. According to Graham this statement can be extended to *sea states with the same probability of exceedance have the same persistence duration characteristics*. Figure 8.3 gives examples of observed durations of calm periods in the summer and winter seasons in the North Sea. In both seasons there are a few calm periods of very long duration, while most of the calm periods have very short duration.

Average Duration of Calms

From Figure 2.36, it is observed that $P(H_s) = 0.7$ for $H_s \simeq 3.0$ m. Using [8.1] the average duration of the calm periods is found to be:

$$\bar{\tau}_c = 20h[-\ln(P(H_s))]^{-\frac{1}{1.3}} = 44.2h$$

This is an all-year average. During the summer season $P(H_s) = 0.7$ for $H_s \simeq 2.0$ m, and during the winter season $P(H_s) = 0.7$ for $H_s \simeq 4.2$ m. Thus, the calm periods have an average duration of 44.2 h for wave heights below 2.0 meters in the summer season and below 4.2 meters in the winter season.

More accurate data from the North Sea are available; see Haakenstad et al. (2021). *Analyzing ten years of data from a site 56°05" north, 5°0" east, the following mean H_s-values are found with $P(H_s) = 0.7$: $H_s = 2.3$, 1.6 and 3.6 m for the all-year, summer and winter data respectively. From the data, the corresponding mean duration of calm periods are found to be 80, 83, and 62 h. Analyzing $P(H_s)$ in the range 0.6 to 0.9 and estimating the values of A and B in [8.1], the following values are found to fit the data: A = 41, 43, 31 h, and B = 1.58, 1.52 and 1.48 for the all-year, summer and winter data respectively.*[2]

Based upon data fitting, Graham (1982) found that the cumulative probability of the duration of a calm period may also be represented by a two-parameter Weibull distribution:

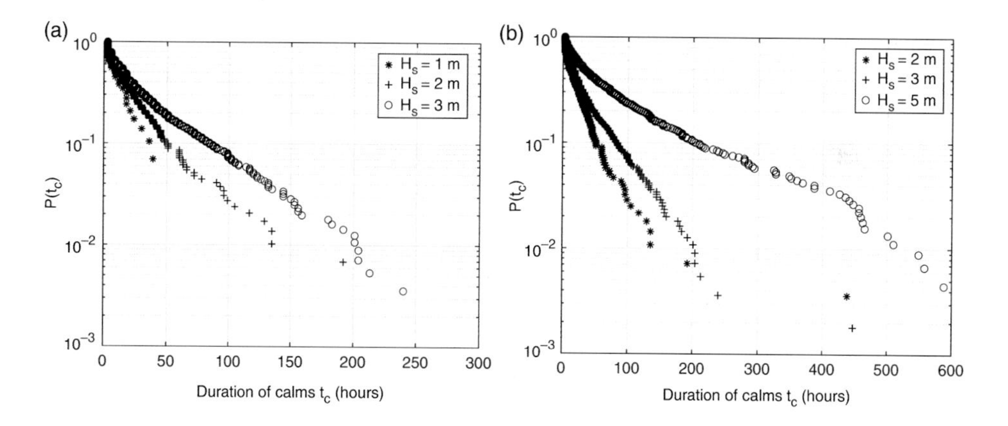

Figure 8.3 Examples of the statistics of duration of the calm periods during winter (January–March) and summer (June–August) in the North Sea. Crossing levels 2, 3 and 5 m for the winter season and 1, 2 and 3 m for the summer season.

[2] See Exercise 8.1 for further analyses of the data.

$$P(t) = 1 - e^{-\left(\frac{t}{t_c}\right)^{\beta_t}}. \quad [8.2]$$

Here, $P(t)$ is the probability that the duration of the calm period is less than t. t_c and β_t are parameters to be determined. The average duration of a calm period may be written:

$$\overline{\tau}_c = \int_0^\infty t \frac{dP(t)}{dt} dt. \quad [8.3]$$

From [8.2], we obtain:

$$\frac{dP(t)}{dt} = \beta \frac{t^{\beta_t - 1}}{t_c^{\beta_t}} e^{-\left(\frac{t}{t_c}\right)^{\beta_t}}. \quad [8.4]$$

By introducing [8.4] into [8.3], the average duration is obtained as:

$$\overline{\tau}_c = t_c \Gamma\left(\frac{1}{\beta_t} + 1\right), \quad [8.5]$$

where $\Gamma()$ denotes the Gamma function. $\overline{\tau}_c$ is now eliminated by combining [8.5] and [8.1], obtaining:

$$t_c = \frac{A[-\ln(P(H_s))]^{-1/B}}{\Gamma(\frac{1}{\beta_t} + 1)}. \quad [8.6]$$

Assuming A, B and $P(H_s)$ are known from observed data, it remains to determine β_t. Vik and Kleiven (1985) determined β_t from measured North Sea wave data and found a relationship as shown in Figure 8.4. It is observed that β_t is a function of the significant wave height. It will certainly also depend upon season and location as well.[3] However, $\beta_t = 0.8$ is a reasonable estimate for wave heights relevant for marine operations.

As most marine operations have limitations related to both the significant wave height as well as the duration of the calm period, it is of interest to establish the probability of $H_s < H_s'$ and duration $t > t'$, i.e., $P(H_s|t > t')$ is to be established. The mean duration of periods with $H_s < H_s'$ and $t < t'$ can be written as:

$$\overline{\tau}_{cs}(H_s < H_s'|t < t') = \frac{1}{P(t')} \int_0^{t'} t \frac{dP(t)}{dt} dt. \quad [8.7]$$

[3] The specific numbers used in the present examples are based upon old wave data. New values of the parameters should therefore be estimated based upon newer observations. Examples on results obtained using newer data are shown in Exercise 8.1.

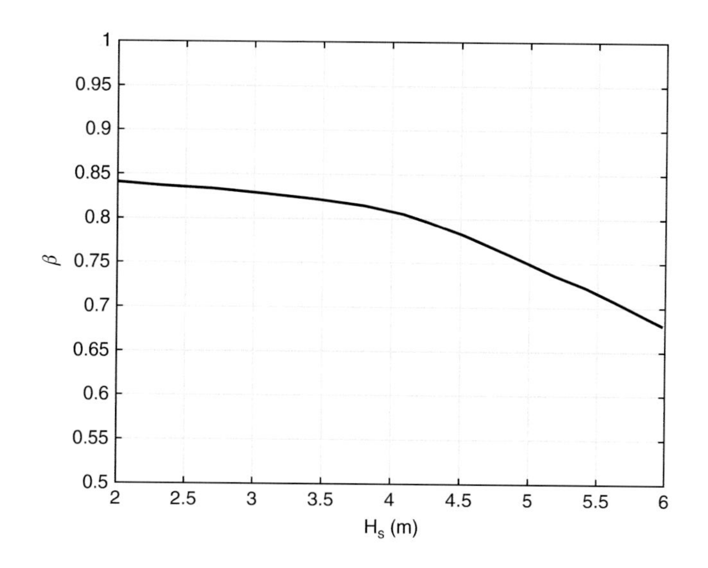

Figure 8.4 β_t versus H_s for North Sea wave data. As obtained by Vik and Kleiven (1985).

The average duration of periods with $H_s < H'_s$ is given above. During a period of duration T_{tot}, the number of calm periods is thus obtained as $N_c = P(H'_s)\frac{T_{tot}}{\overline{\tau}_c}$. Of this total number of calm periods, N'_c will have a duration shorter than t', such that:

$$N'_c(H_s < H'_s | t < t') = P(t')N_c. \qquad [8.8]$$

The total time with $H_s < H'_s$ and $t < t'$ is given by the number of occurrences and the average duration, i.e., $T'_c = N'_c \overline{\tau}_{cs}$. The relations necessary to find the "accumulated available operational time," i.e., periods with $H_s < H'_s$ and $t > t'$, T_{acc} are now established:

$$T_{acc} = T_c - T'_c = T_{tot}P(H'_s) - N'_c\overline{\tau}_{cs} = T_{tot}P(H'_s)(1 - \lambda). \qquad [8.9]$$

Here, λ is given by:

$$\lambda = \frac{1}{\overline{\tau}_c}\int_0^{t'} t\frac{dP(t)}{dt}dt. \qquad [8.10]$$

λ is in the range $(0, 1)$. The cumulative probability of $H_s < H'_s$ and $t > t'$ is now written as:

$$P(H'_s | t > t') = P(H'_s)(1 - \lambda). \qquad [8.11]$$

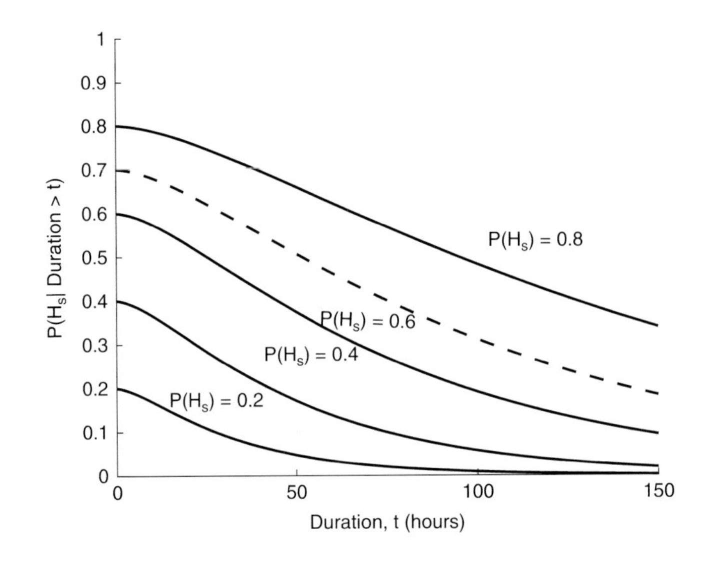

Figure 8.5 The conditional probability of the significant wave height H_s staying below a certain level for a duration of t hours. [8.11] is used with $A = 20$ *hours*, $B = 1.3$ and $\beta_t = 0.8$. The values for $P(H_s)$ are established by using the three-parameter Weibull distribution in Section 2.3.2 with $\alpha = 2.37$, $\beta = 1.425$ and $\gamma = 0.6234$. Dashed line corresponds to $P(H_s) = 0.7$.

It is observed that as $t' \rightarrow 0$ m the above cumulative distribution tends to the cumulative distribution of H_s (see Figures 2.36 and 2.37). In Figure 8.5, $P(H_s|t > t')$ is shown as a function of t' for various $P(H_s)$ levels.

Minimum Duration of Calms

Continued from the previous example, i.e., $P(H_s) = 0.7$. Assume the operation to be performed needs a minimum duration of 50 h. From Figure 8.5, it is obtained that $P(H_s|t > 50hours) = 0.51$. Thus, in an average summer season, the wave conditions will be such that 51% of the time the waves stay below 2.0 m for periods of duration at least 50 h. For the remaining time, the wave height is either above 2.0 m (30% of the time) or below 2.0 m, but for periods shorter than 50 h (19% of the time). Similar results are found for the winter season, but now with a threshold for the wave height of 4.2 m.

In the above examples, $P(H_s)$ for an average season was applied. However, the weather conditions vary significantly from one year to another. If wave measurement over long periods of time, e.g., 20–30 years, are available or such data are established based on hindcast analyses, the variability of the wave conditions may be taken into consideration. Then $P(H_s)$ for a certain return period may be used in the analyses. This can be used to add a safety margin to the estimates of operational availability.

The above derivation is based upon Graham (1982) and the key assumption that sea states with the same probability of exceedance have the same average duration. Mathiesen (1994) states that in general this assumption does not hold. He asserts that to obtain reliable statistics on the mean duration of exceedance, the distribution of the wave height as well as the average rate of change of significant wave height must be known. In Mathiesen's approach A and B in [8.1] depend upon $P(H_s)$ as well as the average rate of change of H_s.

8.3.3 Time Domain Simulation of Marine Operations

In Section 8.3.2 a statistical method for estimating "available time for operation" was presented. The method is fast and easy to use and suited for obtaining estimates for feasibility studies at an early design stage. However, the method is approximate and averages out much of the information in the real wave records. If long wave records are available, time domain simulation of the operations is an attractive approach that will produce more reliable estimates.

Depending upon the operation considered, various vessels are involved and various weather limitations are present. Dinwoodie et al. (2015) uses the limitations shown in Table 8.2 for the analysis of weather-limited marine operations. In Table 8.3 typical durations for the operations are given. Similar data for simulation are given in, e.g., Dalgic et al. (2015). In addition to the operations at the field, mobilization time and transit time must be accounted for. These have different weather restrictions.

In simulation marine operations using historic records of waves and wind, one must consider how decisions are made during real time operations. In the real operational case, decisions must be made based upon the expected duration of the

Table 8.2 *Typical weather limitation criteria for vessels involved in installation and maintenance of offshore wind turbines. Based on Dinwoodie et al. (2015).*

	Crew Transfer Vessel (CTV)	Field Support Vessel (FSV)	Heavy-Lift Vessel (HLV)
Governing weather criteria	Wave	Wave	Wave/Wind
Limitation (H_s / U_{mean})	1.5 m	1.5 m	2.0 m/10 m/s
Comments			Wave height: jack-up (duration approx. 3 h) Wind: crane operation

Table 8.3 *Some operations considered with vessels involved and corresponding average operational time. From Dinwoodie et al. (2015).*

	Minor Repair	Major Repair	Major Replacement
Vessel used	CVT	FSV	HLV
Typical duration of operation (h)	7.5	26	52

operation as well as uncertainties in the weather forecasts. A typical procedure is described here.

8.3.4 Simulation of Weather-Restricted Marine Operations

The following procedure is a short summary of the procedure as described by DNV (2021d). Assume a marine operation has an estimated or planned duration of T_{pop}. Due to uncertainties in this estimate, a contingency time T_c should be added to the estimate. Depending upon how detailed the assessment of the time estimates are, the ratio T_c/T_{pop} may vary. If no detailed assessment is made, $T_c/T_{pop} > 1$. $T_c/T_{pop} = 0.5$ is normally accepted if the operation considered is:

- an operation with an extensive experience basis from similar operations;
- a towing operation with properly assessed towing speed;
- repetitive, where T_{pop} has been accurately defined based on experience with the actual operation and vessel.

The total time to be used for establishing the weather window, the reference period T_R, is thus $T_R = T_{pop} + T_c$. Thus, to increase the number of available weather windows, marine operations should be designed so that they can be interrupted if the weather conditions deteriorate. That will shorten T_R.

If $T_R < 96$ *hours* or $T_{pop} < 72$ *hours*, the operation may be considered to be "weather-restricted." In such cases the operation may be started based upon a reliable weather forecast. If the duration is longer than the above limits, the operation will normally be considered to be an "unrestricted" operation. That implies that weather forecasts cannot be used for starting the operation. In such cases, the operation must be designed for extreme weather conditions for the location and season considered.

For weather-restricted operations, uncertainties in the weather forecasts must be accounted for. These uncertainties increase as the length of the weather forecasts increases. It is also assumed that the forecasts for mild weather conditions (low wave heights and low wind speeds) are relatively more uncertain than estimates for

Table 8.4 *α_H-factor for significant wave height according to Offshore Standard DNV-ST-N001 (DNV, 2021d). Assuming use of public domain weather forecast and no environmental monitoring during operation.*

Operational Period (h)	Design Significant Wave Height H_{sD} (m)			
	1	2	4	6
$T_{pop} \leq 12h$	0.65	0.76	0.79	0.80
$T_{pop} \leq 24h$	0.63	0.73	0.76	0.78
$T_{pop} \leq 36h$	0.62	0.71	0.73	0.76
$T_{pop} \leq 48h$	0.60	0.68	0.71	0.74
$T_{pop} \leq 72h$	0.55	0.63	0.68	0.72

Table 8.5 *α_W-factor for wind according to Offshore Standard DNV-ST-N001 (DNV, 2021d). Assuming use of public domain weather forecast and no environmental monitoring during operation.*

	Design Wind Speed V_{wD}	
Operational Period (h)	$V_{wD} < 0.5V_{10year\ return}$	$V_{wD} > 0.5V_{10year\ return}$
$T_{pop} \leq 24h$	0.80	0.85
$T_{pop} \leq 48h$	0.75	0.80
$T_{pop} \leq 72h$	0.70	0.75

more harsh conditions. If a specific operation has as design limit[4] corresponding to a certain significant wave height H_{sD} and wind speed V_{wD}, then the limiting wave height and wind speed in the weather forecast should be lower to account for the uncertainty in the forecast. This is accounted for by introducing a so-called α-factor on the design wave height and wind speed. The forecasted significant wave height H_{sF} during the reference period should thus not exceed $\alpha_H H_{sD}$ and, similarly, the forecasted wind speed V_{wF} should not exceed $\alpha_W V_{wD}$. If the operation is very sensitive to the wave period, the joint probability of the wave height and, e.g., spectral peak period must be considered. The "alpha"-factor is a function of the duration of the operation considered, the operational period. For operations of duration a few hours, the factor is close to 1, while for longer durations, the factor is reduced.

In Table 8.4, the α_H-factor for typical marine operations according to Det Norske Veritas' Offshore Standard (DNV, 2021d) is given. It is observed that for operations with low design wave height and long durations, the α_H-factor is below 0.7. In Table 8.5 similar values for α_W are given. Again, it is observed that the factor

[4] A design limit in this context is the maximum allowable wave height, wind speed etc. or combinations of these that are acceptable due to, e.g., strength or deformation criteria. In establishing these design limits, proper safety factors are included on load estimates and material capacities.

has the lowest values for long durations and low wind speeds. Linear interpolation is to be used for wave heights between the tabulated values. The specific examples are valid while using publicly available weather forecasts and no environmental monitoring during the operation. The load and resistance factor design (LRFD) approach is assumed to be used in the structural design. If other design principles or if other weather forecasts are applied, the α- factors should be modified. The same is the case if environmental monitoring is applied during the operation.

Use of α-Factor in Marine Operations

Assume an operation with estimated duration of 26 h and a limiting design significant wave height of 1.5 m (see Tables 8.2 and 8.3). From Table 8.4 we obtain that the α-factor to be used is 0.67 ($T_{pop} \leq 36$ h, linear interpolation between 1 and 2 m wave height). With $T_c = 0.5T_{pop}$ we must thus require a weather forecast of $H_{sF} < 0.67 \cdot 1.5\ m = 1.0\ m$ for a period of $T_R = 1.5 \cdot 26\ h = 39\ h$ before the operation is started. This is a very strict requirement; the probability distribution of H_s for the North Sea used in Chapter 2 gives $P(H_s < 1\ m) = 0.07$ on a yearly basis. Adding the requirement of a duration of 39 h gives a probability of approximately 0.01. Thus, such an operation should preferably be designed to be less sensitive to the wave height and/or the operation should be interruptible, so that shorter weather windows can be considered.

8.4 Dynamics of Lifting Operations

Lifting operations are an important part of installation and maintenance operation related to offshore wind. In this section, the lifting of objects by a floating crane vessel will be discussed in some detail. Several simplifications are made to better illustrate some main features of the dynamics. For planning and dimensioning analyses, more complete multibody computer programs for dynamic analysis should be applied.

It may be useful to distinguish between "light-lift" operations and "heavy-lift" operations. In the present context the two categories are described as follows. For light-lift operations it is assumed that the load is very light compared to the vessel. It may be assumed that the motion characteristics of the vessel (at the top of the crane) are unaffected by the presence of the load. The order of magnitude of the load will be less than 1–2% of the vessel displacement. For such loads, heave compensation is possible.

For heavy-lift operations the coupled dynamics of the vessel-load system must be considered as there will be mutual interaction. For a system comprising a crane vessel, a barge and a lifted object, in principle one has to consider an 18-degree-of-freedom dynamic system with hydrodynamic as well as structural interaction. The structural interaction is due to, e.g., mooring lines between crane vessel and barge, and hoisting wire and tugger lines between load and crane vessel. The order of magnitude of the load will be more than 1–2% of the vessel displacement and typically more than a few hundred tons. Heave compensation may not be possible in such cases.

The lifting operations in the offshore wind industry imply higher lifting heights than in other marine applications. Therefore, wind loads and vessel-lifted object interaction effects may be more severe than elsewhere.

8.4.1 Coupled Dynamics for a Simple Lifting Operation

To illustrate the effect of coupling between vessel motions and the motion of a lifted load, the system shown in Figure 8.6 is considered. This is a two-body system where the bodies are connected via a single lifting wire. Each body is rigid with six degrees of freedom. However, as only a single wire connects the two bodies, the rotational motions of the lifted object are disregarded. If the rotations should be controlled, extra "tugger wires" have to be included. Only three degrees of freedom are thus included for the lifted object, i.e., the linear motions in the horizontal plane and the vertical motion. The following considerations are based upon a linear analysis, i.e., small motions are assumed.

A Cartesian earth-fixed coordinate system with the z axis positive upward and zero at the mean free water surface is used. The x-axis points aft along the vessel and $y = 0$ is assumed to be a plane of symmetry. All motions $\eta_i, i = 1, 9$ are referred to this coordinate system. The motions are given as deviations from the static equilibrium position. The initial position of the load is given as:

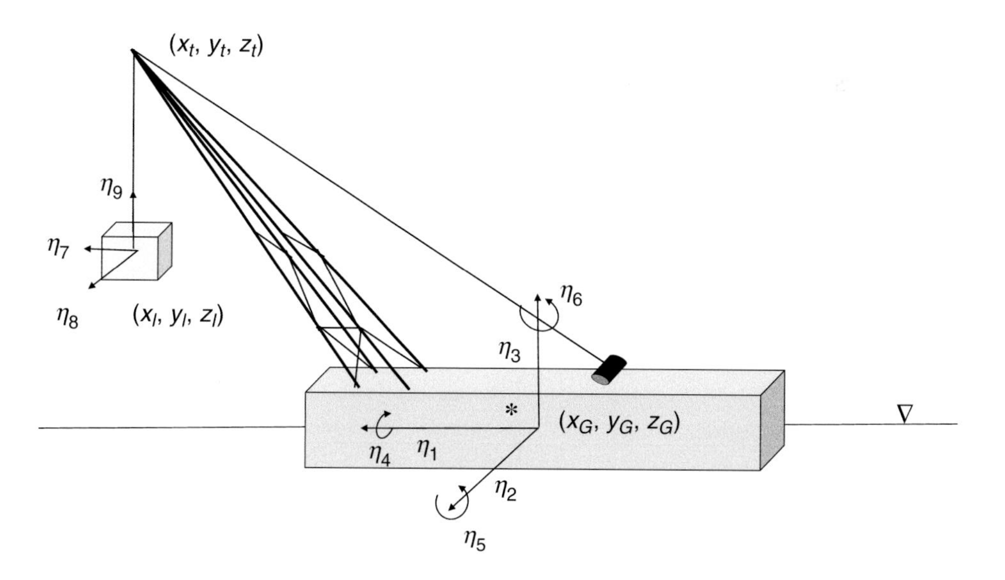

Figure 8.6 Illustration of a simple lifting operation. The star indicates the center of gravity of the vessel.

$$\begin{aligned} x_l &= x_t \\ y_l &= y_t \\ z_l &= z_t - l_s. \end{aligned} \quad [8.12]$$

Here, (x_t, y_t, z_t) is the static equilibrium position of the top of the crane and l_s is the length of the lifting wire from the crane top to the load. To perform the dynamic analysis, mass, damping and inertia matrices as well as excitation loads must be established. As a start, the mass and stiffness matrices are established. The undamped eigenmodes and eigenfrequencies can thereby be found. These are important in order to understand the coupled dynamics.

The mass matrix is obtained as follows:

$$\mathbf{M} = \begin{bmatrix} m + A_{11} & 0 & A_{13} & 0 & mz_G + A_{15} & -my_G & 0 & 0 & 0 \\ & m + A_{22} & 0 & -mz_G + A_{24} & 0 & mx_G + A_{26} & 0 & 0 & 0 \\ & & m + A_{33} & my_G & -mx_G + A_{35} & 0 & 0 & 0 & 0 \\ & & & I_{44} + A_{44} & I_{45} & I_{46} + A_{46} & 0 & 0 & 0 \\ & & & & I_{55} + A_{55} & I_{56} & 0 & 0 & 0 \\ & & & & & I_{66} + A_{66} & 0 & 0 & 0 \\ & & & Sym & & & m_l + a_{11} & 0 & 0 \\ & & & & & & & m_l + a_{22} & 0 \\ & & & & & & & & m_l + a_{33} \end{bmatrix}. \quad [8.13]$$

Here, m and m_l are the dry masses of the vessel and lifted load respectively. (x_G, y_G, z_G) is the location of the center of gravity (CG) of the vessel. Note that even if $y = 0$ is a plane of geometric symmetry, the CG of the vessel may be offset from the symmetry plane as the load is not necessarily located at $y_l = 0$. The moments of inertia of the vessel mass are given from:

$$\begin{aligned} I_{44} &= \int (y^2 + z^2) dm \\ I_{55} &= \int (z^2 + x^2) dm \\ I_{66} &= \int (x^2 + y^2) dm \\ I_{45} &= -\int xy \, dm = 0 \\ I_{46} &= -\int xz \, dm \\ I_{56} &= -\int yz \, dm = 0. \end{aligned} \quad [8.14]$$

A_{ij} are the components of the hydrodynamic added mass matrix for the vessel. Note that several of these components are zero due to the assumption that the vessel geometry is symmetric about $y = 0$. The a_{ii} components are the added mass for linear translations of the lifted load in direction i. These are zero when the lift is performed in air. Note that there are no coupling terms between the inertia terms for the vessel and the lifted load. In principle, for a submerged load, hydrodynamic coupling terms will exist, even if in most cases these are negligible. As discussed before, the added mass terms in general are frequency-dependent.

The stiffness matrix for the system may conveniently be split into three contributions: the hydrostatic effects, the effects of the positioning system and the coupling between the vessel and load due to the lifting line. The vessel may either be kept in position by mooring lines or by a dynamic positioning system. For a conventional catenary mooring system, the horizontal restoring effects are the most important ones; the vertical components are frequently ignored as they are small compared to the hydrostatic effects. The restoring effects of the mooring system may thus be approximated by the following stiffness matrix:

$$\mathbf{C}_m = \begin{bmatrix} C_{m11} & C_{m12} & 0 & 0 & 0 & C_{m16} & 0 & 0 & 0 \\ & C_{m22} & 0 & 0 & 0 & C_{m26} & 0 & 0 & 0 \\ & & 0 & 0 & 0 & 0 & 0 & 0 & 0 \\ & & & 0 & 0 & 0 & 0 & 0 & 0 \\ & & & & 0 & 0 & 0 & 0 & 0 \\ & & & & & C_{m66} & 0 & 0 & 0 \\ & & Sym & & & & 0 & 0 & 0 \\ & & & & & & & 0 & 0 \\ & & & & & & & & 0 \end{bmatrix}. \quad [8.15]$$

If the mooring system is symmetric, only the diagonal terms are non-zero. Note, however, that due to the nonlinear force-displacement characteristics of catenary mooring systems, a mean force on the vessel, due to wind, waves and current effects, will make the tension in the lines unequal and thus create coupling effects even if the layout pattern is symmetric. More on mooring line characteristics can be found in Section 7.5. If the vessel is kept in position by a dynamic positioning system, the system may be approximated by a linear spring-damper system and the restoring effects modeled similarly as above.

The hydrostatic stiffness matrix may be written as:

$$
\mathbf{C}_h = \begin{bmatrix}
0 & 0 & 0 & 0 & 0 & 0 & 0 & 0 & 0 \\
 & 0 & 0 & 0 & 0 & 0 & 0 & 0 & 0 \\
 & & C_{h33} & C_{h34} & C_{h35} & 0 & 0 & 0 & 0 \\
 & & & C_{h44} & C_{h45} & C_{h46} & 0 & 0 & 0 \\
 & & & & C_{h55} & C_{h56} & 0 & 0 & 0 \\
 & & & & & 0 & 0 & 0 & 0 \\
 & & Sym & & & & 0 & 0 & 0 \\
 & & & & & & & 0 & 0 \\
 & & & & & & & & 0
\end{bmatrix} . \qquad [8.16]
$$

Here, it is assumed that the lifted load either is fully submerged or hanging in air. If the load is partly submerged, additional terms will appear ($C_{h39}, C_{h49}, C_{h59}, C_{h99}$). The above hydrostatic coefficients are given by:

$$
\begin{aligned}
C_{h33} &= \rho g A_{wl} \\
C_{h34} &= \rho g \int_{A_{wl}} y dA = 0 \\
C_{h35} &= -\rho g \int_{A_{wl}} x dA \\
C_{h44} &= \rho g \int_{A_{wl}} y^2 dA + \rho g V z_B - m g z_G \\
C_{h45} &= -\rho g \int_{A_{wl}} xy dA = 0 \\
C_{h46} &= -\rho g V x_B + m g x_G \\
C_{h55} &= \rho g \int_{A_{wl}} x^2 dA + \rho g V z_B - m g z_G \\
C_{h56} &= -\rho g V y_B + m g y_G \, .
\end{aligned} \qquad [8.17]
$$

Here, ρ is the density of water, A_{wl} is the area of the water plane of the vessel and (x_B, y_B, z_B) is the center of buoyancy (CB) of the vessel. For a freely floating vessel, static equilibrium requires $\rho V = m$, $x_B = x_G$ and $y_B = y_G$. This is not necessarily the case in the present lifting operation. C_{h34} and C_{h45} are zero due to the assumed geometric symmetry of the vessel.

To establish the stiffness matrix for the coupling between the vessel and the load, the load is given a unit displacement in each of the three modes of motion (direction 7–9). The corresponding reaction forces in all modes of motion are then identified. The force in each of the nine directions may be written as:

$$
\begin{aligned}
F_1 &= -\frac{w}{l_s}(\Delta x - \eta_7) \\
F_2 &= -\frac{w}{l_s}(\Delta y - \eta_8) \\
F_3 &= -\frac{AE}{l_e}(\Delta z - \eta_9) \\
F_4 &= -F_2 z_t + F_3 y_t \\
F_5 &= -F_3 x_t + F_1 z_t \\
F_6 &= -F_1 y_t + F_2 x_t \\
F_7 &= \frac{w}{l_s}(\Delta x - \eta_7) = -F_1 \\
F_8 &= \frac{w}{l_s}(\Delta y - \eta_8) = -F_2 \\
F_9 &= \frac{AE}{l_e}(\Delta z - \eta_9) = -F_3.
\end{aligned} \qquad [8.18]
$$

Here, $(\Delta x, \Delta y, \Delta z)$ is the displacement of the top of the crane, given by:

$$
\begin{aligned}
\Delta x &= \eta_1 + z_t\eta_5 - y_t\eta_6 \\
\Delta y &= \eta_2 + x_t\eta_6 - z_t\eta_4 \\
\Delta z &= \eta_1 + y_t\eta_4 - x_t\eta_5.
\end{aligned} \qquad [8.19]
$$

Further, AE is the modulus of elasticity of the lifting wire multiplied by the cross-sectional area of the wire. l_e is the “elastic” length of the wire, i.e., the length of the wire that is stretched due to the wire force. Here, one may include the effect of a possible crane deformation as well.

By associating forces and corresponding displacements, the elements of the stiffness matrix due to the coupling effects are found as $C_{vij} = -\partial F_i/\partial \eta_j$ and the complete matrix thus becomes:

$$
\mathbf{C}_v = \begin{bmatrix}
\frac{w}{l_s} & 0 & 0 & 0 & \frac{w}{l_s}z_t & -\frac{w}{l_s}y_t & -\frac{w}{l_s} & 0 & 0 \\
 & \frac{w}{l_s} & 0 & -\frac{w}{l_s}z_t & 0 & \frac{w}{l_s}x_t & 0 & -\frac{w}{l_s} & 0 \\
 & & \frac{AE}{l_e} & \frac{AE}{l_e}y_t & -\frac{AE}{l_e}x_t & 0 & 0 & 0 & -\frac{AE}{l_e} \\
 & & & C_{v44} & -\frac{AE}{l_e}x_t y_t & -\frac{w}{l_s}z_t x_t & 0 & \frac{w}{l_s}z_t & -\frac{AE}{l_e}y_t \\
 & & & & C_{v55} & -\frac{w}{l_s}z_t y_t & -\frac{w}{l_s}z_t & 0 & \frac{AE}{l_e}x_t \\
 & & & & & C_{v66} & \frac{w}{l_s}y_t & -\frac{w}{l_s}x_t & 0 \\
 & & & Sym & & & \frac{w}{l_s} & 0 & 0 \\
 & & & & & & & \frac{w}{l_s} & 0 \\
 & & & & & & & & \frac{AE}{l_e}
\end{bmatrix}. \quad [8.20]
$$

The terms C_{vii} are given as:

$$
\begin{aligned}
C_{v44} &= \frac{w}{l_s}z_t^2 + \frac{AE}{l_e}y_t^2 \\
C_{v55} &= \frac{w}{l_s}z_t^2 + \frac{AE}{l_e}x_t^2 \\
C_{v66} &= \frac{w}{l_s}y_t^2 + \frac{w}{l_s}x_t^2 .
\end{aligned} \quad [8.21]
$$

The total stiffness matrix is now given as $\mathbf{C} = \mathbf{C_m} + \mathbf{C_h} + \mathbf{C_v}$. Given the mass matrix and the stiffness matrix, the undamped eigenvalue problem may be solved, providing the undamped natural frequencies and modes. Assuming the solution for each mode can be written as $\eta_i = \eta_{Ai}e^{i\omega t}$, where η_{Ai} is a complex amplitude, the homogeneous undamped coupled equation may be written as:

$$
\left(-\omega^2\mathbf{M} + \mathbf{C}\right)\boldsymbol{\eta} = 0. \quad [8.22]
$$

The eigenvalues and eigenmodes are thus obtained from the eigenvalue problem: $\lambda\boldsymbol{\psi} = \mathbf{M}^{-1}C\boldsymbol{\psi}$. To each eigenvalue, $\lambda_i = \omega_{0i}^2$, corresponds an eigenvector, $\boldsymbol{\psi}_i$, defining the contribution from each of the degrees of freedom to that specific resonant mode of motion. ω_{0i} is the undamped natural frequency for eigenmode i. For rigid bodies one frequently talks about "natural frequency in heave or pitch," etc. That implies that the eigenmodes are almost equal to a pure displacement along the z-axis, or a rotation about the y-axis. For a coupled system the eigenmodes may be composed of several modes of motion, making it difficult to given them proper names. An example of this is given here.

If external forces from wind or waves are acting on the system, it is may still be convenient to solve the problem in the frequency domain. A forcing vector is then written as:

$$\mathbf{F} = \mathrm{Re}\{\mathbf{F_A} e^{i\omega t}\}, \qquad [8.23]$$

where $\mathbf{F_A} = (F_1, F_2, \ldots \;, F_9)^T$ is the complex forcing vector and ω is the frequency of excitation. The wave loads can be obtained as discussed in Chapter 7. In addition, wind excitation may be included. Including a damping matrix, $\mathbf{B}$, the motion response of the vessel-load system is obtained from:

$$\boldsymbol{\eta} = \left(-\omega^2\mathbf{M} + i\omega\mathbf{B} + \mathbf{C}\right)^{-1}\mathbf{F}_A e^{i\omega t}. \qquad [8.24]$$

$\boldsymbol{\eta}$ is now a 9 x 1 complex vector containing amplitudes and phases for each degree of freedom.

Crane Load on a Floating Vessel

Consider a semisubmersible vessel with the following simplified properties.

Mass:	$m = 5.17E7\ kg$
Center of gravity:	(x_g, y_g, z_g) = *(0, 0, -0.02L), L= 100 m*
Radii of inertia:	(r_{44}, r_{55}, r_{66}) =*(.33L,.32L,.35L)*, $r_{ij} = 0,\ i \neq j$

Added mass matrix for vessel:

$$\frac{\mathbf{A}}{m} = \begin{bmatrix} .7 & 0 & 0 & 0 & 0 & 0 \\ & 1 & 0 & 0 & 0 & 0 \\ & & 1 & 0 & 0 & 0 \\ & & & .1L^2 & 0 & 0 \\ & Sym & & & .1L^2 & 0 \\ & & & & & .1L^2 \end{bmatrix}.$$

(cont.)

The vessel restoring matrix is assumed to be diagonal and is given as:

$$\frac{\mathbf{C}}{C_{33}} = \begin{bmatrix} .025 & 0 & 0 & 0 & 0 & 0 \\ & .025 & 0 & 0 & 0 & 0 \\ & & 1 & 0 & 0 & 0 \\ & & & .05L^2 & 0 & 0 \\ & Sym & & & .05L^2 & 0 \\ & & & & & .002L^2 \end{bmatrix}.$$

Here, $C_{33} = \rho g A_{wl} = 7.85\ MN/m$.

The load and lifting line have the following properties.

Mass of load:	$m_l = 0.02m$
Position of top of crane:	$(x_t, y_t, z_t) = (.6L, 0, .5L)$
Length of wire (from top of crane to load):	$l_s = 0.5L$
Elasticity of line:	*AE=3.96E09 N*
Added mass of load:	*zero*

I.e., in this case the whole system is symmetric about y = 0. The vessel mass and restoring matrices are both pure diagonal except for a minor coupling as the CG has a vertical offset from origin. Considering the eigenvalues and eigenvectors for the vessel alone and the load alone, the following is obtained.

Eigenperiods for the load alone: 14.19 14.19 2.87 s.
Corresponding eigenvectors (the columns of the matrix):

$$\xi_{load} = \begin{bmatrix} 1 & 0 & 0 \\ 0 & 1 & 0 \\ 0 & 0 & 1 \end{bmatrix}.$$

The two first eigenvectors correspond to displacement in horizontal direction, i.e., a pendulum motion, while the third value corresponds to vertical motion of the load, i.e., stretching of the wire.

Eigenperiods for the vessel alone: 132.99 32.43 144.25 32.99 22.81 170.10 s.
Corresponding eigenvectors (the columns of the matrix):

(cont.)

$$\psi_{vessel} = \begin{bmatrix} -1.00 & 0.0218 & 0 & 0 & 0 & 0 \\ 0 & 0 & 1.00 & 0.0184 & 0 & 0 \\ 0 & 0 & 0 & 0 & 1.00 & 0 \\ 0 & 0 & 0.003 & -0.9998 & 0 & 0 \\ 0.0036 & 0.9998 & 0 & 0 & 0 & 0 \\ 0 & 0 & 0 & 0 & 0 & 1.00 \end{bmatrix}.$$

In the above matrix, each column represents one eigenvector corresponding to the eigenvalues listed above. In these results coupling effects are observed. Thus, the eigenmodes are not corresponding to the pure vessel modes as defined by the coordinate system. In naming the eigenmodes, the name of the dominating vessel mode is frequently used.

Note that the eigenvectors are normalized so the sum of the squares of the components equals one. I.e., the eigenvector will look different if we define rotations in degrees or radians. As a rotation of one radian is very large, the eigenvectors usually are easier to interpret if the rotations are defined in degrees. Degrees are used in the present example, even if the mass and restoring matrix are referred to radians.

The first eigenvector above is dominated by the first degree of freedom, i.e., it is dominated by surge with a small component of pitch. The second mode is dominated by the fifth degree of freedom, i.e., the pitch motion with a minor coupling to surge. Similarly, the third and fourth modes are sway and roll eigenmodes with small coupling effects between the two. The last two vectors correspond to heave and yaw, which in this case appear without any coupling terms. The undamped heave and yaw eigenperiods are 22.8 and 170.1 s respectively.

Solving for the complete nine-degree-of-freedom system, the following eigenperiods are obtained.

2.80 13.90 13.69 22.91 33.57 33.37 133.77 172.99 144.89 s.

The corresponding eigenvectors are obtained as:

$$\xi_{vessel} = \begin{bmatrix} -0.0007 & 0.0124 & -0.0023 & 0.0002 & -0.0039 & 0.0033 & 0.7044 & -0.0020 & -0.0009 \\ -0.0001 & 0.0019 & 0.0105 & 0.0000 & -0.0025 & -0.0041 & -0.0006 & 0.0166 & -0.7407 \\ 0.0101 & 0.0006 & -0.0000 & 0.7226 & 0.0047 & -0.0028 & 0.0000 & 0.0000 & -0.0000 \\ 0.0055 & -0.0059 & -0.0333 & -0.0068 & 0.3335 & 0.5342 & -0.0000 & -0.0007 & -0.0012 \\ -0.0343 & 0.0334 & -0.0063 & 0.0431 & -0.4634 & 0.3840 & -0.0012 & -0.0001 & -0.0000 \\ 0.0000 & 0.0005 & 0.0315 & 0.0000 & 0.0085 & 0.0198 & 0.0089 & 0.6744 & 0.0749 \\ 0.0012 & -0.9823 & 0.1800 & 0.0613 & -0.4988 & 0.4089 & 0.7097 & -0.1207 & -0.0141 \\ 0.0002 & -0.1800 & -0.9825 & 0.0097 & -0.3466 & -0.5486 & 0.0088 & 0.7283 & -0.6675 \\ -0.9993 & -0.0370 & 0.0008 & 0.6871 & 0.5523 & -0.3140 & 0.0013 & 0.0000 & -0.0002 \end{bmatrix}.$$

(cont.)

In this case the coupling effects are much more involved and it is more difficult to interpret the eigenmodes. However, using the uncoupled results as a starting point, it is observed that the different eigenmodes have their major contribution to motion from the following modes of motion.

Eigenmode no.	1	2	3	4	5	6	7	8	9
Dominating mode of motion/ direction	9	7	8	3	5	4	1	6	2
Name	Load vertical	Load x-dir.	Load y-dir.	Heave	Pitch	Roll	Surge	Yaw	Sway

The "load eigenperiods" are all somewhat shorter in the coupled case than in the uncoupled case. This reflects that the corresponding motion of the vessel is in opposite phase to the load motion, e.g., vessel heave and vertical load motion have opposite sign in mode 1. Similarly, the negative pitch motion component for mode 1 moves the crane top upward, while the load moves downward. The two pendulum modes (2 and 3) have different natural periods in the coupled case. All the vessel natural periods become a little longer in the coupled case, due to the increased inertia in the system.

Next, the dynamic response of the system is demonstrated by exciting the system with a simplified forcing:

$$\mathbf{F} = \mathrm{Re}\left\{ \begin{bmatrix} iC_{11} & 0 & C_{33} & 0 & 0 & 0 & 0 & 0 & 0 \end{bmatrix}^T e^{i\omega t} \right\}.$$

This is a frequency-independent excitation in surge and heave with a magnitude so the response should converge toward unity in the two degrees of freedom for large periods of oscillation. The surge and heave forces are 90 deg out of phase, which to some extent resembles wave load excitation. The results are illustrated in Figure 8.7. The solid black lines are for the case defined above, while the dotted line is obtained using $m_{l1} = m_l/1000$, i.e., almost no coupling from the load to the vessel. It is observed that the surge and heave motions show no coupling effects. These are both controlled by the external excitation. The pitch motion, however, shows a significant effect of the load. The vertical force in the load causes a pitch motion via the force in the crane, even if the motion in absolute terms is small. The horizontal motion of the load, mode 7, follows the horizontal vessel motion for long periods of excitation, However, clear tops in the response are observed for periods corresponding to the eigenperiods for pendulum motion, vessel heave motion and vessel pitch motion. The vertical motion of the load follows the motion of the crane top, which has a resonant peak at the heave natural period. A small effect of the pitch natural period is observed as well.

(cont.)

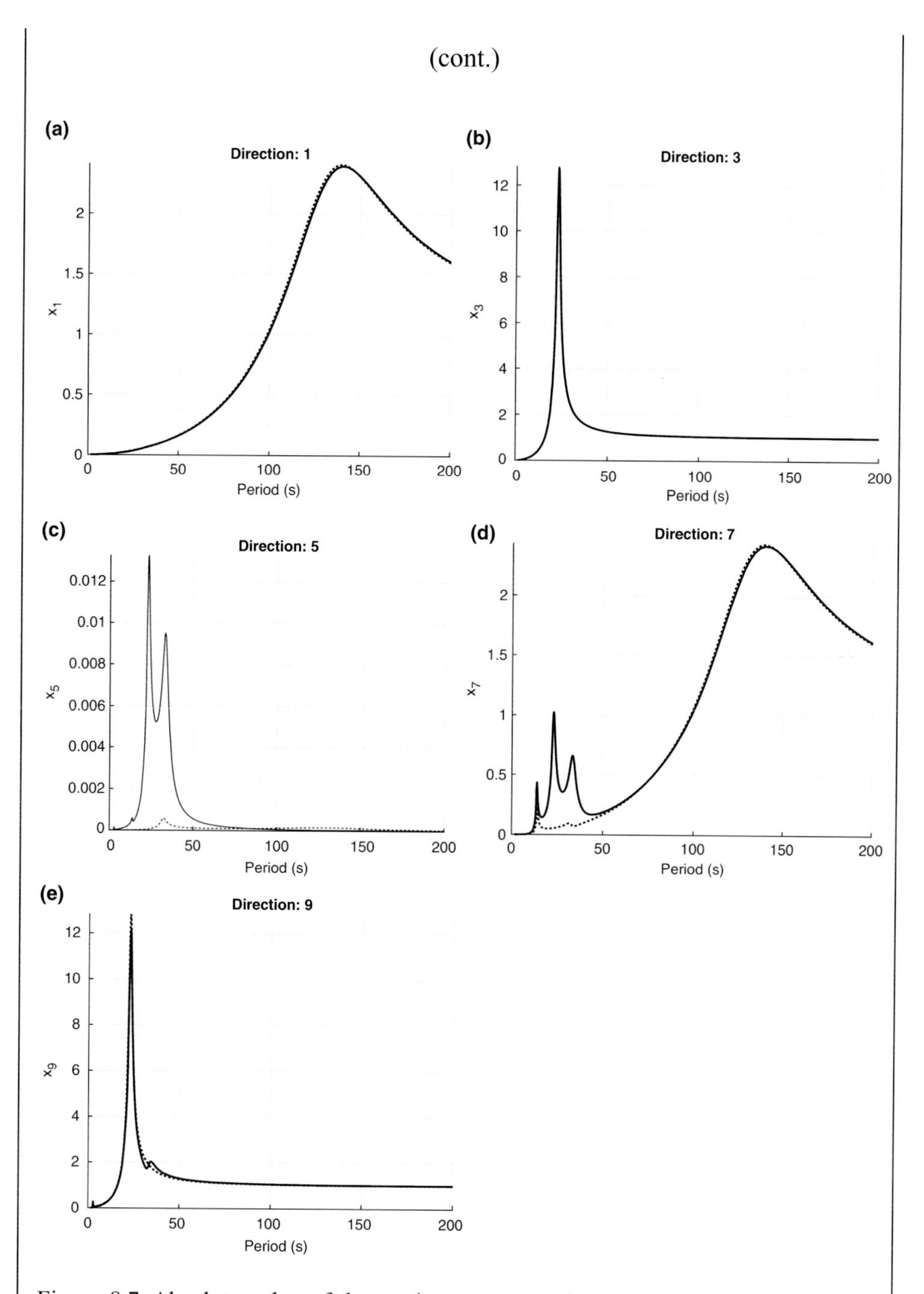

Figure 8.7 Absolute value of the motion response of the vessel and load when exposed to a harmonic force in surge in heave. Solid line: $m_l = 0.02m$. Dotted line: $m_{l1} = m_l/1000$.

8.4.2 Mathieu Instability

Consider a situation with a load with mass m hanging in a crane. The mass is modeled as a point mass. The top of the crane is performing a pure heave motion and the lifting line is assumed to be without mass. No other external forces are acting on the load and the load is assumed to move in one horizontal direction only. The following two equations for dynamic are obtained:

$$\begin{aligned} m\ddot{\eta}_1 + B_{11}\dot{\eta}_1 + \frac{T}{l}\eta_1 &= 0 \\ m\ddot{\eta}_3 + B_{33}\dot{\eta}_3 + k\eta_3 &= k\eta_{3T}. \end{aligned} \tag{8.25}$$

Here, η_1 and η_3 are the horizontal and vertical motion respectively; B_{ii} is the damping; T is the tension in the line of length l; k is the elastic stiffness; and $\eta_{3T} = \eta_{3Ta}e^{i\omega t}$ is the harmonic vertical motion of the top of the crane. The vertical motion of the load is obtained as:

$$\eta_3 = \frac{k}{-\omega^2 m + i\omega B_{33} + k}\eta_{3Ta}e^{i\omega t}. \tag{8.26}$$

The dynamic tension in the line is thus given by:

$$T_d = (\eta_{3T} - \eta_3)k = T_{dA}\cos(\omega t + \psi), \tag{8.27}$$

where

$$\begin{aligned} T_{dA} &= \frac{\sqrt{\left[-\omega^2 m(k - \omega^2 m) + \omega^2 B_{33}^2\right]^2 + (\omega B_{33}k)^2}}{(k - \omega^2 m)^2 + \omega^2 B_{33}^2}k\eta_{3Ta} \\ \psi &= \arctan\left(\frac{\omega B_{33}k}{-\omega^2 m(k - \omega^2 m) + \omega^2 B_{33}^2}\right). \end{aligned} \tag{8.28}$$

For zero damping and a very stiff line, $\omega^2 \ll k/m$, this simplifies to:

$$T_d = -\omega^2 m\eta_{3Ta}e^{i\omega t}. \tag{8.29}$$

The equation for the horizontal motion of the load is thus obtained as:

$$m\ddot{\eta}_1 + B_{11}\dot{\eta}_1 + \frac{1}{l}[T_0 + T_{dA}\cos(\omega t + \psi)]\eta_1 = 0. \tag{8.30}$$

Here, $T_0 = mg$ is the static tension. Rearranging this equation and introducing the natural frequency for the pendulum motion, $\omega_0 = \sqrt{g/l}$ [8.30] may be written as:

$$\ddot{\eta}_1 + 2\zeta\omega_0\dot{\eta}_1 + \omega_0^2\left[1 + \frac{T_{dA}}{T_0}\cos(\omega t + \psi)\right]\eta_1 = 0. \qquad [8.31]$$

The last term in this equation may be considered an excitation term. As the excitation is due to a time-dependent variation of the stiffness term, it is called "parametric excitation." This particular equation is called the Mathieu equation. Depending upon the damping ratio, $\zeta = B_{11}/(2m\omega_0)$, the frequency ratio, ω/ω_0, and the excitation ratio, T_{dA}/T_0, the equation has a stable or unstable behavior. In Figure 8.8, approximate ranges of the stable and unstable regions are given for $\zeta = 0$ and $\zeta = 0.05$.

For zero damping there is a large region of instability close to $\omega/\omega_0 = 2$ and a more narrow one at $\omega/\omega_0 = 1$. As the T_{dA}/T_0 ratio increases, the instability region widens with respect to the excitation frequencies. For $\omega/\omega_0 = 2$ the instability region starts from zero force amplitude ratio. For zero damping there exist instability regions for $\omega/\omega_0 = 2/n$, $n = 1, 2, 3, \ldots$. As the damping increases, a finite force amplitude ratio is required to excite the instability. This is illustrated by the dashed line in Figure 8.8. The zones of instability in the figure are approximate. For a thorough discussion of the Mathieu instability, see, e.g., Nayfeh and Mook (1979).

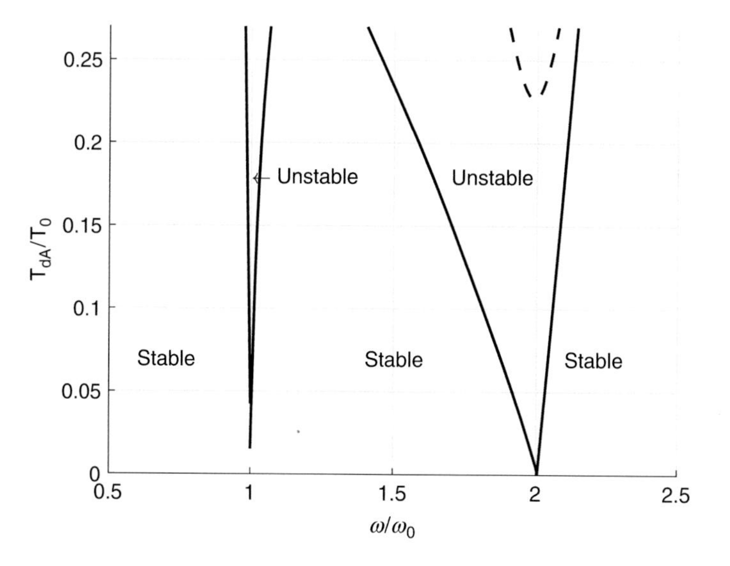

Figure 8.8 Illustration of regions of Mathieu instability according to [8.31]. Solid lines: $\zeta = 0$; dashed line: $\zeta = 0.05$. Approximate results obtained by time-domain simulations.

For practical marine operations, the region in the proximity of $\omega/\omega_0 = 2$ is the most important. That is, if the crane top, due to wave-induced motions, moves vertically almost harmonically with a period T_{ex}, then large pendulum motions of the crane load may occur if the length of the pendulum is such that the natural period is $2T_{ex}$.

Another example is a vessel moving forward in head waves. In this case the vessel may experience excessive roll motion if the heave natural period is excited by the wave-induced forces and the heave natural period is half the roll natural period. The roll motion usually has low damping, and the roll stiffness depends upon the water plane area of the vessel. Thus, the water plane area and the roll-restoring force change as the vessel heaves.

8.4.3 Simulation of Impact during Mating Operations

Many of the lifting operations that take place during the installation and maintenance of offshore wind turbines involve a mating operation, i.e., one heavy body is to be installed with high precision on top of another. Simulation of such operations involves nonlinear multibody simulations with time- and position-dependent properties. Such an operation is discussed here using a very simplistic approach. For real design purposes special-purpose time domain simulation programs have to be used (see, e.g., Reinholdtsen, Mo and Sandvik, 2003).

The following system is considered: a mass m_1 is hanging on a wire from a crane. The mass is lowered with a speed v_1 (negative downward). The crane wire system is modeled by a spring-damper system, as illustrated in Figure 8.9. The mass is to be mated on top of a floating substructure. The substructure has mass m_2 including the effect of added mass. The hydrostatic effects are modeled by a spring-damper system, k_2 and b_2. At the mating area a spring-damper arrangement is used to reduce the impact forces. This arrangement is modeled by k_c and b_c. The spring-damper arrangement has a finite stroke length. Here, only the vertical motions are considered, i.e., the vertical motion of the load and the vertical motion of the substructure.

The coordinate system is such that z is vertical, positive upward and zero at the water plane. The time-dependent position of the lower end of m_1 is denoted $z_1(t)$. The initial gap between m_1 and the top of m_2 is d_0. This is with zero static elongation of the lifting wire. It is assumed that m_1 is hanging in the air. The static displacement of m_1 due to the weight in air is thus $\Delta z_{1s} = -m_1 g/k_1$. The position of the top end of m_2 is denoted $z_2(t)$. m_2 is connected to the "earth" by a spring with stiffness k_2 and a damper b_2. These model the hydrostatic stiffness and damping of the floater. It is assumed that m_2 is floating in water and is neutrally buoyant prior to on-loading of m_1.

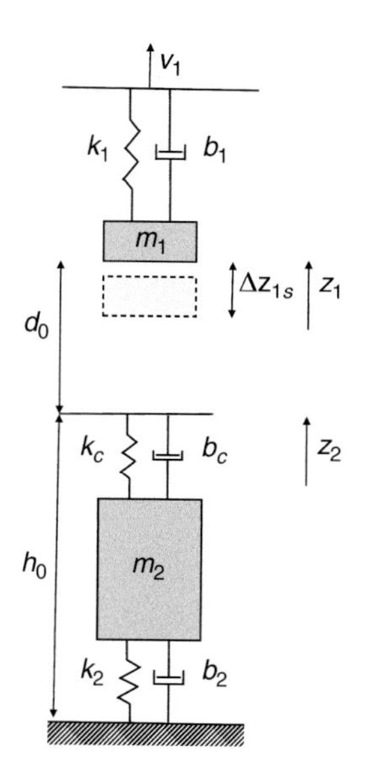

Figure 8.9 Illustration of the dynamic system considered for the mating operation.

The initial position at the start of dynamic simulations of m_1 is located at $z_1(t = 0) = h_0 + d_0 - m_1 g/k_1$. The initial position of the top of the substructure (m_2) is $z_2(t = 0) = h_0$. The initial position of the top of the "crane" (spring k_1) is $z_0 = h_0 + d_0 + l_0$. l_0 is the unstretched length of the lifting line.

Prior to contact, m_1 and m_2 are parts of two independent dynamic systems. The undamped natural periods are in this case given by:

$$T_{0i} = 2\pi\sqrt{\frac{m_i}{k_i}}, \quad i = 1, \ 2. \qquad [8.32]$$

After contact, and with no slack in the springs, the coupled dynamic system has mass and stiffness matrices given by:

$$\mathbf{M} = \begin{pmatrix} m_1 & 0 \\ 0 & m_2 \end{pmatrix} \qquad [8.33]$$

$$\mathbf{K} = \begin{pmatrix} k_1 + k_c & -k_c \\ -k_c & k_2 + k_c \end{pmatrix}. \qquad [8.34]$$

The corresponding undamped natural periods are obtained by solving for the eigenvalues λ_i from:

$$(\boldsymbol{\lambda} - \mathbf{M}^{-1}\mathbf{K})\boldsymbol{\psi} = 0$$
$$T_{0i} = \frac{2\pi}{\sqrt{\lambda_i}}. \quad [8.35]$$

After contact, and with slack in the lifting line, the natural periods will be given as in [8.35], but with $k_1 = 0$ in [8.34].

Assume the top end of spring 1 is moving with a constant speed v_0, negative when lowering the load. The wire will be taut on the condition that there is no contact between m_1 and the springs on top of m_2, i.e., $z_1 > z_2$. Contact between the bodies is present if $z_2 \geq z_1$. The spring force between the bodies is given by $(z_2 - z_1)k_c$. The equation of motion can be written as:

$$\mathbf{M}\ddot{\mathbf{z}} + \mathbf{B}\dot{\mathbf{z}} + \mathbf{K}\mathbf{z} = \mathbf{F}. \quad [8.36]$$

Here, $\mathbf{z} = (z_1, z_2)^T$. The damping and stiffness matrices become case-dependent and may be written as:

$$\mathbf{B} = \begin{pmatrix} b_{11} + b_{12} & -b_{12} \\ -b_{12} & b_{22} + b_{12} \end{pmatrix} \quad [8.37]$$

$$\mathbf{K} = \begin{pmatrix} k_{11} + k_{12} & -k_{12} \\ -k_{12} & k_{22} + k_{12} \end{pmatrix}. \quad [8.38]$$

The constants can be derived from the forces acting on the two masses. In the linear case the forces acting upon the two bodies are:

$$F_1 = k_{11}(h_0 + d_0 + v_0 t - z_1) + k_{12}(z_2 - z_1) + b_{11}(v_0 - \dot{z}_1) + b_{12}(\dot{z}_2 - \dot{z}_1) - m_1 g$$
$$F_2 = k_{22}(h_0 - z_2) + k_{12}(z_1 - z_2) - b_{22}\dot{z}_2 + b_{12}(\dot{z}_1 - \dot{z}_2). \quad [8.39]$$

The weight and buoyancy forces on m_2 in the initial position are assumed to balance each other and are omitted from the equations.

The initial conditions become:

$$\begin{aligned} z_1(t = 0) &= h_0 + d_0 - m_1 g / k_1 \\ z_2(t = 0) &= h_0 \\ v_1(t = 0) &= v_{10} \\ v_2(t = 0) &= 0. \end{aligned} \quad [8.40]$$

As observed from [8.40], the initial velocity for m_1 may differ from the lowering velocity. This makes it possible to study, e.g., zero lowering velocity but with a

Table 8.6 *Damping and restoring coefficients for the four cases considered (see [8.39]).*

		b_{11}	b_{12}	b_{22}	k_{11}	k_{12}	k_{22}
No slack, no contact	$z_1(t) < v_0 t + h_0 + d_0$ $z_2 < z_1$	b_1	0	b_2	k_1	0	k_2
No slack, contact	$z_1(t) < v_0 t + h_0 + d_0$ $z_2 \geq z_1$	b_1	b_c	b_2	k_1	k_c	k_2
Slack, no contact	$z_1(t) \geq v_0 t + h_0 + d_0$ $z_2 < z_1$	0	0	b_2	0	0	k_2
Slack, contact	$z_1(t) \geq v_0 t + h_0 + d_0$ $z_2 \geq z_1$	0	b_c	b_2	0	k_c	k_2

Table 8.7 *Forcing terms in [8.39].*

	F_1	F_2
No slack, no contact	$k_1(h_0 + d_0 + v_1 t) + b_1 v_1 - m_1 g$	$k_2 h_0$
No slack, contact	$k_1(h_0 + d_0 + v_1 t) + b_1 v_1 - m_1 g$	$k_2 h_0$
Slack, no contact	$-m_1 g$	$k_2 h_0$
Slack, contact	$-m_1 g$	$k_2 h_0$

finite initial velocity. The contributions from k_1 and b_1 are accounted for if the line is taut according to the above criterion. Similarly, the contributions from k_c and b_c are accounted for if there is contact between the bodies. The damping and stiffnesses to be used in [8.36] are listed in Table 8.6. Similarly, the forcing terms are listed in Table 8.7.

In a linear implementation, the vertical hydrostatic force may be written as $F_h = k_2(h_0 - z_2) = \rho g A_{wl}(h_0 - z_2)$. In the case of, e.g., a conical vertical cylinder, the total hydrostatic force must be considered, i.e., the buoyancy force at the displaced position $z_2(t)$ is used. $\mathbf{K}(2,2)$ in [8.36] is thus zero and the corresponding force term in Table 8.7 becomes $F_2 = \rho g(V(z_2) - V(0))$. $V(z_2)$ is the actual submerged volume of the substructure.

Numerical Simulation of Mating Operation

A simplified mating operation with the following system data is considered.

Mass of body 1, m_1	*1000 Mg*		
Mass of body 2, m_2	*10000 Mg*		
Stiffnesses, k_1, k_c, k_2	*10000*	*7998.4*	*789.4 kN/m*

(cont.)

Damping, b_1, b_c, b_2	*700*	*1000*	*500 kN/(m/s)*
Initial gap, D_0	*1.5 m*		
Initial position of m_2, z_{20}	*10 m*		
Static displacement of m_1, dz_{s1}	*-0.98 m*		
Static position of m_1, z_{10}	*10.52 m*		
Gap after static elongation	*0.52 m*		
Lowering velocity, V_0	*-0.2 m/s*		

Properties of springs at interconnection (different coefficients are used depending upon stroke) are as follows.

Number of springs	*2*
Stroke length	*0.5 m*
Restoring coeff. in compression at half-stroke	7998.4 kN/m
Damping coeff. in compression at 0.1 m/s	1000 kN/(m/s)
Restoring coeff. fully engaged damper (kc_max)	10000 kN/m
Damping coeff. fully engaged damper (bc_max)	500 kN/(m/s)

Wave:
Wave period 10 s
Wave amplitude 1.5 m
Water depth 150 m

Computed undamped natural periods:		
Prior to contact (s):	*22.36*	*1.99*
With contact and kc at half-stroke (s):	*1.47*	*8.77*
With kc_max (s):	*1.39*	*8.36*
With contact (kc_max) and slack line (s):	*1.89*	*23.46*

The time domain simulation is performed using the fourth-order Runge–Kutta method with a time step 0.01 s. Selected time histories of the results are shown in Figure 8.10. From the plots, the following observations are made. After about 6.5 s, contact between the masses is established and a gradual load transfer begins. However, due to the wave motion, contact is lost after about 12.7 s, causing a more forceful and impact-like contact after about 18 s. The maximum force in the lifting line exceeds the static value, but gradually decreases until the line finally is unloaded after 77 s. The high-frequency variations during the first phase of the load transfer are due to the stiff springs between the two bodies.

(cont.)

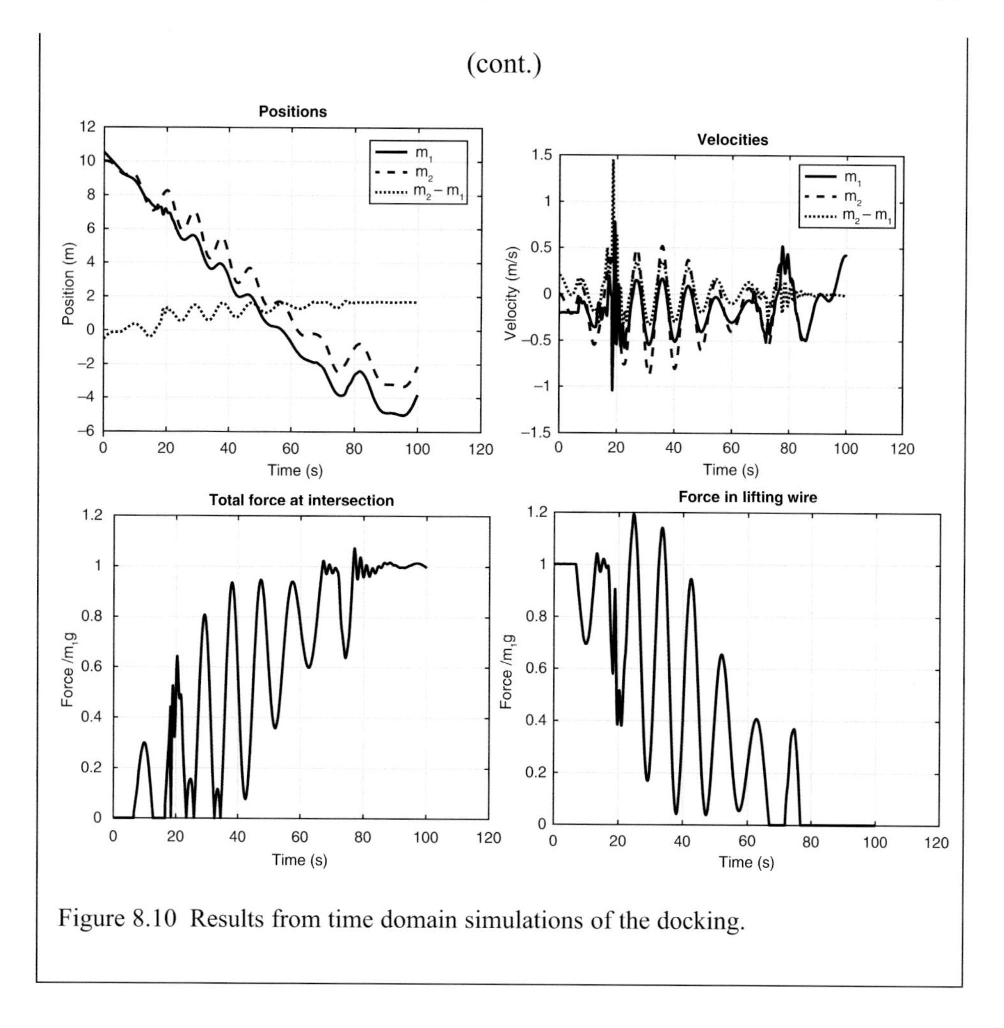

Figure 8.10 Results from time domain simulations of the docking.

8.5 Statistics of Impacts and Snatch Loads during Mating Operations

During mating operations, the distance between the two objects to be mated is gradually reduced. Thus, if some relative motions due to, for example, waves are present, the probability of an unintended impact increases as the average distance between the two bodies is reduced. Similarly, just after contact is established, there is a probability that the slack in the lifting wire is too small to avoid lift-off and corresponding snatch loads in the wire due to the wave-induced motions. Examples of such events are illustrated above in "Numerical Simulation of Mating Operation". Here, an approach to estimate the probability of impacts and snatch loads will be outlined. The approach was first published by Vinje, Kaalstad and Daniel (1991).

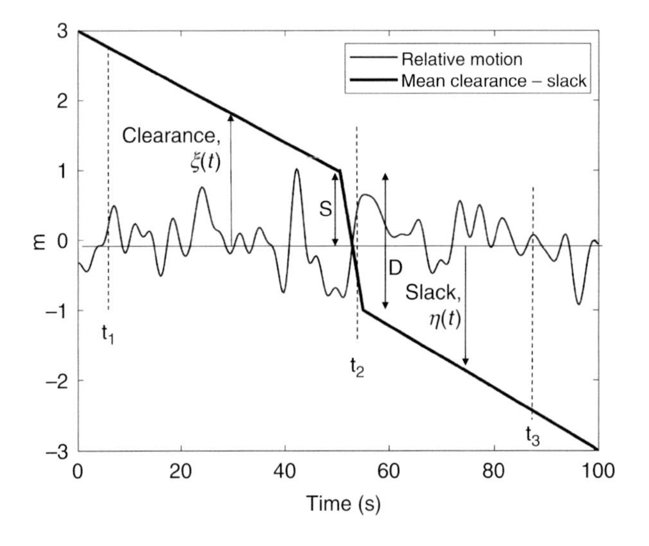

Figure 8.11 Illustration of clearance, slack and relative motion during a mating operation.

Consider the situation illustrated in Figure 8.11. The oscillating line represents the relative vertical motion between the two objects. For simplicity, it may be thought of as the vertical motion of the crane top of a floating crane vessel, while the substructure is at rest. At time t_1 the clearance between the load and the substructure is $\xi(t_1)$. The clearance is the distance, ignoring the wave-induced motions. The clearance is reduced linearly with a speed v_ξ. The clearance at time t is thus $\xi(t) = \xi(t_1) - v_\xi(t - t_1)$. At time t_2 a sudden load transfer is assumed to take place, i.e., the clearance is reduced much faster than the lowering speed. The load transfer is supposed to take place during a time interval much shorter than a wave period. This load transfer may be studied by the approach described in the previous paragraph. The clearance prior to the load transfer is $\xi(t_2) = S$. The load transfer can be performed by, for example, a rapid filling of ballast tanks in the crane vessel or by mobilizing special-purpose hydraulic cylinders. The total "stroke length" during the load transfer is denoted D. After the load transfer, there is a slack in the lifting wire $\eta(t_2) = D - S$. By continuing to give out wire, the slack will increase linearly after the load transfer, and thus the slack in the wire is given by:

$$\eta(t) = \eta(t_2) + v_\eta(t - t_2). \tag{8.41}$$

Here, the wire spool-out velocity v_η may differ from the lowering velocity.

The probability of impacts prior to the load transfer and of snatch loads after the load transfer are considered. The considerations are similar for the two processes. The impact probability prior to load transfer is considered first. It is assumed that the relative motion process, denoted $x(t)$, is a stationary, narrow-banded Gaussian process with standard deviation σ_x. The narrow-band assumption implies that there is only one extremum between each crossing of the mean value. The amplitude of the relative motion is given by a Rayleigh distribution, i.e., by:

$$p(A) = \frac{A}{\sigma_x^2} e^{-\frac{A^2}{2\sigma_x^2}}. \qquad [8.42]$$

The corresponding expected frequency of positive crossings of level ξ is given by:

$$f_x = \frac{1}{T_x} e^{-\frac{\xi^2}{2\sigma_x^2}}. \qquad [8.43]$$

Here, T_x is the expected zero up-crossing period of $x(t)$. A positive crossing of level ξ implies an impact. [8.43] is valid for ξ constant. In the present case ξ is a function of time. However, ξ may be assumed to vary at a much longer time scale than x. The principle of multiple-scale analysis may thus be applied, which implies that within each cycle of $x(t)$, ξ may be considered constant. Under these assumptions, the expected number of positive crossings of level $\xi(t)$ during the time interval from t_1 to t_2 can be written as:

$$N_x(t_1, t_2) = \int_{t_1}^{t_2} f_x(t)dt = \int_{t_1}^{t_2} \frac{1}{T_x} e^{-\frac{[\xi(t)]^2}{\sigma_x^2}} dt. \qquad [8.44]$$

Under the assumption that ξ varies linearly with time and observing that the integrand in [8.44] is similar as for the probability density of a zero-mean Gaussian process, the number of positive-level crossings, or impacts, can be written as:

$$N_x(t_1, t_2) = \frac{\sigma_x}{v_\xi T_x} \sqrt{2\pi} \left[F_N\left(\frac{\xi_1}{\sigma_x}\right) - F_N\left(\frac{\xi_2}{\sigma_x}\right) \right]. \qquad [8.45]$$

Here, $F_N(z) = P(Z<z)$ is the probability function in the standard normal distribution, $N(0, 1)$. To simplify, $\xi(t_i)$ is written as ξ_i. The probability density and cumulative probability for the standard normal distribution are shown in Figure 8.12. If it is assumed that at $t = t_1$ the clearance is very large, i.e.,

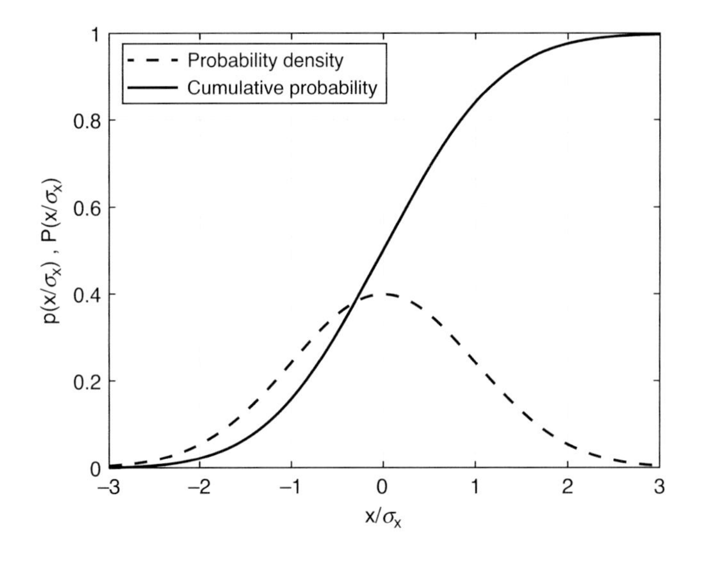

Figure 8.12 The standard normal distribution.

$\xi_1/\sigma_x \gg 1$, then $F_N\left(\frac{\xi_1}{\sigma_x}\right) \simeq 1$ and the number of impacts prior to the load transfer is approximately:

$$N_x(t_1, t_2) \simeq \frac{\sigma_x}{v_\xi T_x}\sqrt{2\pi}\left[1 - F_N\left(\frac{\xi_2}{\sigma_x}\right)\right]. \qquad [8.46]$$

The minimum value of ξ_2 is zero, and thus an upper limit of the number of impacts prior to load transfer is:

$$N_{x,\max}(t_1, t_2) = 0.5\frac{\sigma_x}{v_\xi T_x}\sqrt{2\pi}. \qquad [8.47]$$

$v_\xi T_x$ is the lowering distance during an average period of oscillation. The relation is shown in Figure 8.13. From [8.47] it is observed that the ratio between σ_x and the average lowering distance during one wave period is the key parameter for the expected number of impacts. For example, if it is required that the expected number of impacts should be less than 0.1, one must require that $\sigma_x/v_\xi T_x < 0.08$. Note that the above expression only gives information about the number of impacts, not their magnitude.

In a given sea state and with a constant lowering velocity, the expected number of impacts is reduced by increasing ξ_2. However, with a limited "load transfer distance" D available, a large ξ_2 will cause a small slack η_2 just after load

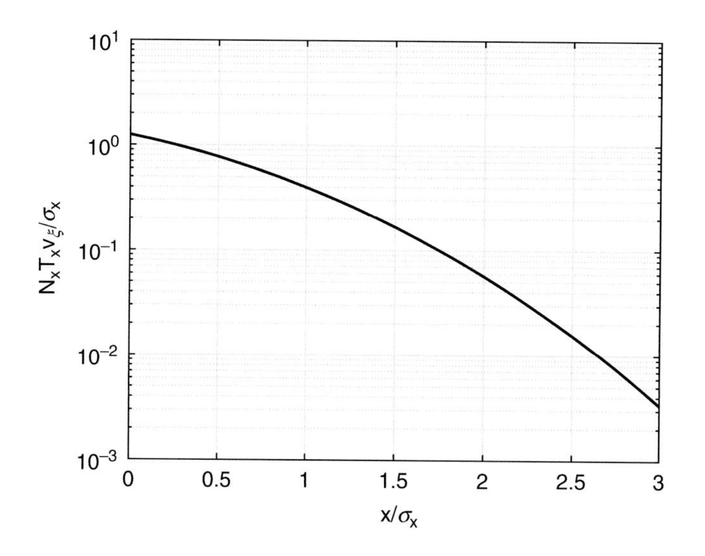

Figure 8.13 Expected number of impacts N_x as a function of the clearance $\xi(t_2)$.

transfer. Thus, the probability of snatch loads in the lifting wire after load transfer will increase. The expected number of snatches is found similarly to the expected number of impacts. Thus, the expected number of snatches may be written as:

$$N_y(t_2, t_3) \simeq \frac{\sigma_y}{v_\eta T_y} \sqrt{2\pi} \left[1 - F_N\left(\frac{D - \xi_2}{\sigma_y}\right)\right]. \quad [8.48]$$

Here, it is assumed that the standard deviation of the relative motion as well as the zero up-crossing period may differ after the mating, i.e., σ_x and T_x are replaced by σ_y and T_y.

If a certain load transfer distance D is available, what is the optimum clearance S to aim for to minimize the number of impacts plus the number of snatches? As the consequences or severity of impacts and snatches may differ, one should give different weight to the two events and aim at minimizing:

$$N_E = \alpha N_x + \beta N_y \ . \quad [8.49]$$

The minimum value of N_E is obtained by finding $\partial N_E / \partial \xi_2 = 0$. For the simplest case of $T_x = T_y$, $v_x = v_y$ and $\alpha = \beta$, the minimum value is obtained as:

$$N_{E,\min} \simeq \frac{\sqrt{2\pi}}{v T_x} (\sigma_x + \sigma_y) \left[1 - F_N\left(\frac{D}{\sigma_x + \sigma_y}\right)\right]. \quad [8.50]$$

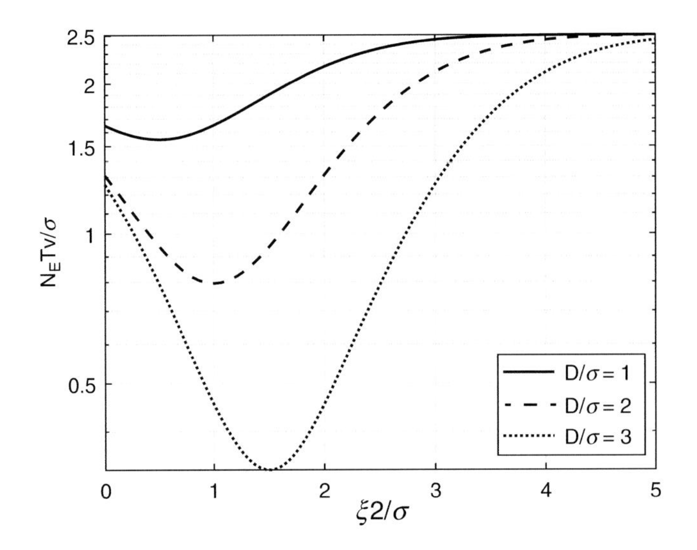

Figure 8.14 Expected number of impacts plus snatches N_E as a function of ξ_2/σ. Three values of D/σ are applied. The following simplifications are used: $T = T_x = T_y$, $\sigma = \sigma_x = \sigma_y$, $\alpha = \beta = 1$.

In Figure 8.14, the expected number of impacts plus snatches are plotted for various clearances prior to load transfer. Three values of relative transfer distance D/σ are used. It is observed that the minima of the curves correspond to the values given by [8.50] using $\sigma = \sigma_x = \sigma_y$.

Exercises Chapter 8

1. Section 8.4 describes that "sea states with the same probability of exceedance have the same persistence duration statistics." This statement shall be checked by using the wave data in the file "Wave_data.txt."
   a. Compute the cumulative probability of H_s during the winter and summer seasons for the ten years of data in the wave data file.
   b. Select some probability levels, e.g., probability $P(H_s) = 0.9, 0.8, 0.7, 0.6$. Find the corresponding H_s values for the summer and winter seasons.
   c. Compute the mean duration of periods with H_s below the found thresholds. You may use linear interpolation to find time intervals within fractions of an hour.
   d. Check the above results on the mean duration toward the above statement about duration statistics.

e. If the wave height limit for a certain operation is $H_s < 2.5$ m, on average, what fraction of the time is available during the summer season versus the winter season?
f. Based upon the wave data file, create a graphic similar to Figure 8.5, showing the conditional probability of the significant wave height.

2. A floating crane has a load hanging on a wire of a length of 35 m. The crane tip performs a pure heave motion.
   a. Which periods of the heave motion may excite large pendulum motions?
   b. Which measures can be taken to avoid the pendulum motions?
3. A fixed crane is lowering an object down towards the sea surface. The lowering process is performed with a low velocity, $v_z = 0.1\,m/s$. The wave condition is characterized by $H_s = 5\,m$ and $T_p = 10\,s$. When the object has a distance s above the mean surface, a sudden and fast lowering takes place.
   a. What is the expected number of impacts on the body from the waves prior to the final fast lowering? Make a plot as function of s.
   b. How large should s be to have the expected number of impacts less than 0.1?
   c. Create a time history of the wave elevation and simulate the lowering process by several realizations. Do the simulation results fit the above results?

9

Offshore Wind Farms

Offshore wind farms consist of many wind turbines connected to a common power system. The earliest offshore wind farm, Vindeby, was installed in 1991 and decommissioned in 2017. It was installed in shallow waters, at a depth of about 4 m, in Danish waters only 2 km from shore. The wind farm consisted of eleven 0.45 MW turbines. This contrasts with recent projects such as those at Dogger Bank. The Dogger Bank Offshore Development Zone is located in the North Sea in UK waters at a distance between 125 km and 290 km east of England. The water depths in this area are in the range of 18 to more than 60 m. Three wind farms are under development: Dogger Bank A, B and C. Each of the projects is planned for an installed power of 1200 MW, covering areas in the range of 515–599 km^2. The wind farms are to be developed in the shallowest area of the Dogger Bank Offshore Development Zone, with water depths ranging from approximately 20 to 35 m. It may be expected that the wind farms will be developed using state-of-the-art offshore wind turbines with rated power in the range 12 to 15 MW. This means that in the order of 100 turbines are needed in each of the wind farms.

For bottom-fixed offshore wind farms, the turbines represent about 35% of the capital cost (CAPEX), while substructures represent about 13%, electrical infrastructure 18% and assembly and installation about 10% (Global Wind Energy Council, 2022). Monopiles are the most frequently used substructure. Floating wind farms are still in their infancy, the largest being the Hywind Tampen wind farm off the west coast of Norway, consisting of 11 turbines each of 8.6 MW capacity on a spar substructure.

In the design of wind farms, several partly conflicting interests must be handled. The license conditions for a specific concession area may contain conditions related to the maximum and minimum installed capacity as well as conditions related to turbine size and interaction with other activities in the area, for example, fishery, military activity and shipping lanes. Further, conditions related to protection of the environment, avoidance of bird collisions etc. may be imposed. In particular, noise related to piling may be a problem to the marine life and noise-reducing measures may be required. As part of the planning process, an environmental impact assessment (EIA) is normally required. The requirements vary from country to country. Ideally the EIA should involve environmental mapping and measurements for a period prior to the development as well as follow-on registrations during the construction and operational phases to confirm the assessments in the EIA.

The cost of electricity delivered from an offshore wind farm depends both on the investment costs, CAPEX and operational costs (OPEX), as well as the amount of energy produced, which depends on the interaction between the wind turbines and thus the layout of the wind farm. Until recently, the cost of electricity from offshore wind farms was too high to be attractive from a pure market perspective. Various support schemes have thus been introduced to stimulate the development of energy from offshore wind. Such schemes have varied between countries and over time. One group of support schemes is related to investment support, either in the form of investment grants or discount on loans etc. Another main group is support during the operational phase. Feed-in tariffs have been used to guarantee the operator of a wind farm a long-term price on the energy delivered. Another system is the premium principle, where the operator sells the energy in the market but receives a price supplement on top. To stimulate cost reduction in the industry, many countries use a premium-based support scheme where the licenses are awarded via an auction system. Details about and an evaluation of the various support schemes are provided by the European Commission Directorate-General for Energy (2022).

9.1 Cost of Electricity

A concession for developing a wind farm is limited to a certain development area. Within the terms given in the license, the developer normally will aim for a maximum revenue of the invested capital. In simple terms, this may be thought of as minimizing the levelized cost of electricity (LCOE), which is frequently used as a measure for comparing various energy projects. In its simplest form, the LCOE may be expressed as:

$$LCOE = \frac{\text{Investments} + \text{Operational costs}}{\text{Energy delivered}} = \frac{I_0 + \sum_{k=1}^{N} \frac{A_k}{(1+i_r)^k}}{\sum_{k=1}^{N} \frac{M_k}{(1+i_r)^k}}. \quad [9.1]$$

Here, I_0 denotes the initial investments in the wind farm or power station, assumed to take place in year zero; N is the expected operational lifetime in years of the wind farm; k is the year of operation; A_k is the operational costs in year k, including, e.g., fuel costs if a fuel-based power plant is considered; M_k is the energy delivered in year k; i_r is the discount rate to be used in present value calculations. As can be seen from [9.1], the LCOE is a cost per energy unit computed as a present cost in year zero. It is thus well suited for comparison of the economy of various projects. The above simplistic electricity cost estimate does not account for factors such as taxes, difference in discount rate using equity versus bank loan, inflation etc. i_r is frequently denoted as "weighted average cost of capital" (WACC). From [9.1] it is observed that the initial investments have a large weight on LCOE, and the operational costs during the first years are also important cost drivers, but the importance diminishes as time increases. The same is valid for power production, but with the opposite effect; a high energy production during the first years of operation contributes to a reduced LCOE. The discount rate is important to the LCOE. A low discount rate favors projects with high initial investments and long expected lifetimes.

When comparing published data for LCOE one should be aware which investments are included, for example, costs of power transmission to shore and grid connections. Such costs may represent a significant cost element. For some projects the cost of power transmission is part of the operational cost, not the investment. Also, the decommissioning (removal) of the wind farm at end of operation should be included in calculating the LCOE.

Renewable energy systems such as wind and solar involve high investment costs and relatively low operational costs. This is in contrast to fossil fuel-based electricity generation such as gas power plants. For a gas power plant, the yearly operational costs, including gas costs, are the major contribution to LCOE. Thus, in the LCOE calculation the discount rate and the estimated lifetime is of less importance to a gas power plant than it is to a renewable energy system. A low discount rate generally favors renewable energy systems.

9.2 Wind Farm Layout

An offshore wind farm consists of four main components: wind turbines, inter-array cables, a substation and an export cable. An illustration of a possible layout of an offshore wind farm is shown in Figure 9.1. The solid lines illustrate a radial layout of the inter-array cables. Adding the cables, shown as dashed lines, a ring layout is obtained. Using a radial layout, a failure in one cable may cause loss of power from several turbines. The ring layout is less vulnerable.

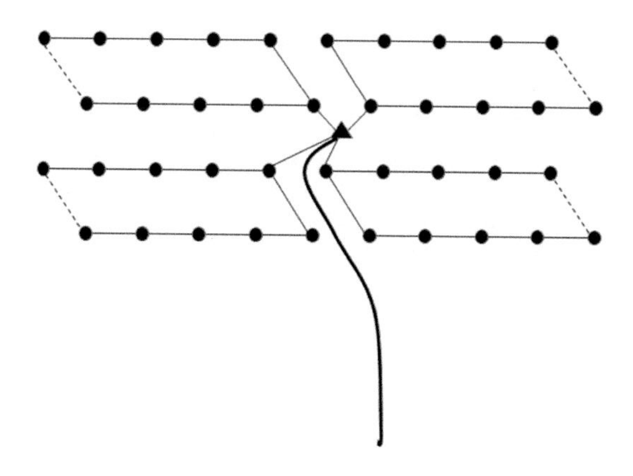

Figure 9.1 Illustration of a wind farm layout with a radial (solid lines) and a ring (solid lines plus dashed lines) inter-array cable configuration. Circles indicate turbine locations, the triangle indicates the substation and the thick line is the export cable.

Each wind turbine delivers power into the inter-array cables, which are connected to the substation, where the voltage is transformed to a higher level. Depending upon the distance to shore and the power capacity of the wind farm, the current may be transformed from alternating current (AC) to direct current (DC) at the substation. This is done to reduce inductive losses in the export cable to shore. Typical voltages in the inter-array cable are in the range of 30–36 kV and in the export cable in the range of 110–220 kV. For newer and larger wind farms, even higher voltages may be preferred.

To compute a preliminary estimate of the yearly energy produced by each turbine, the yearly distribution of wind speeds together with the turbine's power curve are used. An example of a yearly distribution of the wind speed is shown in Figure 9.3, together with a power curve. The observed wind statistics are fitted to a two-parameter Weibull distribution. Using the wind data and the power curve shown in Figure 9.3, a distribution of the power production over the year is obtained. The cumulative distribution is shown. It is observed that for about 18% of the year, there is no production, mainly due to wind speeds being below cut-in. In addition downtime due to failures and maintenance must be accounted for. For the case shown in Figure 9.3, the turbine runs at rated power approximately 17% of the time. Most of the time the turbine runs at partial load. This implies that in a wind farm setting, for most of the time the power delivered from the turbine is very sensitive to changes in the wind speed due to, for example, wake effects.

At sites with comparatively high mean wind speeds, the rotor diameter may be reduced relative to sites with lower average wind speeds and keeping the same capacity factor. This may reduce investment costs. However, the present trend has been to keep a large rotor diameter and take advantage of the higher yearly energy yield.

The placement of the turbines within the concession area is an optimization challenge. An important factor for optimization is the distribution of wind directions. Examples of the directional distribution of the mean wind speed are shown in Figure 9.2. This way of presenting the directional distribution of the mean wind speed is called a wind rose. The wind rose shows the frequency of occurrence of various wind speed ranges as a function of wind direction. In the example shown in Figure 9.2, directional intervals of 10 deg are used. Within each directional interval the frequency of occurrence of wind speeds in 2 m/s intervals are shown. The example in Figure 9.2 (left) is from a site not far from the west coast of Norway. The coastline runs roughly from south to north. It is observed that the prevailing wind directions follow the coastline. For the site further offshore a greater variation in wind directions is observed, with prevailing winds in the directions from south-west to north-west. The difference in the directional distribution of the mean wind speed for the two sites has implications on the optimum layout of the wind farms for the two sites. To avoid wind turbines operating in the wake of upstream turbines one should avoid placement of turbines along lines in the prevailing wind direction. If a wind turbine does operate in the wake of an upstream wind turbine, reduced power production will frequently occur for the downstream turbine, in particular for wind velocities below rated wind speeds. Also, an increased turbulence level is experienced in the wake, causing increased fatigue loads. Turbine wakes are discussed in more detail in Section 9.3, but for the present considerations the wake may be thought of as a conically shaped area behind a wind turbine within

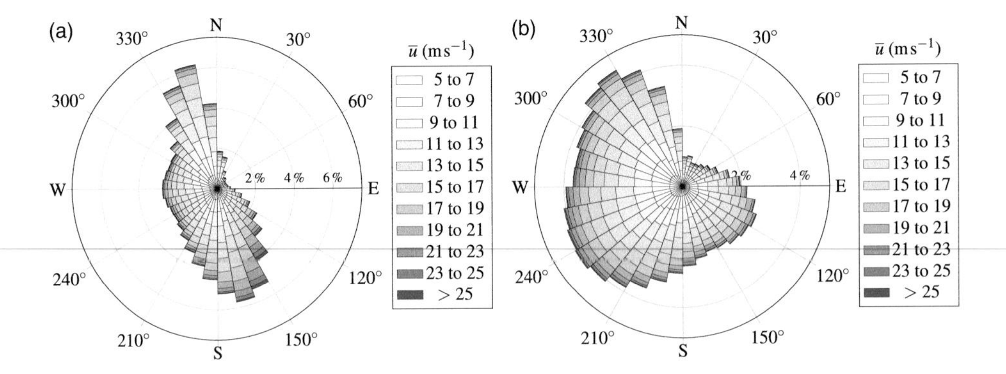

Figure 9.2 Wind roses as obtained from a site close to the coast (left) and a site further offshore (right). The "close to coast" case is the Utsira Nord area off the west coast of Norway. The "further offshore" case is from Sørlige Nordsjø II in the North Sea. The data are obtained from the NORA3 (Haakenstad et al., 2021) at 150 m above sea level. The results are based upon hourly data over a period of 27 years. The wind directions are given as "coming from," e.g., "West" is wind blowing from the west. Courtesy Etienne Cheynet, University of Bergen.

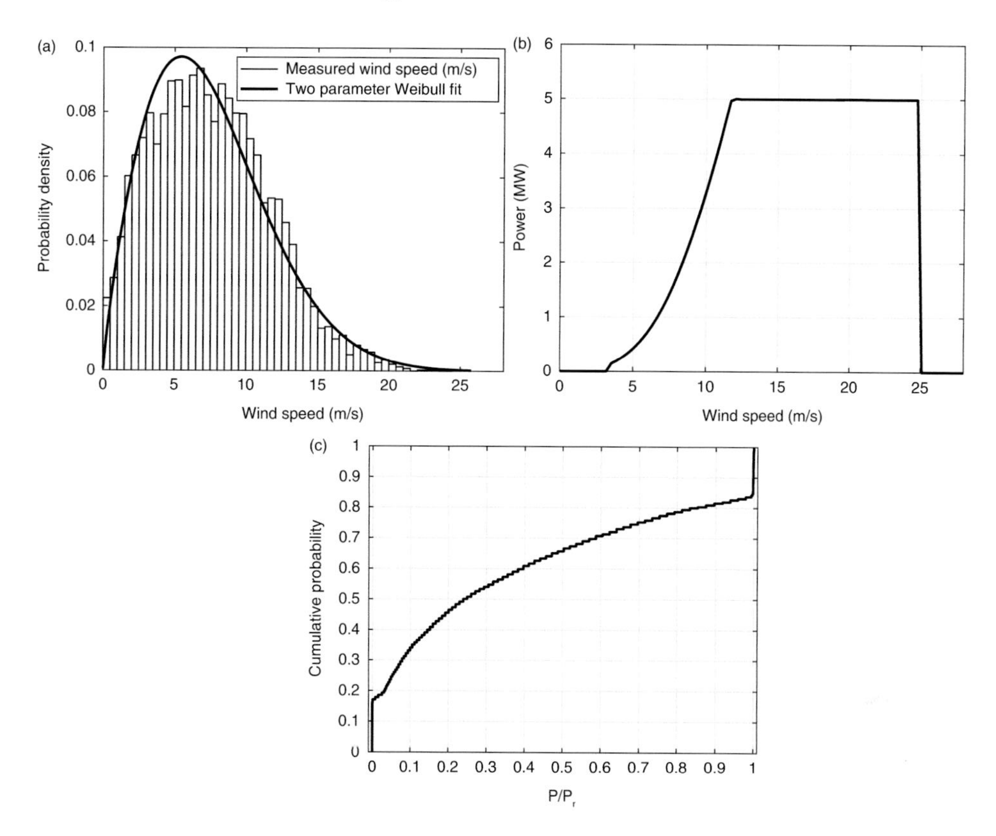

Figure 9.3 Distribution of wind speeds and corresponding power production. Upper-left: distribution of wind speeds for one year and a corresponding two-parameter Weibull distribution fitted to the data. Upper-right: schematic power curve for a 5 MW turbine. Bottom: cumulative distribution of delivered power over one year.

which the wind velocity is reduced due to the momentum loss at the turbine and the turbulence level is increased due to the large velocity shear. The wake broadens and the velocity deficit is reduced as the distance from the turbine increases, as illustrated in Figure 9.4.

For floating wind turbines, with very low natural frequencies in surge and sway, operation partially in a wake or in a meandering wake may cause unwanted and severe low-frequency motions with corresponding large dynamic loads in the mooring system.

As discussed in Section 9.3, the recovery of the wind speed in the wake region depends upon several parameters such as the turbulence level in the incident wind field and the stability of the atmospheric boundary layer. As the turbulence level in general is lower offshore than onshore, the wake effect persists for a longer distance

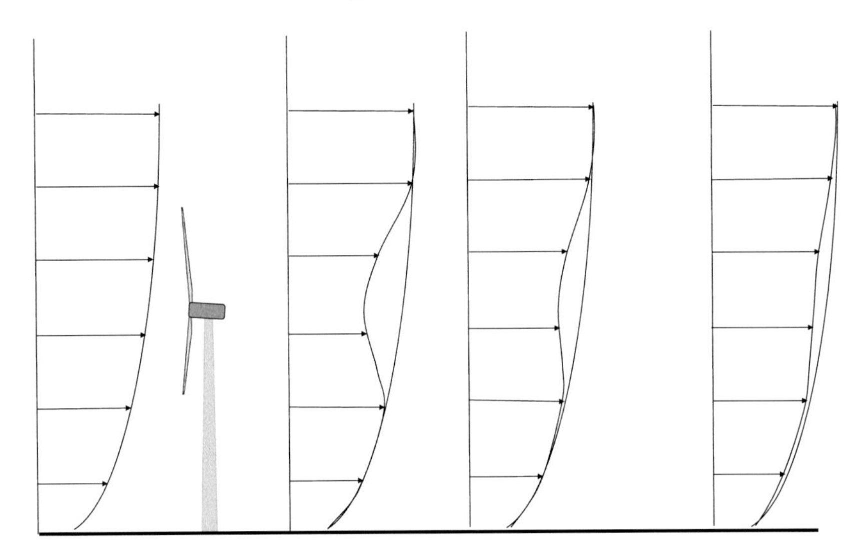

Figure 9.4 Schematic illustration of a wind turbine wake. The velocity deficit is reduced downstream while the wake width increases.

offshore than onshore. In general, the preferred distance between turbines is thus longer offshore than onshore. A distance between the turbines in the range of six to more than ten rotor diameters is used. Longer distances reduce the wake effects but increase inter-array cable costs and limit the total installed power within a certain area.

Other issues that must be considered in the planning of the wind farm layout include the following.

- Water depth. Costs, e.g., for monopiles and jacket substructures, increase considerably as the water depth increases. An illustration of the costs as a function of water depth is given in Figure 4.9.
- Bottom sediment characteristics. The choice of foundation and length of monopiles or piles depends upon the sediment characteristics. If the characteristics vary considerably over the farm area, each turbine foundation may need to be designed individually.
- For floating wind turbines with catenary mooring, common anchor points for several mooring lines may be considered. This will reduce the cost of anchors but restrict possible turbine positions.
- Neighbor wind farms. In areas where several wind farms are planned within a close proximity, such as the southern North Sea, the wake behind one wind farm may cause reduced inflow wind speeds for a downstream wind farm (see Section 9.3.2).

- Environmental issues related to marine life and birds must have attendance in the planning phase and may impact the layout of the wind farm.
- Ship traffic, fishing activities and military considerations may restrict the layout of the wind farm.
- As offshore wind farms occupy huge areas, combined use of the areas should be considered. This may involve fish farming or seaweed farming, wave energy or solar power. Such activities may use common infrastructure and coordinate operational activities and thus reduce costs.

Area Needs for Offshore Wind Farms

Assume an offshore wind farm of $P_{tot} = 1200\ MW$ *in a quadratic layout with equal distance between the turbines in both directions. The distance between the turbines is denoted* αD *and* D *is the rotor diameter. The rated power of each turbine is* P_r *and the capacity factor* C_c. *The number of turbines needed is* $N = P_{tot}/P_r$ *and the yearly electric energy production* $E = 8760\ P_{tot}C_c[MWh]$. *The length of each side of the wind farm becomes* $L = \alpha D(\sqrt{N} - 1)$ *and the area needed* $A = L^2 = (\alpha D)^2(\sqrt{N} - 1)^2 \simeq (\alpha D)^2 N$ *if* N *is large. The rated power is given from* $P_r = \frac{1}{2}\rho\frac{\pi D^2}{4}C_{Pr}U_r^3$, *where* C_{Pr} *is the power coefficient at rated wind speed,* U_r. *Inserting in the approximate expression for the farm area, one obtains* $A \simeq \frac{8}{\pi}\frac{\alpha^2 P_{tot}}{\rho C_{Pr}U_r^3}$. *Thus, the area needed is approximately independent of the size of the turbines (rotor diameter) but sensitive to the rated wind speed and the power coefficient at rated power.*

Using some characteristic values (see Table 3.3): $P_r = 10\ MW$, $D = 178.3\ m$, $U_r = 11.4\ m/s$, $C_{Pr} = 0.4414$ *and, choosing* $\alpha = 8$, *the area needed for the 1200 MW wind farm is obtained as* $244\ km^2$. *This in less than half the area given for the Dogger Bank projects (515–599 km²). The difference may be explained by several factors: non-quadratic layout, larger distance between turbines, bottom conditions, prevailing wind direction and in particular the rated wind speed. Reducing the rated wind speed to 10 m/s, the needed area increases to* $361\ km^2$.

Using a yearly capacity factor $C_c = 0.45$, *the yearly energy production for the wind farm becomes* $E = 4730\ GWh$, *corresponding to an average power production per unit wind farm area of* $P_A = E/(8760\ A) = 2.21\ W/m^2$.

9.3 Wakes

As discussed in Chapter 3, the power offtake by the wind turbine causes a reduced wind speed behind the turbine. This flow field with reduced flow velocity is called the turbine wake. Close to the turbine, the flow field is controlled by local details such as flow around the nacelle, the lift distribution along the length of the blade, the number of blades and the tip vortices. As the flow progresses downstream, shear and turbulence cause mixing of the flow and the local effects diminish. The near field may be assumed to extend to two to four rotor diameters downstream (Porté-Agel, Bastankhah and Shamsoddin, 2020). Further downstream, in the far field, most of the local flow effects may be disregarded and the flow may be modeled more generically. As the distance between turbines in an offshore wind farm normally exceeds five rotor diameters, the far field models are used in analyses of interaction between wind turbines in a wind farm. As obtained from the momentum theory in Chapter 3, the incident flow is slowed down in front of the turbine; this upstream zone is called the induction zone. The induction effect may influence the efficiency of wind turbines in the front row of a wind farm (Nygaard et al., 2020).

To fully understand and model the wakes behind wind turbines in a wind farm, the flow must be considered in a large range of time and spatial scales, from meso-scale wind, via flow at the wind farm scale, to the rotor scale and down to blade cross-sectional scale. The longest scales are of hours and hundreds of kilometers, while the shortest are in fractions of seconds and centimeters. It is not possible to handle this within the framework of a single numerical model. Efficient computational methods are required in the planning and study of the flow within a wind farm. Therefore, simplified models based on combined analytical and empirical methods are needed. Validation of the wake models is challenging as detailed mapping in time and space of the full-scale wake flow is difficult. However, wind tunnel tests have been used to study the flow behind single turbines as well as several turbines with interaction. Also, flow modeling using LES and RANS are frequently applied to study the wake flow. These may, however, be too computationally demanding for use in, e.g., wind farm layout optimization.

9.3.1 Simple Wake Models

Over the years, several simple wake models suitable for the far wake have been developed. Sanderse (2009) gives a review of some basic wake models, and in Krutova et al. (2020) a review of some simple wake models are given and compared to results from LES.

A 1D wake model is obtained using the actuator disk model (Chapter 3). The velocity deficit in the wake is directly related to the thrust on the turbine. Half of the velocity deficit is obtained at the rotor plane; the remaining half is obtained in the downstream far field. Thus, to fulfil the continuity requirements, the wake area must expand downstream. In the simplest version of wake models, the velocity deficit is assumed to be constant over the wake area, as illustrated in Figure 9.5. These are known as "top hat" models. In the "Jensen model" (Jensen, 1983; Katic,

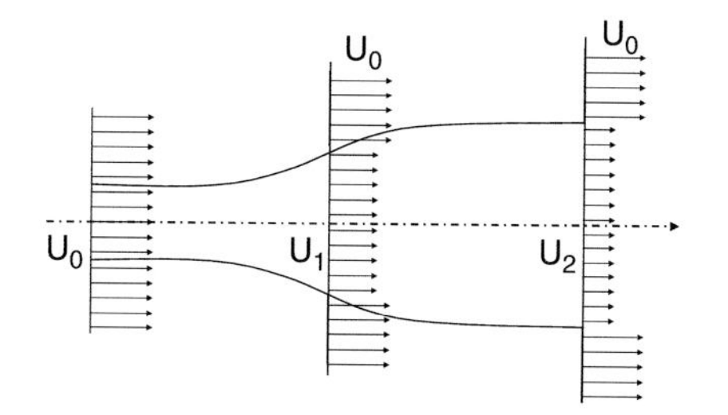

Figure 9.5 Illustration of 1D "top hat" wake model based upon actuator disk approach.

Højstrup and Jensen, 1987; Krutova et al., 2020), the diameter of the wake is assumed to expand linearly with the distance downstream from the rotor plane:

$$D_w(\xi) = D(1 + 2k\xi). \tag{9.2}$$

Here, $\xi = x/D$ is the nondimensional downstream distance from the rotor plane. The assumption of a linearly expanding wake is supported by wind tunnel tests (Porté-Agel, Bastankhah and Shamsoddin, 2020). From potential flow theory we should expect a slower expansion, proportional to $\xi^{1/3}$ (Frandsen et al., 2006). The linear expansion and the constant k are thus assumed to be related to the turbulence level. As the wake expands, a mixing of the wake and the ambient wind will take place, mainly due to turbulence. From continuity, the velocity in the wake $U_w(x)$ may be written as:

$$\pi \frac{D_w^2(\xi)}{4} U_w(\xi) = \pi \frac{D^2}{4} U_w(0) + \left[\pi \frac{D_w^2(\xi)}{4} - \pi \frac{D^2}{4} \right] U_0. \tag{9.3}$$

$U_w(0)$ is given from the induction factor. The momentum theory states $U_1 = U_w(0) = (1 - a)U_0$, while far downstream the factor is $(1 - 2a)$. In the Jensen model this far field factor is used for $U_w(0)$. Under the assumption of a linearly expanding wake and continuity according to [9.3], the wake velocity is obtained as:

$$\frac{U_w(\xi)}{U_0} = 1 - \frac{2a}{(1 + 2k\xi)^2}. \tag{9.4}$$

Utilizing the relation between the axial induction factor and the thrust coefficient, the velocity deficit $\Delta U_w(\xi) = U_0 - U_w(\xi)$ is obtained as:

$$\frac{\Delta U(\xi)}{U_0} = \frac{(1 - \sqrt{1 - C_T})}{(1 + 2k\xi)^2}. \tag{9.5}$$

The relation between k and the turbulence intensity of the ambient flow is proposed as $k \simeq 0.4I_0$.

Based upon analysis of an inviscid spiraling vortex system behind the rotor, an alternative expression for the velocity deficit was derived by McCormick (1999); see also Sanderse (2009) and Werle (2008):

$$\frac{\Delta U(\xi)}{U_0} = \frac{1 - \sqrt{1 - C_T}}{2}\left(1 + \frac{2\xi}{\sqrt{1 + 4\xi^2}}\right). \quad [9.6]$$

As seen from this expression, the velocity deficit rapidly approaches the far field limit corresponding to the Betz theory, $\Delta U(\xi \to \infty)/U_0 = 2a$. The model predicts no further velocity decrease beyond that value.

The top hat assumption for the velocity deficit is simple but obviously unphysical. Therefore, a self-similar Gaussian velocity profile has been proposed (Bastankhah and Porté-Agel, 2014):

$$\frac{\Delta U}{U_0} = C(\xi) f\left(\frac{r}{R_*(\xi)}\right). \quad [9.7]$$

Here, $C(\xi)$ is the maximum normalized velocity deficit at $x = \xi D$, $f(r/R_*(\xi))$ is the radial distribution of the velocity deficit, $R_*(\xi)$ is a characteristic wake radius at the actual position and r is the radius from the center line of the wake. f is assumed to be independent of the downstream position, and thus the formulation is self-similar. It is further assumed that f has a bell shape described by a Gaussian function. The function is assumed to be valid for the velocity deficit, and thus the shape of the wake velocity will not satisfy the self-similarity due to the shear of the mean wind speed, as illustrated in Figure 9.4. The radial distribution of the velocity deficit can be written as:

$$f\left(\frac{r}{R_*(\xi)}\right) = \exp\left(-\frac{r^2}{2R_*^2}\right). \quad [9.8]$$

Thus, $r = R_*$ corresponds to the radius where the wake deficit is 0.6065 times the maximum value at that x-position. Invoking the momentum equation and using the turbine thrust coefficient, the following expression is obtained for the maximum velocity deficit along the x-axis:

$$C(\xi) = 1 - \sqrt{1 - \frac{C_T}{8(R_*/D)^2}}. \quad [9.9]$$

As in the Jensen model, Bastankhah and Porté-Agel (2014) assume a linear expansion of the wake:

$$\frac{R_*(\xi)}{D} = k_*\xi + \varepsilon. \qquad [9.10]$$

Here, εD is the characteristic radius of the wake at the rotor position. The expressions are not valid in the near wake, but this initial value is needed for the far field expression. Combining the above expressions, the following is obtained for the velocity deficit:

$$\frac{\Delta U}{U_0} = \left[1 - \sqrt{1 - \frac{C_T}{8(k_*\xi + \varepsilon)^2}}\right] \cdot \exp\left[-\frac{(r/D)^2}{2(k_*\xi + \varepsilon)^2}\right]. \qquad [9.11]$$

The parameter ε is obtained by equating the deficit in the mass flow obtained by the present method at $\xi = 0$ and that obtained by the top hat model by Frandsen (Frandsen et al., 2006; Bastankhah and Porté-Agel, 2014):

$$\varepsilon = \frac{1}{8}\left[\frac{1 + \sqrt{1 - C_T}}{\sqrt{1 - C_T}}\right]^{1/2}. \qquad [9.12]$$

Comparisons with LES simulations indicate that the above value of ε is slightly too large and better agreement is obtained using a factor of 1/10 instead of 1/8 (Bastankhah and Porté-Agel, 2014).

9.3.2 Summation of Wakes

In a wind farm, the wakes from the individual turbines are to be added together. A commonly used principle (Katic, Højstrup and Jensen, 1987) is to assume a linear expansion of the wakes and estimate the velocity deficit at a downstream position from the first wind turbine in a row by a summation of squares of the velocity deficits. Thus, for a row of turbines where all downstream turbines are inside the wakes of the upstream turbines and a top hat distribution is assumed, the total velocity deficit behind i turbines may be approximated by:

$$\Delta U(\xi_*) = U_0 - U(\xi_*) \simeq U_0 - \left[\sum_i \left(\Delta U_i^2(\xi_*)\right)^2\right]^{1/2}. \qquad [9.13]$$

Here, $\Delta U_i(\xi_*) = U_0 - U_i(\xi_*)$ is the difference between the wind velocity incident to the wind farm and the wake velocity from turbine i at the position considered, disregarding the other turbines. The summation is to be taken over all the turbines upstream of the position considered. It is assumed that the downstream turbines are fully immersed in the wake from the upstream turbines. Other models have been proposed; see an overview in Porté-Agel, Bastankhah and Shamsoddin (2020). An alternative to the square-root sum of squares above is a linear summation of the velocity deficits. Also, an alternative velocity deficit has been proposed, where the difference between the incident velocity to turbine i and the wake behind that turbine is considered, $\Delta U_i(\xi) = U_{in,\,i} - U_i(\xi)$. A result of [9.13] is that most of the velocity loss in a row of wind turbines takes place behind the first few turbines. This has been confirmed by measurements in real wind farms (Nygaard et al., 2020; Niayifar and Porté-Agel, 2016; Barthelmie et al., 2010). When the wind direction aligns with a row of turbines in a wind farm, there may be a significant difference in power between the first and the second turbine in the row. For turbines further down in the row, a more gradual decay in power output is observed, see Figure 9.6. This is the case when the first turbine is operating below rated wind speed. If it is operating above rated wind speed, downstream turbines may also operate above rated wind speed and thus at rated power.

As mentioned above, if the inflow is laminar, a wake expansion proportional to $\xi^{1/3}$ should be expected, while experiments in turbulent flow indicate a linear expansion. Thus, there is a relation between the wake expansion and the level of turbulence in the flow. From the theory of turbulent boundary layers, it is known

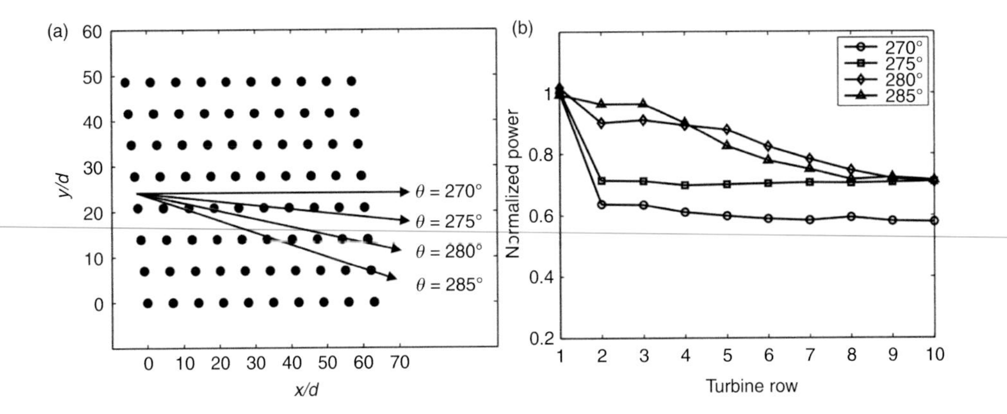

Figure 9.6 Left (a): layout of the Horns Rev offshore wind farm. Right (b): normalized average power production turbines in a row for various wind directions. Wind speed 8.0±0.5 m/s. Figure taken from Porté-Agel, Bastankhah and Shamsoddin (2020) under the terms of the Creative Commons Attribution 4.0. International data obtained from field measurements (Barthelmie et al., 2010).

that turbulence is generated by shear. The shear in the wake is largest close to the rotor, and thus a lot of turbulence is expected to be generated in this region, while the generation of turbulence diminishes further downstream. The turbulence generated in the wake region is superposed to be the ambient turbulence. The most common superposition principle used is a quadratic summation, i.e., the total turbulence intensity is obtained as:

$$I = \sqrt{I_0^2 + I_w^2}. \qquad [9.14]$$

Here, I_0 is the turbulence intensity in the ambient flow and I_w is the turbulence intensity generated in the wake. For the linearly expanding wake, it was assumed that the expansion rate is proportional to the ambient turbulence intensity. If it is assumed that the local expansion rate $dD_w(x)/dx$ is proportional to the local turbulence intensity I, the wake expansion will be largest close to the rotor and then diminish as the turbulence intensity diminishes downstream (Nygaard et al., 2020).

The ambient turbulence level depends upon atmospheric stability; thus, the wake development also depends upon atmospheric stability. A more efficient mixing of the ambient flow into the wake is expected during unstable atmospheric conditions than during stable conditions. Comparisons of various Gaussian wake models versus LES results under various atmospheric stability conditions can be found in Krutova et al. (2020).

9.3.3 Other Wake Issues

9.3.3.1 Wake Meandering

The ambient turbulent wind field contains low-frequency large turbulent eddies. Such eddies will interact with the turbine wake and result in a slowly oscillating trajectory of the wake. This is in particular the case in the horizontal plane, but to some extent it also takes place in the vertical plane. These oscillations are called wake meandering. It may cause downstream turbines to partly or fully move in and out of the wakes of upstream turbines with corresponding variations in power output and structural loads. For floating turbines, meandering may represent a challenge, as the low natural frequencies may be close to the frequencies of the meandering. Jacobsen and Godvik (2020) analyzed motion responses of floating wind turbines operating in free wind and in a wake. For the turbine operating in a wake, increased motion responses were observed as compared to the motion responses while operating in free wind. This was the case in stable atmospheric conditions; during neutral and unstable conditions, the effect of the wake was more difficult to observe. It is not obvious from the measurements by Jacobsen and Godvik (2020) whether the increased motion responses were due to increased turbulence level or due to wake meandering.

9.3.3.2 Wind Veering

Due to the earth's rotation the combined effect of the Coriolis force and the vertical shear of the mean wind speed causes the mean direction of the wind to change with height, called wind veering. The veer exists in the free wind and will also modify the wake flow from individual turbines as well as the wake of wind farms (Englberger and Lundquist, 2020; Murphy, Lundquist and Fleming, 2020).

9.3.3.3 Wind Farm Wakes

The collective effect of the turbine wakes in a wind farm creates a wind farm wake (Porté-Agel, Bastankhah and Shamsoddin, 2020). The development of the wind farm wake is closely related to the atmospheric stability condition and turbulence. The downstream extent of the wind farm wake is expected to be much longer than the wakes of the individual turbines. Examples of wind farm wakes are shown in Figure 9.7. Here, the mean wind speed 10 m above sea level in the German Bight is

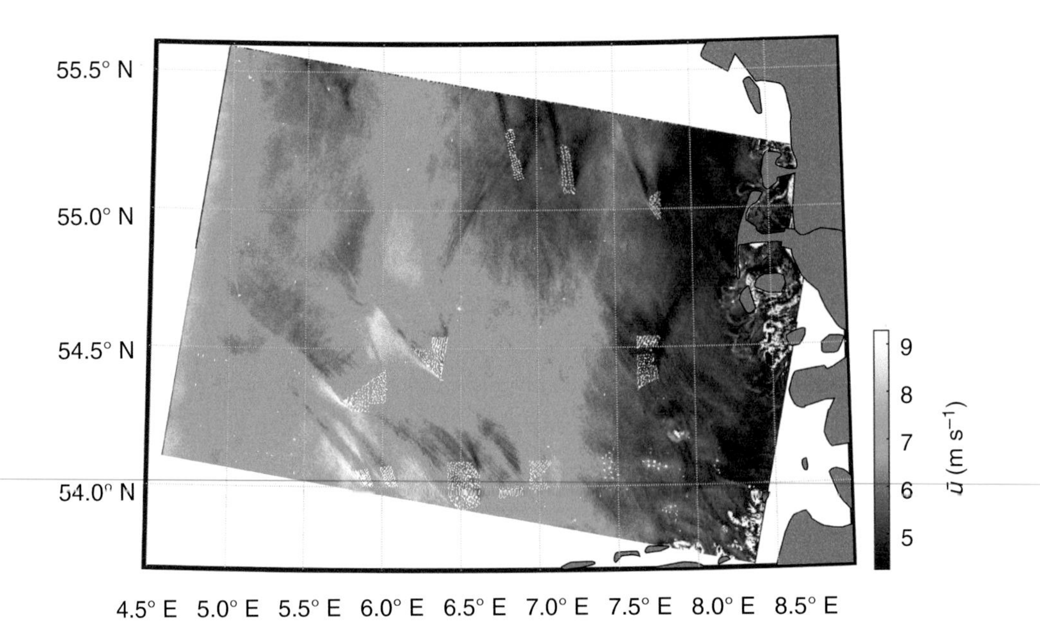

Figure 9.7 Illustration of wakes behind wind farms in the German Bight as observed by the Sentinel-1A satellite, April 17, 2022. Individual wind turbines can be seen as white dots. Along the lower edge of the picture the Borkum-Riffgrund, Nordsee and Gode wind farms are located. The color bar indicates the mean wind speed 10 m above sea level. The map is produced and used courtesy of Etienne Cheynet, University of Bergen.

shown. The map is derived from satellite measurements from April 17, 2022. The wind is blowing from the southeast. Reduced wind speed behind the wind farms is clearly observed. The extent of the visible speed reduction is several times the characteristic dimension of the wind farm. The wind farm wake may reduce the wind speed incident to a downstream wind farm, with a reduced power production as a possible consequence. The extent of the wind farm wake is sensitive to the stability conditions and thus the turbulence level in the atmospheric boundary layer.

9.3.3.4 Effect of the Induction Zone

The wind speed is slowed down in front of an operating wind turbine. This is called an induction effect. The extent of the induction zone for an individual turbine is of the order of the rotor diameter. However, the collective action of the turbines located in the front row of a wind farm may have a considerably larger extent. The induction effect causes a slow-down of the wind speed in the middle section of the front row and a speed-up along the edges. It will also cause a speed-up of the flow above the wind farm (Porté-Agel, Bastankhah and Shamsoddin, 2020).

9.3.3.5 Wind Farm Control/Turbine Yaw

To avoid the turbine wake hitting a downstream turbine, the upstream turbine may be yawed relative to the free wind direction. The yaw angle will result in a transverse component of the thrust force, causing a transverse component of the wake. Thus, the wake may be guided away from the downstream turbine. The advantage of this approach is increased production and reduced fatigue loads on downstream turbines, while the disadvantage may be reduced power production of the upstream turbine. Together with power control of the individual turbines in a wind farm, turbine yaw and wake steering may be parts of a wind farm control strategy aiming at optimizing the power production from the farm and at the same time minimizing the fatigue loads on the turbines.

Exercises Chapter 9

1. We will compare the cost of energy for an onshore and an offshore wind farm using the following data. (MW-data refers to installed capacity.)

	Offshore	Onshore
Discount rate (%)	6	6
Lifetime (years)	27	27
Initial investments (year zero) (million euro/MW)	2.37	1.40
O&M costs (thousand euro/MW year)	76	43
Energy produced (GWh/year)	4.47	3.73

a. What is the LCOE for each of these wind farms?
b. How will the LCOE change if the discount rate is doubled?
c. A life extension to 30 years is considered. For which of the above two discount rates is this most favorable? Compute the new LCOEs.
d. If the investment costs take place two years prior to start-up of production, how will this change the LCOE?

2. Consider the 5 MW turbine as given by Jonkman et al. (2009) (see Section 3.8). The wind conditions at the site considered are modeled by a Weibull distribution, with $\alpha = 12.0$, $\gamma = 0$ and $\beta = 2$. (With $\gamma = 0$ and $\beta = 2$ the Weibull distribution becomes a Rayleigh distribution; see Sections 2.1.4 and 2.3.1) Compute:
   a. the annual average wind speed
   b. the number of hours per year that the wind is below the cut-in velocity
   c. the number of hours per year that the turbine will be shut down due to wind speeds above the cut-out velocity
   d. the total energy production per year
   e. the fraction of the total energy produced when the turbine is running at below rated power
3. Consider the two wind roses in Figure 9.2. Assume you shall construct a wind farm within a rectangular domain with the long side twice the length of the short side.
   a. How will you orient the rectangle in the two cases? List your arguments.
   b. How will you place the wind turbines within the two rectangles? List your arguments.
4. In Section 9.4.1 a top-hat and a self-similar Gaussian velocity profile for the wake deficit are presented.
   a. Make plots of the two profiles as a function of downstream distance from the turbine. Use the same turbine thrust and rotor diameter. Present the results in the range of 2–15 diameters downstream.
   b. Discuss how the presence of the sea surface and the vertical shear profile of the wind field modify the wake field.

Appendix A

2D Boundary Layer Equations

A 2D channel flow is considered. The $x-$ axis is pointing in the flow direction and the $z-$ axis is vertical with $z = 0$ at the bottom surface, the wall.[1] Incompressible flow is assumed. The continuity equation becomes in this case:

$$\frac{\partial u}{\partial x} + \frac{\partial w}{\partial z} = 0. \qquad [A.1]$$

The equation for momentum conservation in the horizontal direction becomes:

$$\frac{\partial u}{\partial t} + u\frac{\partial u}{\partial x} + w\frac{\partial u}{\partial z} = -\frac{1}{\rho_a}\frac{\partial p}{\partial x} + \nu\frac{\partial^2 u}{\partial z^2} . \qquad [A.2]$$

Here, $u = \overline{u} + u'$ and $w = \overline{w} + w'$ are the velocities in horizontal and vertical direction respectively. ν is the kinematic viscosity of the fluid. For stationary flow, the first term of [A.2], the acceleration term is zero. Inserting for the mean and turbulent contributions to the velocities and considering the mean quantities, the second term on the left-hand side of [A.2] may be written as:

$$u\frac{\partial u}{\partial x} = (\overline{u} + u')\left(\frac{\partial \overline{u}}{\partial x} + \frac{\partial u'}{\partial x}\right) = \overline{u}\frac{\partial \overline{u}}{\partial x} + \overline{u}\frac{\partial u'}{\partial x} + u'\frac{\partial \overline{u}}{\partial x} + u'\frac{\partial u'}{\partial x} . \qquad [A.3]$$

Invoking that the mean of u' is zero, the second and third term vanish. Further, it may be assumed that $u'\frac{\partial u'}{\partial x} = \frac{1}{2}\frac{\partial}{\partial x}(u'u')$. The mean of [A.3] is thus obtained as:

$$\overline{u\frac{\partial u}{\partial x}} = \overline{u}\frac{\partial \overline{u}}{\partial x} + \frac{1}{2}\frac{\partial}{\partial x}\left(\overline{u'u'}\right) . \qquad [A.4]$$

Using the same considerations for the last term on the left-hand side of [A.2], the mean of the term may be written as:

$$\begin{aligned} \overline{w\frac{\partial u}{\partial z}} &= \overline{w}\frac{\partial \overline{u}}{\partial z} + \overline{w'\frac{\partial u'}{dz}} = \overline{w}\frac{\partial \overline{u}}{\partial z} + \frac{\partial}{\partial z}\left(\overline{u'w'}\right) - \overline{u'\frac{\partial w'}{dz}} \\ &= \overline{w}\frac{\partial \overline{u}}{\partial z} + \frac{\partial}{\partial z}\left(\overline{u'w'}\right) + \frac{1}{2}\frac{\partial}{\partial x}\left(\overline{u'u'}\right). \end{aligned} \qquad [A.5]$$

[1] This appendix provides information relevant to Chapter 2 in this volume.

In the last step, the continuity equation [A.1] has been invoked. The equation for conservation of the mean momentum in $x-$ direction in steady flow thus becomes:

$$\overline{u}\frac{\partial\overline{u}}{\partial x}+\overline{w}\frac{\partial\overline{u}}{\partial z}=-\frac{1}{\rho_a}\frac{\partial\overline{p}}{\partial x}+\nu\frac{\partial^2\overline{u}}{\partial z^2}-\frac{\partial}{\partial z}\left(\overline{u'w'}\right)-\frac{\partial}{\partial x}\left(\overline{u'u'}\right). \quad [A.6]$$

Here, the last term may be ignored as the gradient of the square of the $x-$ component of the horizontal velocity fluctuations is assumed to be small. The vertical gradient of horizontal shear stress is given from the two middle terms on the right-hand side, i.e.:

$$\frac{\partial\tau}{\partial z}=\mu\frac{\partial^2\overline{u}}{\partial z^2}-\rho_a\frac{\partial}{\partial z}\left(\overline{u'w'}\right), \quad [A.7]$$

or

$$\tau=\mu\frac{\partial\overline{u}}{\partial z}-\rho_a\overline{u'w'}. \quad [A.8]$$

$\mu=\rho_a\nu$ is called the dynamic viscosity, while ν is the kinematic viscosity.[2]

In the laminar sublayer, very close to the boundary, $\overline{w}$ as well as w' will tend to zero. The wall stress is thus approximately given by:

$$\tau_w\simeq\mu\frac{\partial\overline{u}}{\partial z}\bigg|_{z\to 0}. \quad [A.9]$$

Similarly, above the laminar sublayer, in the surface layer, where the effect of viscosity is small, the shear stress is given from the turbulent velocities only:

$$\tau\simeq-\rho_a\overline{u'w'}. \quad [A.10]$$

A quantity denoted the *friction velocity* is introduced. In turbulent flow the friction velocity is defined by:

$$u_*\equiv\sqrt{\frac{\tau}{\rho_a}}=\sqrt{-\overline{u'w'}}\ . \quad [A.11]$$

Considering the laminar sublayer, the friction velocity is obtained by integrating [A.9] in vertical direction. The velocity in this layer may thus be written as:

$$\overline{u}(z)=u_*^2\frac{z}{\nu}. \quad [A.12]$$

Thus, a linear increase in mean velocity with the distance from the wall is expected in the laminar sublayer.

In the surface layer, but still close to the ground level, Prandtl's mixing length theory may be invoked (see, e.g., Curle and Davies, 1968). The assumption behind Prandtl's mixing length theory implies that close to a wall but inside the turbulent boundary layer, the turbulent shear stress is related to the vertical gradient of the mean flow, i.e.:

[2] At an air temperature of 15°C and 1 atm pressure, the density of air is $\rho_a=1.225kg/m^3$ and the kinematic viscosity $\nu=1.470\cdot10^{-5}m^2/s$.

$$\overline{u'w'} \propto -l^2\left(\frac{\partial\overline{u}}{\partial z}\right)^2. \qquad [A.13]$$

The constant l is thus assumed to be proportional to the vertical distance from the wall, $l = kz$. The shear stress close to wall but outside the laminar sublayer is thus obtained by combing [A.10] and [A.13] as:

$$\tau \simeq \rho_a k^2 z^2 \left(\frac{\partial\overline{u}}{\partial z}\right)^2. \qquad [A.14]$$

Rearranging and introducing the friction velocity:

$$\frac{\partial\overline{u}}{\partial z} = \frac{1}{kz}\sqrt{\frac{\tau}{\rho_a}} = \frac{1}{kz}u_*. \qquad [A.15]$$

Integrating, the horizontal mean velocity is obtained as:

$$\overline{u} = \frac{u_*}{k}\ln\frac{z}{z_0} + C. \qquad [A.16]$$

This is equivalent to [2.3], where $k = k_a \simeq 0.40$ is recognized as the von Kármán constant. z_0 is a roughness length scale given by $z_0 = \nu/u_*$. According to Curle and Davies (1968), the range of validity of [A.16] is typically $20 < z/z_0 < 400$, while C is approximately 5.5. For $z/z_0 < 10$ the velocity profile is according to the laminar flow expression ([A.12]). For atmospheric boundary layers, the logarithmic velocity profile is assumed valid throughout the surface layer during neutral atmospheric conditions and with a constant surface roughness upwind of the section considered. That is, the turbulent fluxes are assumed constant in this region (Lee, 2018).

Appendix B

Asymptotic Behavior of the Theodorsen Function

The Theodorsen function is defined by:

$$C(k_f) = F(k_f) + iG(k_f) = \frac{H_1^{(2)}(k_f)}{H_1^{(2)}(k_f) + iH_0^{(2)}(k_f)}. \quad \text{[B.1]}$$

Here, $H_n^{(2)}(k_f) = J_n(k_f) - iY_n(k_f)$ is the Hankel function of second kind and order n; J_n and Y_n are the Bessel functions of first and second kind and order n.[1] To find the limiting values as $z \to 0$ and $z \to \infty$, the following asymptotic values for the Bessel functions are employed (see Abramowitz and Stegun, 1970):

$$\begin{aligned} J_n(z) &\simeq \frac{\left(\frac{1}{2}z\right)^n}{\Gamma(n+1)}, \quad z \to 0 \\ Y_n(z) &\simeq -\frac{1}{\pi}\Gamma(n)\left(\frac{1}{2}z\right)^{-n}, \quad z \to 0, \ \text{n} > 0 \\ Y_0(z) &\simeq \frac{2}{\pi}\ln(z), \quad z \to 0, \end{aligned} \quad \text{[B.2]}$$

i.e., for $n = 0$ and $n = 1$ the following expressions are obtained:

$$\begin{aligned} J_0(z) &\simeq \frac{\left(\frac{1}{2}z\right)^0}{\Gamma(1)} = 1, \quad z \to 0 \\ J_1(z) &\simeq \frac{\left(\frac{1}{2}z\right)^1}{\Gamma(2)} = \frac{1}{2}z, \quad z \to 0 \\ Y_0(z) &\simeq \frac{2}{\pi}\ln(z), \quad z \to 0 \\ Y_1(z) &\simeq -\frac{1}{\pi}\Gamma(1)\left(\frac{1}{2}z\right)^{-1} = -\frac{2}{\pi z}, \quad z \to 0. \end{aligned} \quad \text{[B.3]}$$

[1] This appendix provides information relevant to Chapter 3 in this volume.

Inserting these asymptotic values in the expression for the Theodorsen function, the following expression is obtained:

$$C(z)=\frac{H_1^{(2)}(z)}{H_1^{(2)}(z)+iH_0^{(2)}(z)}=\frac{J_1(z)-iY_1(z)}{J_1(z)-iY_1(z)+i[J_0(z)-iY_0(z)]}$$

$$\simeq\frac{\frac{1}{2}z-i\left(-\frac{2}{\pi z}\right)}{\frac{1}{2}z-i\left(-\frac{2}{\pi z}\right)+i\left[1-i\frac{2}{\pi}\ln(z)\right]}$$

$$=\frac{\frac{1}{2}z^2+i\frac{2}{\pi}}{\frac{1}{2}z^2+\frac{2z}{\pi}\ln(z)+i\left(z+\frac{2}{\pi}\right)}$$

$$=\frac{\left[z^4+\frac{4z^3}{\pi}\ln(z)+\frac{8z}{\pi}+\frac{16}{\pi^2}\right]+i\left[-2z^3+\frac{16}{\pi^2}z\ln(z)\right]}{\left[z^2+\frac{4z}{\pi}\ln(z)\right]^2+\left[\frac{4}{\pi}+2z\right]^2}\simeq\frac{1+\frac{\pi z}{2}+i\cdot z\ln(z)}{1+\pi z}\quad\text{as } z\rightarrow 0. \tag{B.4}$$

As $z\rightarrow 0$ it is observed that the asymptotic values become $\mathrm{Re}\{C(0)\}=1$ and $\mathrm{Im}\{C(0)\}=0$. For large values of z the following asymptotic form is used for the Hankel functions (Abramowitz and Stegun, 1970):

$$H_n^{(2)}(z)\simeq\sqrt{\frac{2}{\pi z}}\;e^{-i\left[z-\left(n+\frac{1}{2}\right)\frac{\pi}{2}\right]}\quad\text{as } z\rightarrow\infty \text{ and } n \text{ is fixed.} \tag{B.5}$$

Using this form, the Theodorsen function for large arguments becomes:

$$C(z)\simeq\frac{e^{-i\left[\pi-\frac{3}{2}\frac{\pi}{2}\right]}}{e^{-i\left[\pi-\frac{3}{2}\frac{\pi}{2}\right]}-ie^{-i\left[\pi-\frac{1}{2}\frac{\pi}{2}\right]}}=\frac{e^{-i\frac{\pi}{4}}}{e^{-i\frac{\pi}{4}}-e^{-i\frac{5\pi}{4}}}=\frac{1-i}{1-i-(-1+i)} \tag{B.6}$$

$$=\frac{1-i}{2(1-i)}=\frac{1}{2}\quad\text{as } z\rightarrow\infty.$$

Appendix C
The Laplace Transform

Fourier transform is a useful tool to analyze stationary dynamic processes.[1] Similarly, Laplace transform is useful for studying transient response and control systems. Frequently, time domain differential equations are easier to manipulate and solve in the Laplace space and then transform back to the time domain. Details about Laplace transformation are given in standard mathematical textbooks, e.g., Butkov (1973). The Laplace transform of a time-dependent function $y(t)$ is given by:

$$Y(s) = \mathscr{L}\{y(t)\} = \int_0^\infty e^{-st} y(t) dt. \qquad [C.1]$$

$y(t)$ is defined on the interval $[0, \infty)$ and is real. $s = \sigma + i\omega$ is complex. The condition for the transform to be valid is that:

$$\int_0^\infty e^{-\sigma_a t} y(t) dt < \infty. \qquad [C.2]$$

Here, σ_a is real. The lowest value of σ_a that satisfies [C.2] is denoted the convergence axis of $y(t)$. Here, a simple second-order linear differential equation is considered as an illustrative example; i.e., consider the equation:

$$m\ddot{x} + b\dot{x} + kx = f(t), \qquad [C.3]$$

with initial conditions at $t = 0$: $x(0) = x_0$ and $\dot{x}(0) = \dot{x}_0$. Further, $f(t) = 0$ for $t < 0$. The equation may be split into a homogeneous and a particular part:

$$\begin{aligned} \ddot{x}_h + 2\lambda\dot{x}_h + \omega_0^2 x_h &= 0 \\ x_h(0) &= x_0 \\ \dot{x}_h(0) &= \dot{x}_0 \end{aligned} \qquad [C.4]$$

[1] This appendix provides information relevant to Chapter 3 in this volume.

and

$$\ddot{x}_p + 2\lambda\dot{x}_p + \omega_0^2 x_p = f(t)/m.$$
$$x_h(0) = 0 \qquad [C.5]$$
$$\dot{x}_h(0) = 0.$$

Here, the parameters $2\lambda = b/m$ and $\omega_0^2 = k/m$ have been introduced.

The Laplace transformation in [C.1] satisfies the following relations for derivatives and integration:

$$\mathscr{L}\left\{\frac{dy(t)}{dt}\right\} = sY(s) - y(0).$$

$$\mathscr{L}\left\{\frac{d^2y(t)}{dt^2}\right\} = s^2Y(s) - y'(0) - sy(0)$$

$$\mathscr{L}\left\{\int_0^t y(t)dt\right\} = \frac{Y(s)}{s}. \qquad [C.6]$$

Other useful Laplace transforms are given in Table C.1 and more transforms may be found in e.g., Abramowitz and Stegun (1972).

Some useful Laplace transforms.

$y(t)$	$\mathscr{L}\{y(t)\}$
$\delta(0)$ unit impulse at $t = 0$	1
$H(0)$	$\frac{1}{s}$
$\frac{t^{n-1}}{(n-1)!}$, $n = 1,\ 2,\ \ldots$	$\frac{1}{s^n}$, $n = 1, 2, \ldots$
e^{-at}	$\frac{1}{s+a}$
$\frac{1}{(n-1)!}t^{n-1}e^{-at}$, $n = 1, 2, \ldots$	$\frac{1}{(s+a)^n}$ $n = 1, 2, \ldots$
$\frac{1}{a}(1 - e^{-at})$	$\frac{1}{s(s+a)}$
$\frac{1}{a^2}(e^{-at} + at - 1)$	$\frac{1}{s^2(s+a)}$
$\cos(at)$	$\frac{s}{s^2+a^2}$
$\cosh(at)$	$\frac{s}{s^2-a^2}$
$\frac{1}{a}\sin(at)$	$\frac{1}{s^2+a^2}$
$\frac{1}{a}\sinh(at)$	$\frac{1}{s^2-a^2}$
$\frac{t}{2a}\sin(at)$	$\frac{s}{(s^2+a^2)^2}$
$t\cos(at)$	$\frac{s^2-a^2}{(s^2+a^2)^2}$
$\frac{1}{b}e^{-at}\sin(bt)$	$\frac{1}{(s+a)^2+b^2}$
$e^{-at}\cos(bt)$	$\frac{s+a}{(s+a)^2+b^2}$

Applying the Laplace transform to $x(t) = x_h(t) + x_p(t)$ above, the following relation is obtained:

$$s^2X(s) - \dot{x}_0 - sx_0 + 2\lambda sX(s) - 2\lambda x_0 + \omega_0^2X(s) = F(s)/m. \qquad \text{[C.7]}$$

Here, $X(s)$ and $F(s)$ are the Laplace transform of $x(t)$ and $f(t)$ respectively. Rearranging [C.7], the following expression is obtained:

$$\begin{aligned} X(s) = X_h(s) + X_p(s) &= \frac{2\lambda x_0 + \dot{x}_0 + sx_0}{s^2 + 2\lambda s + \omega_0^2} + \frac{\frac{1}{m}F(s)}{s^2 + 2\lambda s + \omega_0^2} \\ &= G(s)[2\lambda x_0 + \dot{x}_0 + sx_0] + \frac{1}{m}G(s)F(s). \end{aligned} \qquad \text{[C.8]}$$

Here, $G(s)$ is the "transfer function" of the dynamic system, i.e.:

$$G(s) = \frac{1}{s^2 + 2\lambda s + \omega_0^2}. \qquad \text{[C.9]}$$

The homogeneous part, $X_h(s)$, can be rearranged and written as:

$$X_h(s) = \frac{x_0(s+\lambda)}{(s+\lambda)^2 + (\omega_0^2 - \lambda^2)} + \frac{\dot{x}_0 + \lambda x_0}{(s+\lambda)^2 + (\omega_0^2 - \lambda^2)}. \qquad \text{[C.10]}$$

The critical damping of the system is given by $b_{cr} = 2m\omega_0$ or $\lambda_{cr} = \omega_0$. Assuming that the system is lightly damped, i.e., $\lambda \ll \lambda_{cr} = \omega_0$, and invoking the inverse Laplace transforms (see Table C.1), the solution of [C.10] may be written as:

$$x_h(t) = x_0e^{-\lambda t}\cos(\omega t) + \frac{\dot{x}_0 + \lambda x_0}{\omega}e^{-\lambda t}\sin(\omega t), \qquad \text{[C.11]}$$

where $\omega^2 = (\omega_0^2 - \lambda^2) \simeq \omega_0^2$ for a lightly damped system.

The solution for the particular part is obtained by invoking the convolution theorem for Laplace transforms, i.e.:

$$\mathscr{L}^{-1}\{x(t)y(t)\} = \int_0^t x(t-\tau)y(t)dt. \qquad \text{[C.12]}$$

The particular solution may thus be written as:

$$\begin{aligned} x_p(t) &= \frac{1}{m}\int_0^t f(\tau)g(t-\tau)d\tau \\ &= \frac{1}{m}\int_0^t f(\tau)\frac{1}{\omega}e^{-\lambda(t-\tau)}\sin\Big(\omega(t-\tau)\Big)d\tau. \end{aligned} \qquad \text{[C.13]}$$

Appendix D

Rotational Terms in the Added Mass Matrix for a Pontoon

Consider the rotational acceleration around each of the axes and the resulting moments about the other axes.[1] Reference is made to Figure 7.8. The following contributions are obtained:

$$\begin{aligned}\Delta M_{11} &= -\Delta F_y z + \Delta F_z y = -\Delta F_h \cos\alpha\ z + \Delta F_z y \\ &= -a_h A_h^{(2D)} \Delta L \cos\alpha\ z + a_v A_v^{(2D)} \Delta L y \\ &= \ddot{\eta}_4 z \cos\alpha\ A_h^{(2D)} \Delta L \cos\alpha\ z + \ddot{\eta}_4 y A_v^{(2D)} \Delta L y \\ &= [z^2 A_h^{(2D)} \cos^2\alpha + y^2 A_v^{(2D)}] \ddot{\eta}_4 \Delta L\end{aligned} \qquad [D.1]$$

Here, ΔM_{11} is the moment around the x-axis from a small section of the pontoon of length ΔL located at (x, y, z) due to an acceleration around the x-axis $\ddot{\eta}_4$. α is the angle between the pontoon axis and the x-axis. a_h and a_v are the horizontal and vertical accelerations of the section considered, a_h being the horizontal component perpendicular to the pontoon axis. $A_h^{(2D)}$ is the 2D added mass for translation in horizontal direction perpendicular to the pontoon axis. Zero added mass is assumed for translation in the axial direction. $A_v^{(2D)}$ is the 2D added mass for translation in vertical direction.

Similarly, the moments around the y-and z-axes become:

$$\begin{aligned}\Delta M_{21} &= -\Delta F_z x + \Delta F_x z = -\Delta F_z x + \Delta F_h \sin\alpha\ z \\ &= -a_v A_v^{(2D)} \Delta L x + a_h A_h^{(2D)} \Delta L \sin\alpha\ z \\ &= -\ddot{\eta}_4 y A_v^{(2D)} \Delta L x + \ddot{\eta}_4 z \cos\alpha\ A_h^{(2D)} \Delta L \sin\alpha\ z \\ &= [-xy A_v^{(2D)} + z^2 A_h^{(2D)} \cos\alpha \sin\alpha] \ddot{\eta}_4 \Delta L.\end{aligned} \qquad [D.2]$$

$$\begin{aligned}\Delta M_{31} &= \Delta F_x y + \Delta F_y x = \Delta F_h \sin\alpha\ y + \Delta F_h \cos\alpha\ x \\ &= a_h A_h^{(2D)} \Delta L \sin\alpha\ y + a_h A_h^{(2D)} \Delta L \cos\alpha\ x \\ &= -\ddot{\eta}_4 z \cos\alpha A_h^{(2D)} \Delta L \sin\alpha\ y - \ddot{\eta}_4 z \cos\alpha\ A_h^{(2D)} \Delta L \cos\alpha\ x \\ &= -[yz \cos\alpha \sin\alpha + zx \cos^2\alpha] \ddot{\eta}_4 \Delta L A_h^{(2D)}.\end{aligned} \qquad [D.3]$$

Considering rotation around the y-axis, the following moments are obtained:

[1] This appendix provides information relevant to Chapter 7 in this volume.

$$\begin{aligned}\Delta M_{22} &= -\Delta F_z x + \Delta F_x z = -\Delta F_v x + \Delta F_h \sin\alpha\, z\\ &= -a_v A_v^{(2D)} \Delta L x + a_h A_h^{(2D)} \Delta L \sin\alpha\, z\\ &= \ddot{\eta}_5 x A_v^{(2D)} \Delta L x + \ddot{\eta}_5 z \sin\alpha\, A_h^{(2D)} \Delta L \sin\alpha\, z\\ &= [x^2 A_v^{(2D)} + z^2 A_h^{(2D)} \sin^2\alpha] \ddot{\eta}_5 \Delta L.\end{aligned} \quad [D.4]$$

$$\begin{aligned}\Delta M_{32} &= -\Delta F_x y + \Delta F_y x = \Delta F_h \sin\alpha\, y + \Delta F_h \cos\alpha\, x\\ &= a_h A_h^{(2D)} \Delta L \sin\alpha\, y + a_h A_h^{(2D)} \Delta L \cos\alpha\, x\\ &= -\ddot{\eta}_5 z \sin\alpha A_h^{(2D)} \Delta L \sin\alpha\, y - \ddot{\eta}_5 z \sin\alpha\, A_h^{(2D)} \Delta L \cos\alpha\, x\\ &= [-yz \sin^2\alpha - zx\cos\alpha \sin\alpha] \ddot{\eta}_5 \Delta L A_h^{(2D)}.\end{aligned} \quad [D.5]$$

$$\begin{aligned}\Delta M_{33} &= -\Delta F_x y + \Delta F_y x = \Delta F_h y \sin\alpha + \Delta F_h x \cos\alpha\\ &= \ddot{\eta}_6 A_h^{(2D)} \Delta L (y\sin\alpha + x\cos\alpha)(y \sin\alpha + x\cos\alpha)\\ &= [x^2 \cos^2\alpha + y^2 \sin^2\alpha + 2xy \sin\alpha \cos\alpha] A_h^{(2D)} \Delta L \ddot{\eta}_6.\end{aligned} \quad [D.6]$$

The above expressions give the moments corresponding to a small 2D section of the pontoon. To get the total moments, integration of the full length of the pontoon is needed. This involves integration of quadratic quantities like x^2, xy, y^2 etc. These integrals are obtained as:

$$\begin{aligned}\int_L x^2 dl &= \int_{-L/2}^{L/2} (x_p + s\cos\alpha)^2 ds = \left[x_p^2 s + \frac{1}{2} 2x_p s^2 \cos\alpha + \frac{1}{3} s^3 \cos^2\alpha\right]_{-L/2}^{L/2}\\ &= x_p^2 L + \frac{1}{12} L^3 \cos^2\alpha.\end{aligned} \quad [D.7]$$

$$\begin{aligned}\int_L y^2 dl &= \int_{-L/2}^{L/2} (y_p + s\sin\alpha)^2 ds\\ &= y_p^2 L + \frac{1}{12} L^3 \sin^2\alpha.\end{aligned} \quad [D.8]$$

$$\int_L z^2 dl = \int_{-L/2}^{L/2} z_p^2 ds = [z_p^2 s]_{-L/2}^{L/2} = z_p^2 L. \quad [D.9]$$

$$\begin{aligned}\int_L xy dl &= \int_{-L/2}^{L/2} (x_p + s\cos\alpha)(y_p + s\sin\alpha) ds\\ &= \int_{-L/2}^{L/2} \left(x_p y_p + x_p s \sin\alpha + y_p s \cos\alpha + s^2 \cos\alpha \sin\alpha\right)\end{aligned}$$

$$= \left[x_p y_p s + \frac{1}{2} x_p s^2 \sin\alpha + \frac{1}{2} y_p s^2 \cos\alpha + \frac{1}{3} s^3 \cos\alpha \sin\alpha \right]_{-L/2}^{L/2} = x_p y_p L + \frac{1}{12} L^3 \cos\alpha \sin\alpha. \quad \text{[D.10]}$$

$$\int_L xz dl = \int_{-L/2}^{L/2} (x_p + s\cos\alpha) z_p ds = \int_{-L/2}^{L/2} \left(x_p z_p + z_p s \cos\alpha \right) ds$$

$$= \left[x_p z_p s + \frac{1}{2} z_p s^2 \cos\alpha \right]_{-L/2}^{L/2} = x_p z_p L. \quad \text{[D.11]}$$

$$\int_L yz dl = \int_{-L/2}^{L/2} (y_p + s\sin\alpha) z_G ds = \int_{-L/2}^{L/2} \left(y_p z_p + z_p s \cos\alpha \right) ds$$

$$= \left[y_p z_p s + \frac{1}{2} z_p s^2 \cos\alpha \right]_{-L/2}^{L/2} = y_p z_p L. \quad \text{[D.12]}$$

Utilizing these results and integrating the above moments, the contributions from the 2D added mass to the rotational added mass matrix $I_{ij(L)}$ are obtained as follows:

$$\begin{aligned} I_{11(L)} &= \int_{-L/2}^{L/2} [z^2 A_h^{(2D)} \cos^2\alpha + y^2 A_v^{(2D)}] dL \\ &= A_h^{(2D)} L z_p^2 \cos^2\alpha + A_v^{(2D)} L \left(y_p^2 + \frac{1}{12} L^2 \sin^2\alpha \right). \end{aligned} \quad \text{[D.13]}$$

$$\begin{aligned} I_{21(L)} &= \int_{-L/2}^{L/2} [-xy A_v^{(2D)} + z^2 A_h^{(2D)} \cos\alpha \sin\alpha] dL \\ &= -A_v^{(2D)} L \left(x_p y_p + \frac{1}{12} L^2 \cos\alpha \sin\alpha \right) + A_h^{(2D)} L z_p^2 \cos\alpha \sin\alpha. \end{aligned} \quad \text{[D.14]}$$

$$\begin{aligned} I_{31(L)} &= -\int_{-L/2}^{L/2} [yz A_h^{(2D)} \cos\alpha \sin\alpha + zx A_h^{(2D)} \cos^2\alpha] dL \\ &= -A_h^{(2D)} L \cos\alpha \sin\alpha \left(y_p z_p \right) - A_h^{(2D)} L \cos^2\alpha \left(x_p z_p \right) \\ &= -A_h^{(2D)} L z_p \cos\alpha \left(y_p \sin\alpha + x_p \cos\alpha \right). \end{aligned} \quad \text{[D.15]}$$

$$I_{22(L)} = \int_{-L/2}^{L/2} [x^2 A_v^{(2D)} + z^2 A_h^{(2D)} \sin^2 \alpha] dL$$
$$= A_v^{(2D)} L \left(x_p^2 + \frac{1}{12} L^2 \cos^2 \alpha \right) + A_h^{(2D)} L z_p^2 \sin^2 \alpha. \quad \text{[D.16]}$$

$$I_{32(L)} = -\int_{-L/2}^{L/2} A_h^{(2D)} [yz \sin^2 \alpha + zx \cos \alpha \sin \alpha] dL$$
$$= -A_h^{(2D)} L \sin^2 \alpha \left(y_p z_p \right) + A_h^{(2D)} L \cos \alpha \sin \alpha \left(x_p z_p \right)$$
$$= -A_h^{(2D)} L z_p \sin \alpha (y_p \sin \alpha + x_p \cos \alpha). \quad \text{[D.17]}$$

$$I_{33(L)} = \int_{-L/2}^{L/2} A_h^{(2D)} [x^2 \cos^2 \alpha + y^2 \sin^2 \alpha + 2xy \sin \alpha \cos \alpha] dL$$
$$= A_h^{(2D)} L \begin{bmatrix} \cos^2 \alpha \left(x_p^2 + \frac{1}{12} L^2 \cos^2 \alpha \right) + \sin^2 \alpha \left(y_p^2 + \frac{1}{12} L^2 \sin^2 \alpha \right) \\ +2 \sin \alpha \cos \alpha \left(x_p y_p + \frac{1}{12} L^2 \sin \alpha \cos \alpha \right) \end{bmatrix} \quad \text{[D.18]}$$
$$= A_h^{(2D)} L \left[\left(x_p \cos \alpha + y_p \sin \alpha \right)^2 + \frac{1}{12} L^2 \right].$$

The effect of the added mass at the end surfaces of the pontoon should be added to the above contributions. Consider an end surface located in (x, y, z_p) and consider an acceleration a_s along the axial direction from end (1) to (2). The accelerations in (x, y, z) direction become:

$$a_x = a_s \cos \alpha$$
$$a_y = a_s \sin \alpha \quad \text{[D.19]}$$
$$a_z = 0.$$

The relations between the axial acceleration and rotations about the global axes are:

$$a_s = -\ddot{\eta}_4 z_p \sin \alpha + \ddot{\eta}_5 z_p \cos \alpha + \ddot{\eta}_6 (x \sin \alpha - y \cos \alpha) \quad \text{[D.20]}$$

The corresponding moments due to an added mass effect at the end surface become then:

$$M_{11(e)} = \ddot{\eta}_4 A_e z_p^2 \sin^2 \alpha$$
$$M_{21(e)} = -\ddot{\eta}_4 A_e z_p^2 \sin \alpha \cos \alpha$$
$$M_{31(e)} = \ddot{\eta}_4 A_e z_p \left(y \sin \alpha \cos \alpha - x \sin^2 \alpha \right)$$
$$M_{22(e)} = \ddot{\eta}_5 A_e z_p^2 \cos^2 \alpha \quad \text{[D.21]}$$
$$M_{23(e)} = \ddot{\eta}_5 A_e z_p \left(x \sin \alpha \cos \alpha - y \cos^2 \alpha \right)$$
$$M_{33(e)} = \ddot{\eta}_6 A_e (x \sin \alpha - y \cos \alpha)^2.$$

To obtain the contribution from both ends, the contributions from (x_1, y_1, z_p) and (x_2, y_2, z_p) are to be added. The contributions to the rotational part of the added mass matrix from the ends thus become:

$$
\begin{aligned}
I_{11(e)} &= 2A_e z_p^2 \sin^2\alpha \\
I_{21(e)} &= -2A_e z_p^2 \sin\alpha \cos\alpha \\
I_{31(e)} &= -2A_e z_p \sin\alpha [x_p \sin\alpha - y_p \cos\alpha] \\
I_{22(e)} &= 2A_e z_p^2 \cos^2\alpha \\
I_{32(e)} &= 2A_e z_p \cos\alpha [x_p \sin\alpha - y_p \cos\alpha] \\
I_{33(e)} &= A_e [(x_1 \sin\alpha - y_1 \cos\alpha)^2 + (x_2 \sin\alpha - y_2 \cos\alpha)^2].
\end{aligned} \quad [D.22]
$$

Appendix E

Notch Filter in Time Domain

A notch filter modifies a stochastic input signal $x(t)$ to a new signal $y(t)$ in which the signal components with frequencies in a narrow neighborhood of the filter frequency f_0 are removed.[1] Normally, filtering of a time series is most conveniently performed in the frequency domain, i.e., the time series considered is converted by Fourier transform to frequency domain. Here, the wanted transfer function is applied and finally the filtered time series is obtained by inverse Fourier transform. This procedure is not feasible if the complete time series is not available at the time of filtering. In real-time application, filtering in time domain is thus required. Here, the procedure for using linear digital filters in time domain as outlined by Press et al. (1989) is summarized.

Consider a time series $x(t)$ sampled at an interval dt, so the sampled time series forms the sequence x_n, $n = [1, n_{\max}]$. A filtered version of x_n may then be obtained by:

$$y_n = \sum_{k=0}^{M} c_k x_{n-k} + \sum_{j=1}^{N} d_j y_{n-j}. \qquad \text{[E.1]}$$

The $M + 1$ coefficients c_k and the N coefficients d_j define the response of the filter. If $N = 0$, the filter is denoted as *non-recursive*, i.e., former values of y do not contribute to y_n . If $N > 0$, the filter is denoted as *recursive*. In this case the values of d_j must fulfil certain requirements to ensure stability of the filter procedure. In the frequency domain the filter transfer function can be written as:

$$H(f) = \frac{\sum_{k=0}^{M} c_k e^{-2\pi ikfdt}}{1 - \sum_{j=1}^{N} d_j e^{-2\pi ijfdt}}. \qquad \text{[E.2]}$$

[1] This appendix provides information relevant to Chapter 7 in this volume.

We normally know the wanted shape of the filter in the frequency domain. The challenge is then to find a set of coefficients, c_k and d_j, that can approximate this shape without having too many terms in the summations in [D.2]. For this purpose, Press et al. (1989) show how a bilinear transformation method may be used. The complex variable z is introduced as:

$$z = e^{2\pi i f dt}. \qquad [E.3]$$

The frequencies are limited by the Nyquist frequency, i.e., $-1 < 2fdt < 1$. A new variable w is introduced. w maps this frequency interval onto the interval $-\infty < w < \infty$ by the transformation:

$$w = i\frac{1-z}{1+z}. \qquad [E.4]$$

Note that w becomes real for real frequencies. Using the above principles, a simple notch filter that filters out the frequencies in the vicinity of f_0 can be written as:

$$H(f) = \frac{w^2 - w_0^2}{(w - i\varepsilon w_0)^2 - w_0^2}. \qquad [E.5]$$

Here, w_0 is obtained from [E.3] and [E.4] by inserting f_0 . ε is a small parameter controlling the bandwidth of the notch filter. Substituting z for w in [E.5] and performing the arithmetic, the filter coefficients are obtained as:

$$c_0 = \frac{1 + w_0^2}{(1 + \varepsilon w_0)^2 + w_0^2}$$

$$c_1 = -2\frac{1 - w_0^2}{(1 + \varepsilon w_0)^2 + w_0^2}$$

$$c_2 = c_0 \qquad [E.6]$$

$$d_1 = 2\frac{1 - \varepsilon^2 w_0^2 - w_0^2}{(1 + \varepsilon w_0)^2 + w_0^2}$$

$$d_2 = -\frac{(1 - \varepsilon w_0)^2 + w_0^2}{(1 + \varepsilon w_0)^2 + w_0^2}.$$

Use of Notch Filter on Wind Spectrum

Consider a wind time series generated according to the Kaimal spectrum with mean wind speed of 10 m/s and turbulence intensity of 10%. The time series has a duration of 1000 s and is sampled at 0.50 s intervals. Figure E.1 shows the theoretical Kaimal spectrum, together with the spectrum obtained after having generated a time history of

(cont.)

the Kaimal spectrum. The random oscillations around the theoretical spectrum are evident and are a result of random phases, finite length of record and finite sampling rate.

A notch filter with filter frequency $f_0 = 0.2$ *Hz and bandwidth* $\varepsilon = 0.1$ *is applied. The characteristic of this filter is shown in Figure E.2. Applying [E.6], the filter constants become:*

$$c_0 = 0.9436, \quad c_1 = -1.5268, \quad c_2 = c_0$$
$$d_1 = 1.5250, \quad d_2 = -0.8891. \qquad [E.7]$$

Figure E.1 shows the spectrum of the time series as obtained after applying the notch filter. The dip at 0.2 Hz is evident.

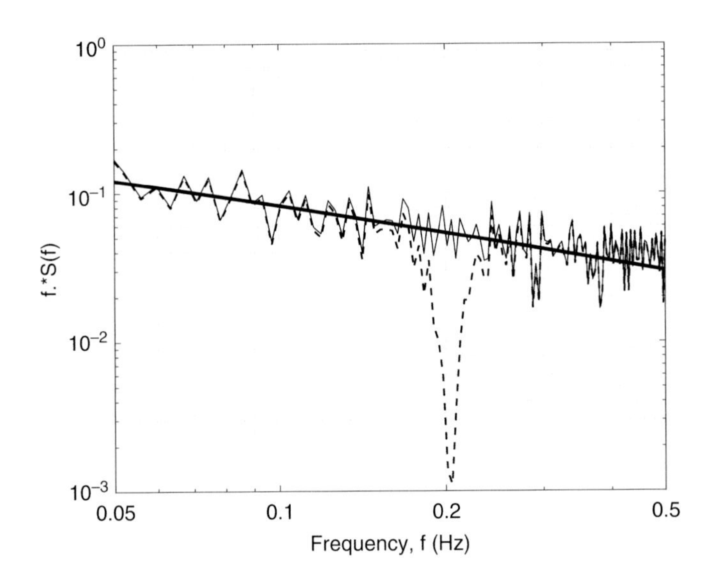

Figure E.1 Solid thick line: the Kaimal wind spectrum; solid thin line: spectrum from generated time series; dashed line: spectrum from time series filtered by notch filter using filter frequency 0.2 Hz and bandwidth 0.1.

(cont.)

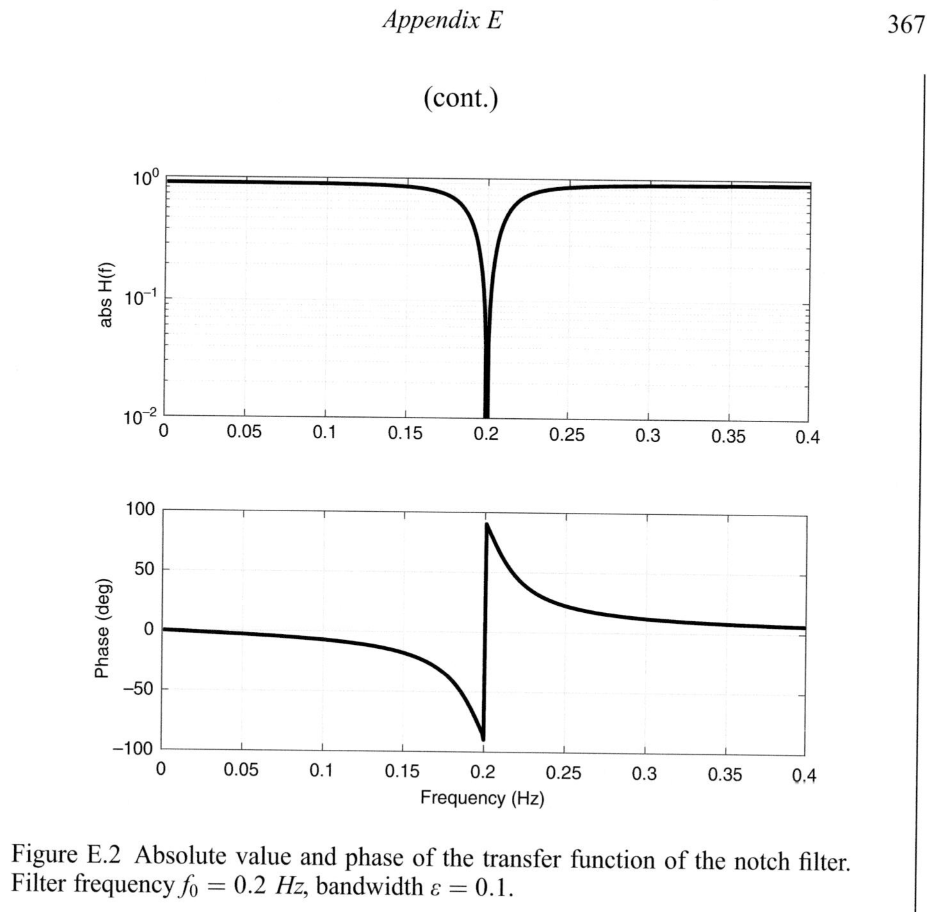

Figure E.2 Absolute value and phase of the transfer function of the notch filter. Filter frequency $f_0 = 0.2\ Hz$, bandwidth $\varepsilon = 0.1$.

References

Aasen, S., A. M. Page, K. S. Skau and T. A. Nygaard. 2017. "Effect of Foundation Modelling on the Fatigue Lifetime of a Monopile-Based Offshore Wind Turbine." *Wind Energy Science* 2 (2): 361–76. https://doi.org/10.5194/wes-2-361-2017.

Abramowitz, M. and I. A. Stegun (ed.). 1970. *Handbook of Mathematical Functions with Formulas, Graphs and Mathematical Tables*. Dover Publications, Inc.

Anaya-Lara, O., J. O. Tande, K. Uhlen and K. Merz. 2018. *Offshore Wind Energy Technology*. Wiley.

Arany, L., S. Bhattacharya, J. Macdonald and S. J. Hogan. 2017. "Design of Monopiles for Offshore Wind Turbines in 10 Steps." *Soil Dynamics and Earthquake Engineering* 92: 126–52. https://doi.org/10.1016/j.soildyn.2016.09.024.

Attari, A. and P. Doherty. 2015. "Gravity Base Foundations in Offshore Wind: Design Drivers." Engineers Journal, April 21. www.engineersireland.ie/Engineers-Journal/More/Renewables/gravity-base-foundations-in-offshore-wind-design-drivers (accessed August 3, 2018).

Bak, C., F. Zahle, R. Bitsche et al. 2013. "Description of the DTU 10 MW Reference Wind Turbine." DTU Wind Energy Report-I-0092, Technical University of Denmark.

Barthelmie, R. J. 1999. "The Effects of Atmospheric Stability on Coastal Wind Climates." *Meteorological Applications* 6 (1): 39–47. https://doi.org/10.1017/S1350482799000961.

Barthelmie, R. J., S. C. Pryor, S. T. Frandsen et al. 2010. "Quantifying the Impact of Wind Turbine Wakes on Power Output at Offshore Wind Farms." *Journal of Atmospheric and Oceanic Technology* 27(8): 1302–17. https://doi.org/10.1175/2010JTECHA1398.1.

Bastankhah, M. and F. Porté-Agel. 2014. "A New Analytical Model for Wind-Turbine Wakes." *Renewable Energy* 70: 116–23. https://doi.org/10.1016/j.renene.2014.01.002.

Benson, T. J. 1996. "Interactive Educational Tool for Classical Aerofoil Theory." NASA Lewis Research Center. www.grc.nasa.gov/WWW/K-12/airplane/FoilTheory.pdf (accessed November 15, 2017).

Betz, A. 1926. *Wind-Energie und Ihre Ausnutzung durch Windmühlen*. Vandenhoeck & Ruprecht.

Beyer, F., T. Choisnet, M. Kretschmer and P. W. Cheng. 2015. "Coupled MBS-CFD Simulation of the IDEOL Floating Offshore Wind Turbine Foundation Compared to Wave Tank Model Test Data." *Proceedings of 25th International Ocean and Polar Engineering Conference*, June 21–26, Kona, 367–74.

Birknes, J., Ø. Hagen, T. B. Johannessen, Ø. Lande and A. Nestegård. 2013. "Second Order Kinematics Underneath Irregular Waves." *Proceedings of the International Conference on Offshore Mechanics and Arctic Engineering – OMAE*, June 8–13, Nantes.

Blevins, R. D. 1977. *Flow-Induced Vibrations*. Van Nostrand Reinhold Company Ltd.

Bramwell, A. R. S., G. Done and D. Balmford. 2001. *Bramwell's Helicopter Dynamics*. 2nd ed. American Institute of Aeronautics and Astronautics, Inc. and Butterworth-Heinemann.

Burton, T., N. Jenkins, D. Sharpe and E. Bossanyi. 2011. *Wind Energy Handbook*. 2nd ed. John Wiley & Sons.

Busch, N. E., Larsen, S. E. and Thomson, D. W. 1978. "Data Analysis of Atmospheric Measurements." In Hansen, B. W. (ed.), *Proceedings of the Dynamic Flow Conference 1978 on Dynamic Measurements in Unsteady Flows*. Springer, 887–908.

Butkov, E. 1973. *Mathematical Physics*. Addison-Wesley Publishing Company.

Campbell, I. M. C. and P. A. Weynberg. 1980. *Measurement of Parameters Affecting Slamming. Rep.440, Tech. Rep. Centre No. OT-R-8042*. Southampton University.

Carswell, W., J. Johansson, F. Løvholt, S. R. Arwade, C. Madshus, D. J. DeGroot and A. T. Myers. 2015. "Foundation Damping and the Dynamics of Offshore Wind Turbine Monopiles." *Renewable Energy* 80: 724–36. https://doi.org/10.1016/j.renene.2015.02.058.

Chaaban, R. 2012. "NREL 5-MW Reference Turbine: CP, CQ, CT Coefficients." NREL Forum. https://wind.nrel.gov/forum/wind/viewtopic.php?t=582 (accessed August 28, 2023).

Charnock, H. 1955. "Wind Stress on a Water Surface." *Quarterly Journal of the Royal Meteorological Society* 81(350): 639–40. https://doi.org/10.1002/qj.49708135027.

Chen, D., K. Huang, V. Bretel and L. Hou. 2013. "Comparison of Structural Properties between Monopile and Tripod Offshore Wind-Turbine Support Structures." *Advances in Mechanical Engineering* 5. https://doi.org/10.1155/2013/175684.

Cheynet, E. 2019. "Influence of the Measurement Height on the Vertical Coherence of Natural Wind." In Ricciardelli, F. and A. M. Avossa (eds.), *Proceedings of the XV Conference of the Italian Association for Wind Engineering. IN VENTO 2018. Lecture Notes in Civil Engineering* 27. Springer International Publishing.

Cheynet, E., J. B. Jakobsen and C. Obhrai. 2017. "Spectral Characteristics of Surface-Layer Turbulence in the North Sea." *Energy Procedia* 137: 414–27. https://doi.org/10.1016/j.egypro.2017.10.366.

Cheynet, E., J. B. Jakobsen and J. Reuder. 2018. "Velocity Spectra and Coherence Estimates in the Marine Atmospheric Boundary Layer." *Boundary-Layer Meteorology* 169(3): 429–60. doi:10.1007/s10546-018-0382-2.

Chougule, A., J. Mann, M. Kelly and G. C. Larsen. 2018. "Simplification and Validation of a Spectral-Tensor Model for Turbulence Including Atmospheric Stability." *Boundary-Layer Meteorology* 167 (3): 371–97. https://doi.org/10.1007/s10546-018-0332-z.

Churchfield, M. J., S. Lee, P. J. Moriarty et al. 2012. "A Large-Eddy Simulation of Wind-Plant Aerodynamics Preprint." January.

Cummins, W. E. 1962. "The Impulse Response Function and Ship Motions." *Schifftechnik* 47 (9): 101–9.

Curle, N. and Davies, H. J. 1968. *Modern Fluid Dynamics, Volume 1: Incompressible Flow*. D. van Nostrand Company.

Dahlquist, G. and Å. Björck. 2003. *Numerical Methods*. Dover Publications, Inc.

Dalgic, Y., I. Lazakis, I. Dinwoodie, D. McMillan and M. Revie. 2015. "Advanced Logistics Planning for Offshore Wind Farm Operation and Maintenance Activities." *Ocean Engineering* 101: 211–26. https://doi.org/10.1016/j.oceaneng.2015.04.040.

Damiani, R., K. Dykes and G. Scott. 2016. "A Comparison Study of Offshore Wind Support Structures with Monopiles and Jackets for U.S. Waters." *Journal of Physics: Conference Series* 753 (9): 092003. https://doi.org/10.1088/1742-6596/753/9/092003.

Davenport, A. G. 1962. "The Response of Slender, Line-Like Structures to a Gusty Wind." *Proceedings of the Institution of Civil Engineers* 23 (3): 389–408. https://doi.org/10.1680/iicep.1962.10876.

De Souza, C. E. S. 2022. "Structural Modelling, Coupled Dynamics, and Design of Large Floating Wind Turbines." PhD thesis, NTNU.

Dinwoodie, I., O.-E. V. Endrerud, M. Hofmann, R. Martin and I. B. Sperstad. 2015. "Reference Cases for Verification of Operation and Maintenance Simulation Models for Offshore Wind Farms." *Wind Engineering* 39 (1): 1–14. https://doi.org/10.1260/0309-524X.39.1.1.

DNV. 2011. "Marine Operations, General, DNV Offshore Standard, DNV-OS-H101." *DNV Offshore Standard*. Oslo.

DNV. 2014a. "Offshore Standard DNV-OS-J101, Design of Offshore Wind Turbine Structures." *DNV Offshore Standard*. Oslo.

DNV. 2021a. "Standard DNV-ST-0126, Support Structures for Wind Turbines." *DNV Offshore Standard*. Oslo.

DNV. 2021b. "Standard DNV-ST-0436 Load and Site Conditions for Wind Turbines." *DNV Offshore Standard*. Oslo.

DNV. 2021c. "Recommended Practice DNV-RP-C205, Environmental Conditions and Environmental Loads." Oslo.

DNV. 2021d. "Standard DNV-ST-N001 Marine Operations and Marine Warranty." *DNV Offshore Standard*. Oslo.

Duarte, T., M. Alves, J. Jonkman and A. Sarmento. 2013. "State-Space Realization of the Wave-Radiation Force within FAST." *Proceedings of the International Conference on Offshore Mechanics and Arctic Engineering – OMAE*, June 8–13, Nantes. https://doi.org/10.1115/OMAE2013-10375.

Englberger, A. and J. K. Lundquist. 2020. "How Does Inflow Veer Affect the Veer of a Wind-Turbine Wake?" *Journal of Physics: Conference Series* 1452 (1): 1–9. https://doi.org/10.1088/1742-6596/1452/1/012068.

European Commission Directorate-General for Energy. 2022. *Study on the Performance of Support for Electricity from Renewable Sources Granted by Means of Tendering Procedures in the Union 2022*. Luxemburg. https://doi.org/10.2833/93256.

Falnes, J. 2002. *Ocean Waves and Oscillating Systems: Linear Interaction Including Wave-Energy Extraction*. Cambridge University Press.

Faltinsen, O. M. 1990. *Sea Loads on Ships and Offshore Structures*. 1st ed. Cambridge University Press.

Faltinsen, O. M. 2005. *Hydrodynamics of High-Speed Marine Vehicles*. Cambridge University Press.

Faltinsen, O. M., J. N. Newman and T. Vinje. 1995. "Nonlinear Wave Loads on a Slender Vertical Cylinder." *Journal of Fluid Mechanics* 289: 179–98.

Fenton, J. D. 1985. "A Fifth Order Stokes Theory for Steady Waves." *Journal of Waterway, Port, Coast and Ocean Engineering* 111 (2). https://doi.org/10.1061/(ASCE)0733-950X(1985)111:2(216).

Frandsen, S., R. Barthelmie, S. Pryor et al. 2006. "Analytical Modelling of Wind Speed Deficit in Large Offshore Wind Farms." *Wind Energy* 9 (1–2): 39–53. https://doi.org/10.1002/we.189.

Furevik, B. R. and H. Haakenstad. 2012. "Near-Surface Marine Wind Profiles from Rawinsonde and NORA10 Hindcast." *Journal of Geophysical Research Atmospheres* 117 (23): 1–14. https://doi.org/10.1029/2012JD018523.

Gaertner, E., J. Rinker, L. Sethuraman et al. 2020. "IEA Wind TCP Task 37: Definition of the IEA 15 MW Offshore Reference Wind Turbine." Technical Report NREL/TP-5000-75698 March. https://doi.org/10.2172/1603478.

Glauert, H. 1935. "Airplane Propellers." In Durand, W. F. (ed.), *Aerodynamic Theory, Vol. IV.* Springer, 169–360.

Global Wind Energy Council. 2022. *Global Wind Report 2022.* https://gwec.net/global-wind-report-2022/ (accessed August 28, 2023).

Graham, C. 1982. "The Parameterisation and Prediction of Wave Height and Wind Speed Persistence Statistics for Oil Industry Operational Planning Purposes." *Coastal Engineering* 6 (4): 303–29. https://doi.org/10.1016/0378-3839(82)90005-9.

Grue, J. and M. Huseby. 2002. "Higher-Harmonic Wave Forces and Ringing of Vertical Cylinders." *Applied Ocean Research* 24 (4): 203–14. https://doi.org/10.1016/S0141-1187(02)00048-2.

Grue, J., J. Kolaas and A. Jensen. 2014. "Velocity Fields in Breaking-Limited Waves on Finite Depth." *European Journal of Mechanics B/Fluids* 47: 97–107. https://doi.org/10.1016/j.euromechflu.2014.03.014.

Haakenstad, H., Ø. Breivik, M. Reistad and O. J. Aarnes. 2020. "NORA10EI: A Revised Regional Atmosphere-Wave Hindcast for the North Sea, the Norwegian Sea and the Barents Sea." *International Journal of Climatology* 40 (10): 4347–73. https://doi.org/10.1002/joc.6458.

Haakenstad, H., Ø. Breivik, B. R. Furevik, M. Reistad, P. Bohlinger and O. J. Aarnes. 2021. "NORA3: A Nonhydrostatic High-Resolution Hindcast of the North Sea, the Norwegian Sea, and the Barents Sea." *Journal of Applied Meteorology and Climatology* 60 (10): 1443–64. https://doi.org/10.1175/JAMC-D-21-0029.1.

Hansen, M. O. L. 2015. *Aerodynamics of Wind Turbines*. 3rd ed. Routledge.

Hansen, M. H., A. D. Hansen, T. J. Larsen, S. Øye, P. Sørensen and P. Fuglsang. 2005. *Control Design for a Pitch-Regulated, Variable Speed Wind Turbine, Control.* Forskningscenter Risoe. Risoe-R No. 1500(EN).

Hansen, M. O. L., J. N. Sørensen, S. Voutsinas, N. Sørensen and Madsen, H. A. 2006. "State of Art in Wind Turbine Aerodynamics and Aeroelasticity." *Progress in Aerospace Sciences* 42 (2006): 285–330.

Haslum, H. and O. M. Faltinsen. 1999. "Alternative Shapes of Spar Platforms for Use in Hostile Areas, OTC 10953." Offshore Technology Conference, May 3–6, Houston.

Haslum, H., M. Marley, S. T. Navalkar, B. Skaare, N. Maljaars and H. S. Andersen. 2022. "Roll–Yaw Lock: Aerodynamic Motion Instabilities of Floating Offshore Wind Turbines." *Journal of Offshore Mechanics and Arctic Engineering* 144 (4): 1–10. https://doi.org/10.1115/1.4053697.

Hess, J. L. and A. M. Smith. 1962. "Calculation of Nonlifting Flow about Arbitrary Three-Dimensional Bodies." *Douglas Aircraft Co. Report No. E.S.* 40622 (also in abbreviated form in the *Journal of Ship Research* 8 (1964)).

Hong, D. C. 1987. "On the Improved Green Integral Equation Applied to the Water-Wave Radiation – Diffraction Problem." *Journal of the Society of Naval Architects of Korea* 24 (1).

Hooft, J. P. 1972. Hydrodynamic Aspects of Semi-Submersible Platforms. H. Veeman on Zonen N.V. PhD thesis, TU Delft.

IEC. 2005. International Standard IEC 61400-1, Wind Turbines Part 1: Design Requirements.

IEC. 2009. International Standard IEC 61400-3, Wind Turbines Part 3: Design Requirements for Offshore Wind Turbines. European Committee for Electrotechnical Standardization.

Irgens, F. 1999. *Dynamikk*. 4th ed. Tapir Forlag.

Jacobsen, A. and M. Godvik. 2020. "Influence of Wakes and Atmospheric Stability on the Floater Responses of the Hywind Scotland Wind Turbines." *Wind Energy* 24 (2): 149–61. https://doi.org/10.1002/we.2563.

Jamieson, P. 2017. "Multi Rotor Solution for Large Scale Offshore Wind Power." EERA DeepWind, 14th Deep Sea Offshore Wind R&D Conference, 18–20 January, Trondheim.

Jensen, N. O. 1983. "A Note on Wind Generator Interaction." Risø-M-2411, Risø National Laboratory Roskilde, 1–16.

Johannessen, T. B. 2011. "Calculations of Kinematics Underneath Measured Time Histories of Steep Water Waves." *Journal of Applied Ocean Research* 32: 391–403. https://doi.org/10.1016/j.apor.2010.08.002.

Jonkman, J., S. Butterfield, W. Musial and G. Scott. 2009. "Definition of a 5-MW Reference Wind Turbine for Offshore System Development." Technical Report, US Department of Energy, February. https://doi.org/10.2172/947422.

Kallehave, D., B. W. Byrne, C. LeBlanc Thilsted and K. K. Mikkelsen. 2015. "Optimization of Monopiles for Offshore Wind Turbines." *Philosophical Transactions of the Royal Society A: Mathematical, Physical and Engineering Sciences* 373 (2035). https://doi.org/10.1098/rsta.2014.0100.

Kalleklev, A. J. and A. Nestegård. 2005. "A Numerical Load Model for Wave Impact on Slender Vertical Cylinders." In C. Soize and G. I. Schuëller (eds.), *Structural Dynamics – EURODYN 2005: Proceedings of the 6th International Conference on Structural Dynamics*, September 4–7, Paris. Millpress, 205–210.

Kalvig, S., E. Manger, B. Hjertager and J. B. Jakobsen. 2014. "Wave Influenced Wind and the Effect on Offshore Wind Turbine Performance." *Energy Procedia* 53: 202–13. http://dx.doi.org/10.1016/j.egypro.2014.07.229.

Katic, I., J. Højstrup and N. O. Jensen. 1987. "A Simple Model for Cluster Efficiency." In W. Palz and E. Sesto (eds.), *EWEC'86: Proceedings* 1, 407–10.

Katz, J. and A. Plotkin. 2001. *Low Speed Aerodynamics*. 2nd ed. Cambridge University Press.

Kiełkiewicz, A., A. Marino, C. Vlachos, F. J. López Maldonado and I. Lessis. 2015. "The Practicality and Challenges of Using XL Monopiles for Offshore Wind Turbine Substructures." University of Strathclyde.

Korotkin, A. I. 2008. *Added Masses of Ship Structures*. Springer.

Korsmeyer, F. T., C.-H. Lee, J. N. Newman and P. D. Sclavounos. 1988. "The Analysis of Wave Effects on Tension Leg Platforms." *Proceedings of the OMAE Conference*, paper 88-611, Houston.

Kramm, G., G. Sellhorst, H. K. Ross, J. Cooney, R. Dlugi and N. Mölders. 2016. "On the Maximum of Wind Power Efficiency." *Journal of Power and Energy Engineering* 4 (1): 1–39. https://doi.org/10.4236/jpee.2016.41001.

Krokstad, J. R., C. Stansberg, A. Nestegård and T. Marthinsen. 1998. "A New Nonslender Ringing Load Approach Verified against Experiments." *Transactions of the ASME Journal of Offshore Mechanics and Arctic Engineering* 120 (1): 20–9.

Krutova, M., M. B. Paskyabi, F. G. Nielsen and J. Reuder. 2020. "Evaluation of Gaussian Wake Models under Different Atmospheric Stability Conditions: Comparison with Large Eddy Simulation Results." *Journal of Physics: Conference Series* 1669: 012016. https://doi.org/10.1088/1742-6596/1669/1/012016.

Lamb, H. 1932. *Hydrodynamics*. 6th ed. Cambridge University Press.

Lang, S. and E. McKeogh. 2011. "LIDAR and SODAR Measurements of Wind Speed and Direction in Upland Terrain for Wind Energy Purposes." *Remote Sensing* 3 (9): 1871–1901. https://doi.org/10.3390/rs3091871.

Larsen, C. M., & Koushan, K. (2005). Empirical Model for the Analysis of Vortex Induced Vibrations of Free Spanning Pipelines. In C. Soize & G. I. Schuëller (Eds.), Proceedings of the 6th International Conference on Structural Dynamics; EURODYN 2005 (pp. 175–180). Millpress, Rotterdam.

Larsen, K. and P. C. Sandvik. 1990. "Efficient Methods for the Calculation of Dynamic Mooring Line Tension." *Proceedings of the First European Offshore Mechanics Symposium*, August 20–22, Trondheim.

Larsen, T. J. and T. D. Hanson. 2007. "A Method to Avoid Negative Damped Low Frequent Tower Vibrations for a Floating, Pitch Controlled Wind Turbine." *Journal of Physics: Conference Series* 75 (1). https://doi.org/10.1088/1742-6596/75/1/012073.

Lee, X. 2018. *Fundamentals of Boundary-Layer Meteorology*. 1st ed. Springer International Publishing. https://doi.org/10.1007/Thompson,1975.

Leishman, J. G. and Martin, G. L. 2002. "Challenges in Modelling the Unsteady Aerodynamics of Wind Turbines." *Wind Energy* 5: 85–132. https://doi.org/10.1002/we.62.

Liu, X, C. Lu, G. Li, A. Godbole and Y. Chen. 2017. "Effects of Aerodynamic Damping on the Tower Load of Offshore Horizontal Axis Wind Turbines." *Applied Energy* 204: 1101–14. https://doi.org/10.1016/j.apenergy.2017.05.024.

MacCamy, R. C. and R. A. Fuchs. 1954. "Wave Forces on Piles: A Diffraction Theory." US Army Coastal Engineering Research Center (Formerly Beach Erosion Board), Technical Memorandum No. 69.

Mann, J. 1994. "The Spatial Structure of Neutral Atmospheric Surface-Layer Turbulence." *Journal of Fluid Mechanics* 273: 141–68. https://doi.org/10.1017/S0022112094001886.

Mann, J. 1998. "Wind Field Simulation." *Probabilistic Engineering Mechanics* 13 (4): 269–82. https://doi.org/10.1016/s0266-8920(97)00036-2.

Manwell, J. F., J. G. McGowan and A. L. Rogers. 2009. *Wind Energy Explained: Theory, Design and Application*. John Wiley & Sons.

Maronga, B., M. Gryschka, R. Heinze et al. 2015. "The Parallelized Large-Eddy Simulation Model (PALM) Version 4.0 for Atmospheric and Oceanic Flows: Model Formulation, Recent Developments, and Future Perspectives." *Geoscientific Model Development* 8 (8): 2515–51. https://doi.org/10.5194/gmd-8-2515-2015.

Marten, D., M. Lennie, G. Pechlivanoglou, C. N. Nayeri and C. O. Paschereit. 2015. "Implementation, Optimization, and Validation of a Nonlinear Lifting Line-Free Vortex Wake Module Within the Wind Turbine Simulation Code QBLADE." *Journal of Engineering for Gas Turbines and Power* 138 (7). https://doi.org/10.1115/1.4031872.

Mathiesen, M. 1994. "Estimation of Wave Height Duration Statistics." *Coastal Engineering* 23 (1–2): 167–81. https://doi.org/10.1016/0378-3839(94)90021-3.

McCormick, B. W. 1999. *Aerodynamics of V/STOL Flight*. Dover Publications, Inc.

Molin, B. 2011. "Hydrodynamic Modelling of Perforated Structures." *Applied Ocean Research* 33: 1–11. https://doi.org/10.1016/j.apor.2010.11.003.

Molin, B. and F. G. Nielsen. 2004. "Heave Added Mass and Damping of a Perforated Disk Below the Free Surface." 20th International Workshop on Water Waves and Floating Bodies, March, Cortona.

Moriarty, P. J. and A. C. Hansen. 2005. *AeroDyn Theory Manual*. NREL/TP-50. National Renewable Energy Laboratory.

Morison, J. R., M. P. O'Brien, J. W. Johnsen and S. A. Schaaf. 1950. "The Force Exerted by Surface Waves on Piles." *Petroleum Transactions AIME* 189: 149–54.

Mork, M. 2010. "Wave Theory." English translation of notes for lecture "Bølgeteori," University of Bergen.

Murphy, P., J. K. Lundquist and P. Fleming. 2020. "How Wind Speed Shear and Directional Veer Affect the Power Production of a Megawatt-Scale Operational

Wind Turbine." *Wind Energy Science* 5 (3): 1169–90. https://doi.org/10.5194/wes-5-1169-2020.

Myren, A. F. 2021. "Sensitivity Analysis in Key Parameters Related to Wind Power Production." Master's thesis, University of Bergen.

Næss, A. 1985. "On the Distribution of Crest to Trough Wave Heights." *Ocean Engineering* 12 (3): 221–34. https://doi.org/10.1016/0029-8018(85)90014-9.

Næss, A. and T. Moan. 2013. *Stochastic Dynamics of Marine Structures*. Cambridge University Press.

Nayfeh, A. H. and D. T. Mook. 1979. *Nonlinear Oscillations*. John Wiley & Sons.

Negro, V., J. S. López-Gutiérrez, M. D. Esteban, P. Alberdi, M. Imaz and J. M. Serraclara. 2017. "Monopiles in Offshore Wind: Preliminary Estimate of Main Dimensions." *Ocean Engineering* 133: 253–61. https://doi.org/10.1016/j.oceaneng.2017.02.011.

Newman, J. N. 1962. "The Exciting Force on Fixed Bodies in Waves." *Journal of Ship Research* 6 (4): 10–17.

Newman, J. N. 1977. *Marine Hydrodynamics*. The MIT Press.

Newman, J. N. 1985. "Algorithms for the Free-Surface Green Function." *Journal of Engineering Mathematics* 19: 57–67.

Newman, J. N. and P. D. Sclavounos. 1988. "The Computation of Wave Loads on Large Offshore Structures." *Proceedings of the BOSS' 88 Conference*, Trondheim.

Niayifar, A. and F. Porté-Agel. 2016. "Analytical Modeling of Wind Farms: A new Approach for Power Prediction." *Energies* 9 (9): 741. https://doi.org/10.3390/en9090741.

Nielsen, F. G., T. H. Søreide and S. O. Kvarme. 2002. "VIV Response of Long Free Spanning Pipelines." *Proceedings of the International Conference on Offshore Mechanics and Arctic Engineering – OMAE*, June 23–28, Oslo. https://doi.org/10.1115/OMAE2002-28075.

Nielsen, F. G., T. D. Hanson and B. Skaare. 2006. "Integrated Dynamic Analysis of Floating Offshore Wind Turbines." *Proceedings of the 25th International Conference on Offshore Mechanics and Arctic Engineering*, June 4–9, Hamburg. https://doi.org/10.1115/OMAE2006-92291.

Nielsen, F. G. 2007. Lecture Notes in Marine Operations. Dept. of Marine Hydrodynamics, Faculty of Marine Technology, Norwegian University of Science and Technology (NTNU), Trondheim.

NORSOK. 2017. Standard N-003:2017. Action and Action Effects. https://handle.standard.no/no/Nettbutikk/produktkatalogen/Produktpresentasjon/?ProductID=873200 (accessed August 28, 2023).

Nybø, A., F. G. Nielsen and J. Reuder. 2019. "Processing of Sonic Anemometer Measurements for Offshore Wind Turbine Applications." *Journal of Physics: Conference Series* 1356: 012006. https://doi.org/10.1088/1742-6596/1356/1/012006.

Nybø, A., F. G. Nielsen, J. Reuder, M. J. Churchfield and M. Godvik. 2020. "Evaluation of Different Wind Fields for the Investigation of the Dynamic Response of Offshore Wind Turbines." *Wind Energy* 23 (9): 1810–30. https://doi.org/10.1002/we.2518.

Nybø, A., F. G Nielsen and M. Godvik, 2021. "Analysis of Turbulence Models Fitted to Site, and Their Impact on the Response of a Bottom-Fixed Wind Turbine." *Journal of Physics: Conference Series* 012028. https://doi.org/10.1088/1742-6596/2018/1/012028.

Nybø, A., F. G. Nielsen and M. Godvik. 2022. "Sensitivity of the Dynamic Response of a Multimegawatt Floating Wind Turbine to the Choice of Turbulence Model." *Wind Energy* 25 (6): 1013–29. https://doi.org/10.1002/we.2712.

Nybø, A., F. G. Nielsen, J. Reuder, M. Churchfield and M. Godvik. 2019. "Evaluation of Different Wind Fields for the Investigation of the Dynamic Response of Offshore Wind Turbines." *Wiley Wind Energy* 23 (9): 1810–30. https://doi.org/10.1002/we.2518.

Nygaard, N. G., S. T. Steen, L. Poulsen and J. G. Pedersen. 2020. "Modelling Cluster Wakes and Wind Farm Blockage." *Journal of Physics: Conference Series* 1618: 062072. https://doi.org/10.1088/1742-6596/1618/6/062072.

Ogilvie, T. F. 1964. "Recent Progress toward the Understanding and Prediction of Ship Motions." *Proceedings of the Fifth Symposium on Naval Hydrodynamics*, September 10–12, Bergen.

Orimolade, A. P., S. Haver and O. T. Gudmestad. 2016. "Estimation of Extreme Significant Wave Heights and the Associated Uncertainties: A Case Study Using NORA10 Hindcast Data for the Barents Sea." *Marine Structures* 49: 1–17. https://doi.org/10.1016/j.marstruc.2016.05.004.

Ormberg, H. and E. E. Bachynski. 2012. "Global Analysis of Floating Wind Turbines: Code Development, Model Sensitivity and Benchmark Study." *Proceedings of the 22nd International Offshore and Polar Engineering Conference*, June 17–22, Rhodes, 1: 366–73.

Paschen, M. and S. Laurat. 2014. "Precision of Cup Anemometers: A Numerical Study." *European International Journal of Science and Technology* 3 (5): 39–45.

Pedersen, M. D. 2017. *"Stabilization of Floating Wind Turbines*." PhD Thesis, NTNU.

Peire, K., H. Nonnemann and E. Bosschen. 2009. "Gravity Base Foundations for the Thornton Bank Offshore Wind Farm." *Terra et Aqua* 115: 20–9.

Perez, T. and T. I. Fossen. 2007. "Kinematic Models for Manoeuvring and Seakeeping of Marine Vessels." *Modeling, Identification and Control* 28 (1): 19 30. https://doi.org/10.4173/mic.2007.1.3.

Porté-Agel, F., M. Bastankhah and S. Shamsoddin. 2020. "Wind-Turbine and Wind-Farm Flows: A Review." *Boundary-Layer Meteorology* 174: 1–59. https://doi.org/10.1007/s10546-019-00473-0.

Press, W. H., B. P. Flannery, S. A. Teukolsky and W. T. Vetterling. 1989. *Numerical Recipes: The Art of Scientific Computing*. Cambridge University Press.

Reinholdtsen, S.-A., K. Mo and P. C. Sandvik. 2003. "Useful Force Models for Simulation of Multibody Offshore Marine Operations." *Proceedings of The Thirteenth International Offshore and Polar Engineering Conference*, May 25–30, Honolulu.

Roddier, D., A. Peiffer, A. Aubault and J. Weinstein. 2011. "A Generic 5 MW Windfloat for Numerical Tool Validation & Comparison against a Generic Spar." *Proceedings of the ASME 2011 International Conference on Ocean, Offshore and Arctic Engineering*, June 19–24, Rotterdam.

Salzmann, D. J. C. and J. van der Tempel. 2005. "Aerodynamic Damping in the Design of Support Structures for Offshore Wind Turbines." *Proceedings of the European Offshore Wind Conference*, October 26–28, Copenhagen.

Sanderse, B. 2009. "Aerodynamics of Wind Turbine Wakes: Literature Review." *Energy Research Centre of the Netherlands*, 1–46.

Sarpkaya, T. and M. Isaacson. 1981. *Mechanics of Wave Forces on Offshore Structures*. Van Nostrand Reinhold Company.

Sathe, A., S.-E. Gryning and A. Peña. 2011 "Comparison of the Atmospheric Stability and Wind Profiles at Two Wind Farm Sites over a Long Marine Fetch in the North Sea." *Wind Energy* 14(6): 767–80. https://doi.org/10.1002/we.456.

Skaare, B., T. D. Hanson and F. G. Nielsen. 2007. "Importance of Control Strategies on Fatigue Life of Floating Wind Turbines." *Proceedings of the International Conference on Offshore Mechanics and Arctic Engineering – OMAE*, San Diego. https://doi.org/10.1115/OMAE2007-29277.

Skaare, B., T. D. Hanson, R. Yttervik and F. G. Nielsen. 2011. "Dynamic Response and Control of the Hywind Demo Floating Wind Turbine." *Proceedings of the European Wind Energy Association (EWEA),* March 14–17, Brussels.

Skaare, B., F. G. Nielsen, T. D. Hanson, R. Yttervik, O. Havmøller and A. Rekdal. 2015. "Analysis of Measurements and Simulations from the Hywind Demo Floating Wind Turbine." *Wind Energy* 18 (6): 1105–22 https://doi.org/10.1002/we.1750.

Skjoldan, P. F. and M. H. Hansen. 2009 "On the Similarity of the Coleman and Lyapunov-Floquet Transformations for Modal Analysis of Bladed Rotor Structures." *Journal of Sound and Vibration* 327 (3–5): 424–39. doi:10.1016/j.jsv.2009.07.007.

Solbrekke, I. M., N. G. Kvamstø and A. Sorteberg. 2020. "Mitigation of Offshore Wind Power Intermittency by Interconnection of Production Sites." *Wind Energy Science* 5 (4): 1663–78. https://doi.org/10.5194/wes-5-1663-2020.

Souza, C. E. S. and E. E. Bachynski. 2019. "Changes in Surge and Pitch Decay Periods of Floating Wind Turbines for Varying Wind Speed." *Ocean Engineering* 180 (April): 223–37. https://doi:10.1016/j.oceaneng.2019.02.075.

Stansberg, C.T. 2011. "Characteristics of Steep Second-Order Random Waves in Finite and Shallow Water." *Proceedings of the ASME 2011 30th International Conference on Ocean, Offshore and Arctic Engineering, OMAE2010*, June 19–24, Rotterdam.

Stull, R. B. 1988. *An Introduction to Boundary Layer Meteorology*. Springer Netherlands.

Suja-Thauvin, L., J. R. Krokstad, E. E. Bachynski and E.-J. de Ridder. 2017. "Experimental Results of a Multimode Monopile Offshore Wind Turbine Support Structure Subjected to Steep and Breaking Irregular Waves." *Ocean Engineering* 146 (October): 339–51. https://doi.org/10.1016/j.oceaneng.2017.09.024.

Theodorsen, T. 1935. "General Theory of Aerodynamic Instability and the Mechanism of Flutter." NACA Report No. 496. 19930090935.

Triantafyllou, M. S. 1990. "Cable Mechanics with Marine Applications, Lecture Notes." Department of Ocean Engineering, Massachusetts Institute of Technology.

University of Strathclyde. 2015. "The Practicality and Challenges of Using XL Monopiles for Offshore Wind Turbine Substructures." www.esru.strath.ac.uk/EandE/Web_sites/14-15/XL_Monopiles/index.html (accessed August 28, 2023).

Van der Laan, M. P., S. J. Anderson, N. Ramos García et al. 2019. "Power Curve and Wake Analyses of the Vestas Multi-Rotor Demonstrator." *Wind Energy Science Discussions* 1932: 1–30. https://doi.org/10.5194/wes-2018-77.

Van Wijk, A. J., A. C. Beljaars, A. A. Holtslag and W. C. Turkenburg, 1990. "Evaluation of Stability Corrections in Wind Speed Profiles over the North Sea." *Journal of Wind Engineering and Industrial Aerodynamics* 33(3): 551–66. https://doi.org/10.1016/0167-6105(90)90007-Y.

Veers, P. S. 1984. "Modeling Stochastic Wind Loads on Vertical Axis Wind Turbines." *Collection of Technical Papers – AIAA/ASME/ASCE/AHS/ASC Structures, Structural Dynamics and Materials Conference*. https://doi.org/10.2514/6.1984-910.

Veers, P. S. 1988. "Three-Dimensional Wind Simulation." Sandia National Laboratories, Albuquerque, New Mexico. Vol. SAND88–015.

Vik, I. and G. Kleiven. 1985. *"Wave Statistics for Offshore Operations."* 8th International Conference on Port and Ocean Engineering under Arctic Conditions (POAC85), September, Narssarssuaq.

Vinje, T. and P. Brevig. 1980. "Breaking Waves on Finite Water Depths: A Numerical Study." Report from "Skip i Sjøgang," Norwegian Institute of Technology, Division of Marine Hydrodynamics and Norwegian Hydrodynamic Laboratories, Division Ship and Ocean Laboratories.

Vinje, T., J. P. Kaalstad and D. W. Daniel. 1991. "A Statistical Method for Evaluation of Heavy Lift Operations Offshore." *Proceedings of the First International Offshore and Polar Engineering Conference (ISOPE)*, August 11–16, Edinburgh.

Von Kármán, T. 1929. "The Impact of Seaplane Floats during Landing." NACA, *Technical Note 321*, Washington, DC.

Wagner, H. 1932. "Über Stoss- und Gleitvorgänge an der Oberfläche von Flüssigkeiten." *Zeitschrift. für Angewandte Mathematik und Mechanic*, 12 (4): 193–235.

WAMIT. 2016. *User Manual Version 7.2. Theory.* http://wamit.com/ (accessed June 5, 2018).

Wehausen, J. V. and E. V. Laitone. 1960. "Surface Waves." In C. Truesdell (ed.), *Fluid Dynamics/Strömungsmechanik*. Springer Verlag, 446–778.

Wenske, J. (ed.). 2022. *Wind Turbine System Design, Volume 1: Nacelles, Drivetrains and Verification.* The Institution of Engineering and Technology, Division Wind Turbine and System Technology.

Werle, M. J. 2008. "A New Analytical Model for Wind Turbine Wakes." Report, FloDesign Inc., 200801, Wilbraham.

Wind Europe. 2020. "Offshore Wind in Europe: Key Trends and Statistics 2019." https://windeurope.org/about-wind/statistics/offshore/european-offshore-wind-industry-key-trends-statistics-2019/ (accessed August 17, 2023).

World Meteorological Organization. 2018. *Guide to Wave Analysis and Forecasting.* WMO-No. 702. Geneva.

Index